ADVANCES IN BIOTECHNOLOGICAL PROCESSES
VOLUME 13

# Bacterial Vaccines

# ADVANCES IN BIOTECHNOLOGICAL PROCESSES

Series Editor **Avshalom Mizrahi**
Israel Institute for Biological Research, Ness Ziona, Israel
TITLES IN THE SERIES

**Advances in Biotechnological Processes**
Volumes 1–5
Avshalom Mizrahi and Antonius L. van Wezel, Editors

**Advances in Biotechnological Processes**
Volume 6
Avshalom Mizrahi, Editor

Volume 7
**Upstream Processes: Equipment and Techniques**
Avshalom Mizrahi, Editor

Volume 8
**Downstream Processes: Equipment and Techniques**
Avshalom Mizrahi, Editor

Volume 9
**Biotechnology in Agriculture**
Avshalom Mizrahi, Editor

Volume 10
**Synthetic Peptides in Biotechnology**
Avshalom Mizrahi, Editor

Volume 11
**Monoclonal Antibodies: Production and Application**
Avshalom Mizrahi, Editor

Volume 12
**Biological Waste Treatment**
Avshalom Mizrahi, Editor

Volume 13
**Bacterial Vaccines**
Avshalom Mizrahi, Editor

ADVANCES IN BIOTECHNOLOGICAL PROCESSES
VOLUME 13

# Bacterial Vaccines

**Editor**
**Avshalom Mizrahi**
Israel Institute for Biological Research
Ness Ziona, Israel



WILEY-LISS

A JOHN WILEY & SONS, INC., PUBLICATION
New York • Chichester • Brisbane • Toronto • Singapore

**Address all Inquiries to the Publisher**
**Wiley-Liss, 41 East 11th Street, New York, NY 10003**



**Printed in United States of America**



**LC 83-644876**
**ISBN 0-471-56219-X**

# Contents

## *Klebsiella* Polysaccharide Vaccines
*S.J. Cryz, Jr.*



## Anthrax Vaccine Development: A Continuing Story
*Peter Hambleton and Peter C.B. Turnbull*



## Production and Control of *Neisseria meningitidis* Vaccines
*Carl E. Frasch*

## Vaccines Against Leprosy
*Duncan E.S. Stewart-Tull*



## Vaccination Against the Cholera/Coli Family of Enterotoxins by Synthetic Peptides
*Chaim O. Jacob*

## Liposomes as Carrier and Immunoadjuvant of Vaccine Antigens
*Nico van Rooijen*



## Surface Chemistry and Classification of Vaccine Adjuvants and Vehicles
*L.F. Woodard*

# Contributors

**Myron Christodoulides,** Department of Pathology, Royal Victoria Infirmary, University of Newcastle-upon-Tyne, Newcastle-upon-Tyne, Tyne and Wear, NE1 4LP, England **[169]**

**Clelia Collotti,** Laboratorio di Immunologia, Istituto Superiore di Sanità, 00161 Rome, Italy **[35]**

**S.J. Cryz, Jr.,** Swiss Serum and Vaccine Institute, 3001 Berne, Switzerland **[87]**

**Carl E. Frasch,** Division of Bacterial Products, Center for Biologics Evaluation and Research, Bethesda, MD 20892 **[123]**

**Giuliano Gentili,** Laboratorio di Immunologia, Istituto Superiore di Sanità, 00161 Rome, Italy **[35]**

**Peter Hambleton,** Division of Biologics, Public Health Laboratory Service, Centre for Applied Microbiology and Research, Porton Down, Salisbury, Wiltshire SP4 0JG, England **[105]**

**Chaim O. Jacob,** Department of Microbiology and Immunology, Stanford University School of Medicine, Stanford, CA 94305 **[223]**

**James L. McDonel,** Department of Biology, Indiana University at South Bend, South Bend, IN 46634 **[1]**

**Paul Nicoletti,** Department of Infectious Diseases, College of Veterinary Medicine, University of Florida, Gainesville, FL 32610 **[147]**

**Jan T. Poolman,** Department of Bacterial Vaccine Development, National Institute of Public Health and Environmental Protection (RIVM), 3720 BA Bilthoven, The Netherlands **[57]**

**Duncan E.S. Stewart-Tull,** Department of Microbiology, University of Glasgow, Bearsden, Glasgow, G12 8QQ, Scotland **[201]**

**Peter C.B. Turnbull,** Division of Biologics, Public Health Laboratory Service, Centre for Applied Microbiology and Research, Porton Down, Salisbury, Wiltshire SP4 0JG, England **[105]**

**Nico van Rooijen,** Department of Histology, Medical Faculty, Free University, Amsterdam, The Netherlands **[255]**

**Maria Wirz,** Laboratorio di Immunologia, Istituto Superiore di Sanità, 00161 Rome, Italy **[35]**

**L.F. Woodard,** Department of Veterinary Sciences, University of Wyoming, Laramie, WY 82070 **[281]**

The numbers in brackets are the opening page numbers of the contributors' articles.

**Bacterial Vaccines, pages 1–33**


# Bacterial Toxin Vaccines

**James L. McDonel**

Department of Biology, Indiana University at South Bend, South Bend, Indiana 46634

## I. INTRODUCTION

Even though the incidence of most infectious diseases and their distribution throughout the world fluctuate with time, it would indeed be a rare infectious disease that simply ceased to be a problem on its own. It is not surprising that medical intervention in one form or another is necessary to improve the health and quality of life for humankind through reduction and eventual elimination of infectious diseases. The two obvious approaches to disease intervention are prevention (prophylaxis) and treatment (therapeutics). While temporary successes have been realized in reversing the epidemic proportions of certain diseases through the application of chemotherapeutic agents, it has become readily evident over the past several decades that these agents alone are insufficient to control and eliminate infectious diseases. Chemotherapeutic agents are clearly indispensable to medical practice, but several problems have limited their overall success. Among these are their propensity for giving rise to resistant strains, the need to administer them repeatedly, their cost, and the difficulty of widespread distribution and administration, especially in underdeveloped countries.

The alternative, vaccination (immunoprophylaxis), certainly has its problems as well, as evidenced by the fact that vaccines haven't eradicated any bacterial diseases either. However, the incidence of disease and death from diphtheria, pertussis, and tetanus, for example, has been reduced to very low rates in the United States, where vaccination levels in children are quite high. Unfortunately, these kinds of optimistic statistics are not to be found in much of the remainder of the world.

Vaccines do have the distinct advantage of having to be administered only once, or infrequently, which places far less of a burden on funding, distribution, and dosage administration facilities. The classic problem, especially with bacterial vaccines, is developing a safe and long-lasting vaccine that actually protects the vast majority of recipients against the target disease.

The general interest in vaccines as a favored approach to the control of infectious disease has had a rebirth in recent years. In part, the lack of interest prior to this rebirth has been the result of several factors, including very high development costs, corporate and even governmental liabilities [1] when side effects or reversion to virulence injure or kill otherwise healthy individuals, and frustrating technical dead ends in developmental programs. However, promising new technologies have become available during the past 15 years, and these are enabling rapid advances in the understanding of molecular aspects of antigenicity and immune response and have opened entirely new avenues of pursuit in the development of safe and efficacious vaccines that were unthinkable only a few years ago.

This chapter will be limited to vaccine developments related to bacteria that cause disease wholly or in part by the production of one or more protein (or peptide) toxins. It has been estimated that more than 220 different bacterial toxins from as many bacterial species have been described to date [2]. Several reviews and books that deal entirely with bacterial toxins have been published recently [3–6]. The exact role of each of these toxins in pathogenesis is not clear in all cases; indeed, the list of toxins known to cause widespread disease in human populations is currently more limited than the large numbers given above [7]. Vaccines based upon the following bacteria and toxins will be reviewed in this chapter: *Corynebacterium diphtheriae* (diphtheria toxin), *Bordetella pertussis* (pertussis toxin), *Clostridium tetani* (tetanus toxin), *Vibrio cholerae* (cholera toxin), *Escherichia coli* (*E. coli* LT and ST enterotoxins), *Pseudomonas aeruginosa* (pseudomonas toxin), *Bacillus anthracis* (anthrax toxin), *Staphylococcus aureus* (toxic shock syndrome toxin-1), and *Streptococcus pneumoniae* (protein M).

Each of the organisms to be discussed has its own characteristics for virulence, invasion, and pathogenesis. A vaccine designed to immunize a host solely against a particular toxin is not necessarily the most effective route of disease prevention and not necessarily sufficient alone in providing total immune protection from one or more disease syndromes that any one organism may be able to cause. It is also important to realize that conventional vaccines based upon whole cells or other nontoxin antigens already exist for some of these diseases [8] and that active research should and does continue along those lines as well as the approaches detailed here.

### A. Subunit Vaccines

Conventional bacterial toxin vaccines, like those derived from tetanus toxin, are prepared by formalin treatment of the protein toxin to render it biologically inactive while retaining immunogenic potential. The formalinized toxoid must then be purified by methods such as salt precipitation, ultrafiltration, or column chromatography to remove extraneous components that can cause untoward side reactions when the vaccine is administered. In the case of tetanus toxoid, some antigenic determinants apparently survive (or result from) the formalin treatment; these can cause hypersensitivity after repeated injections and are not necessarily related or essential to inducing protective immunity (immunogenicity) [9]. One possible solution to this problem is to develop split-type vaccines that use degraded toxin fragments that presumably are free of the unwanted determinants [10].

Formalinized toxoids can also partially revert to toxicity and cause disease symptoms or other painful reactions in recipients. A cholera toxin vaccine that developed this kind of problem has been described [11]. Subunit vaccines are more precisely characterized than whole-toxin derivatives and should have fewer extraneous components associated with them. However, low yields and the need for addition of detergents or other potentially troublesome agents to produce the subunits and separate them from residues have created significant problems with this approach.

### B. Synthetic Peptides

Even more precisely characterized and homogeneous than naturally derived subunit vaccines are peptides that have been artificially synthesized from scratch in the laboratory [12,13]. This can be done by "classical" synthesis [14,15] (less commonly used for peptide vaccine construction) or solid-phase synthesis [16]. Today, determining the exact sequence of amino acids in a toxic protein (or peptide) is routinely done either through sequential analysis of the protein or, more commonly, through decoding the DNA sequence(s) ultimately responsible for production and assembly of the toxin in the microorganism.

In theory at least, peptides can be synthesized essentially free of any unwanted toxin moieties, extraneous bacterial components, growth-medium residues, or other reaction-inducing compounds. Furthermore, a peptide can be synthesized with the exact combination of characteristics and amino acid components desired. As more is learned about complex issues like T- and B-cell–dependent immune response, atomic mobilities ("hot" and "cold" areas) in protein molecules, discontinuous epitopes, and immunodominant epitopes, synthesis of carefully engineered peptides that are more effective even than native immunogens may be possible [17]. Finally, stability, ease of administration, and the cost of large-scale production of synthetic peptides should result in a cost-effective and readily available product.

A short peptide is more readily synthesized (by strictly chemical means) than a long one. Experimental synthetic peptide vaccines thus far have been based upon a limited but antigenically important region of a relevant protective antigen in the target organism or toxin [18]. However, such a peptide alone does not necessarily generate the same immune response as its counterpart from the natural source. Enhancement of immunogenicity of a synthetic peptide may be attained through a number of possible methodologies [19]. One is the utilization of added (e.g., muramyl dipeptide [MDP]) or built-in (e.g., coupling MDP to the peptide) adjuvants [20]. "Divalent" peptides contain one segment that reacts well with T lymphocytes and one that reacts with B lymphocytes [21]. The synthetic peptide can also be bound to protein, polysaccharide, or synthetic and other macromolecular carriers [22] (large molecules often elicit a better immune response than small ones). The peptide can be polymerized, or its ends can be connected to make a cyclic peptide. Finally, immunogenicity can be enhanced through antigen presentation in micelles [23] or with immunostimulating complexes (iscoms) [24,25]. Some theoretical and practical aspects of synthetic peptides in vaccine development have recently been presented in a two-volume series entitled *Synthetic Vaccines* [27].

### C. Recombinant DNA Methodologies

Rapid advances in recombinant DNA methodologies have enabled new and promising approaches to vaccine development. With these technologies, the genetic material that codes for a particular immunity-inducing antigen is isolated from the causative organism and inserted into the DNA of another. Or, genes required for virulence but not antigenicity can be deleted from organisms grown for whole-cell–derived vaccines [27]. The recipient organism can be grown in large quantities to produce the antigen(s) that is then isolated and purified for vaccine production, or the organism can be an otherwise harmless one that can be given to the vaccinee in whom it will propagate and present the antigen(s) to the immune system. In either case extraneous and harmful components associated with the parent organism are avoided. The new producer organism can be a prokaryotic or eukaryotic cell or a virus. As examples, studies have been done to develop a pertussis vaccine [28,29] using recombinant DNA that codes for pertussis toxin in *E. coli* (prokaryote); a hepatitis B vaccine now on the market that is based upon hepatitis B surface antigen (HBsAg) produced in yeast (eukaryote) has proved effective [30]; and vaccinia virus is being used as a cloning and expression vector to carry different or even multiple antigens [31,32] that hopefully will induce a protective immune response in recipients.

### D. Anti-idiotype Antibodies

Idiotypes are recognition sites located at or near the antigen binding site of an antibody molecule. Anti-idiotype antibodies are antibodies against an idiotope (a single determinant) or another antibody molecule [33,34]. It has been demonstrated with bacteria [35,36], as well as viruses, parasites, and tumor cells, that anti-idiotypic antibodies can behave as effective substitutes for antigens. They also can be used as immune modulators in concert with standard antigen challenge. More work has to be done to gain a better understanding of some of the complex interactions and responses in the immune system to these ''anti-antibody'' antibodies, but results reported to date suggest that there may be significant application of these antibodies to vaccine development, including bacterial toxin vaccines.

## II. DIPHTHERIA TOXIN VACCINES

Commercially available diphtheria vaccine is composed of diphtheria toxoid (biologically inactive diphtheria toxin). It is usually administered along with tetanus toxoid and pertussis vaccine in a combined preparation called DTP (diphtheria, tetanus, pertussis). Sometimes diphtheria toxoid is administered just with tetanus toxoid (DT vaccine). Because immunity following diphtheria vaccination in childhood is temporary, serum levels of diphtheria antitoxin in adults have been found to be quite low or nonexistent [37]. With older children and adults, a reduced amount of diphtheria toxoid with the normal tetanus toxoid dose (Td vaccine) has been reported to be sufficient to induce protective levels of immunity [38], although better immune responses occur with higher diphtheria toxoid doses [39,40].

Considerable controversy exists regarding DTP vaccine because of occasional serious side reactions. Of those that are truly attributable to the vaccine, most are believed to originate with the pertussis component (see below). Diphtheria toxoid today is highly purified and is relatively free of extraneous proteins and other reactive components. This, coupled with the reduced dosage of the toxoid in the Td vaccine, causes the frequent but minor local and systemic reactions to diphtheria toxoid to be transient [41] and of no serious or long-term consequence [42]. Nevertheless, new vaccine methodologies would be desirable to eliminate side reactions altogether and hopefully boost immunity in terms of degree and duration.

Traditional diphtheria vaccine is made by formalin inactivation of diphtheria toxin, which is produced and purified from *Corynebacterium diphtheriae* culture by methods that have been reviewed elsewhere [43]. The toxin in crude culture filtrate is inactivated by 0.4–0.5% formalin treatment for several weeks at 37°C until completely inactivated. Purification can be performed after formalin treatment (still done by many producers) or before. The advantage of purification before formalin inactivation is that extraneous bacterial proteins are unavailable to be cross-linked with the toxin by the formalin. Special conditions must be used in treating purified toxin, however, to ensure that reversion of some molecules to toxicity does not occur [43].

Diphtheria toxin is synthesized as a large precursor ($M_r$ 66,000–68,000) and secreted as a single polypeptide into the culture medium [44,45]. The intact protein is a proenzyme that, when "nicked" by a serine protease such as trypsin, is able to directly inhibit cellular protein synthesis. The nicked protein has two fragments (termed A and B) joined by a disulfide bridge. Fragment A is the enzyme that blocks protein synthesis, whereas fragment B is essential for delivery of fragment A to the cytoplasm of a target cell. It has been reported that the B-cell and T-cell responses to determinants on fragments A and B are separate and distinct from one another [46]. Several approaches to the development of antigen substitutes for whole formalin-inactivated toxin are being studied; these include synthetic peptides, cross-reactive materials (CRMs), and rDNA methodologies.

### A. Synthetic Peptides

Investigators have reported [47] the development of synthetic peptides corresponding to fragments of diphtheria toxin such as the 16-amino-acid loop [48,49], which corresponds to two functional segments of the native toxin. These synthetic fragments have been named STDP (based upon amino acid residues 188–201), SHDP (186–201), SODP-SH and SODP-ACM (186–201), and SDP (192–201) [50]. They have been studied as immunogens in their free state, as homopolymers, and coupled to carriers. It was found that the STDP, SHDP, SODP-SH, and SODP-ACM peptides, when coupled to natural or synthetic carriers and used with MDP in saline as adjuvant, could elicit neutralizing anti-DT antibodies [47,50]. More recently, four synthetic peptides based upon fragments of diphtheria toxin, *Streptococcus pyogenes* M protein, HBsAg, and a protein from a parasite were covalently bound within the same construct. When given to mice with saline and murabutide (a form of MDP), higher levels of antibodies against each component species were obtained than when each peptide species was used alone. This demonstrates that a totally synthetic vaccine administered in saline with a synthetic adjuvant can be immunogenic in the absence of a protein carrier [51]. These results are also significant in that they allow for the possibility of constructing multivalent synthetic vaccines that could simultaneously afford protection against numerous diseases.

### B. Cross-Reacting Materials

Cross-reacting materials (CRMs) are nontoxic or very weakly toxic proteins produced by mutant strains of *C. diphtheriae* that cross-react with native toxin. Corynebacteriophage beta carries the structural gene for diphtheria toxin [48,52], and phage mutants have been isolated that produce CRM 197 (class I), which is immunologically indistinguishable from native diphtheria toxin. CRM 45 (class II) is only partially related to native toxin, and class III mutants do not produce detectable *tox* gene products [47,53–55]. When treated with 0.7% formalin, CRM 197 elicits an antibody response in animal models similar to that observed with standard toxoid [56,57]. High yields of CRM 197 can be produced by constructing bacterial strains with multiple copies of the mutant gene responsible for CRM 197 production [58]. Presumably the major advantage of utilization of CRM-derived vaccines would be that the starting material is completely nontoxic and in its native form it competes with diphtheria toxin for receptors on sensitive cells. It would appear that CRM 197 provides a simpler and more practical source of immunizing antigen against diphtheria than native diphtheria toxin [43].

### C. Recombinant DNA Methodologies

Genetic engineering techniques have been used to clone portions of the *tox* gene from *C. diphtheriae* in *E. coli* that are either biologically active [59,60], for purposes of studying the structure–function relationships of the toxin, or inactive [61,62], for possible use as immunogens. The latter could provide a ready and inexpensive source of vaccine without the added steps necessary to inactivate native toxin.

A concept similar to that described above for synthetic peptide vaccine has been studied in which fusion proteins, including diphtheria CRM 228 and HBsAg, can be made by expression of hybrid genes from the PR promotor of bacteriophage lambda. These can be used together as immunogenic carriers of heterologous antigens [63]. Such application of insertion and expression of multiple genes could eventually lead to a genetically engineered organism able to produce a multivalent product that would constitute a vaccine against several diseases.

## III. PERTUSSIS TOXIN VACCINES

While there currently exists a vaccine that is effective in protecting against pertussis (whooping cough) [64,65], it has remained essentially unmodified for several decades [66]. Pertussis cases still occur (though relatively rarely) in countries where use of the vaccine is widespread. Probably of greater significance are the low frequency, potentially severe side effects of vaccine administration, which range from local swelling and redness to fever, encephalitis, and even rare collapse and death [67,68]. Courts have found manufacturers liable for damages to children given the vaccine, and health officials have suspended pertussis vaccination programs at one time or another in Japan, Sweden, and the United Kingdom. Even so, one encounters from time to time the argument that there exists no unequivocal scientific proof of the vaccine's role in causing these untoward side effects [66]. However, this is not unlike similar arguments about oncogenic viruses or other putative causes of cancer in humans. The kinds of experiments that would provide unequivocal proof of these causal relationships are not humanely or ethically possible. One thing is clear; when vaccination programs are suspended or public acceptance of the vaccine drops because of fear of side effects or liabilities, the results in terms of incidence of pertussis are dramatic [69,70]. Without a doubt, a new, more clearly defined, safe and efficacious pertussis vaccine is needed.

The toxin known as "pertussigen" or "pertussis toxin" has been reported to have different activities and properties that have been attributed to more than one protein by different authors [71]. Toxic factors associated with

*Bordetella pertussis* are heat-labile or dermonecrotizing toxin (HLT), lipopolysaccharide (LPS), and pertussigen, which is also called lymphocytosis-promoting factor (LPF). There is also a filamentous hemagglutinin (FHA) that is thought to have a role in the pathogenesis of pertussis as an attachment factor [72,73].

An acellular vaccine has been developed based upon LPF (pertussis toxin) and FHA [74–76]. In Japan this vaccine is manufactured by centrifugation of the supernatant from static cultures of *B. pertussis*. The pellet containing LPF and FHA is precipitated with 33% ammonium sulfate, resuspended, and centrifuged on a sucrose gradient to remove LPS (endotoxin). Toxicity, primarily due to LPF, is inactivated with formalin [69], and these components are then combined and adsorbed on aluminum hydroxide/aluminum phosphate along with diphtheria and tetanus components [74].

This acellular vaccine has been recommended exclusively for pertussis prophylaxis in Japan since 1981 and has been reported to be safe, less reactogenic than whole-cell vaccines, and to have an efficacy estimated at between 78 and 92% [69,74–76]. However, efficacy estimates, even for whole-cell vaccines that have been studied for years, have been reported to range from 4 to 97% [77] and 63 to 95% [69]. These discrepancies appear to be due in part to variables in diagnosis of disease, case definition, and other poorly defined parameters. Immunization with pertussis vaccine does not always prevent the disease, but may allow only mild symptoms of short duration. Therefore, the protection against disease imparted by pertussis vaccines may be greater than that against infection [78]. Clearly, subjective clinical decisions about which vaccinees have or have never gotten the disease will affect reported efficacy rates.

Results of a limited study in the United States have recently been reported in which a vaccine (Wyeth Laboratories) similar to that described above was compared, in 20 infants 6 months of age or younger, to conventional whole-cell pertussis vaccine [79]. The investigators concluded that the acellular vaccine was immunogenic and significantly less reactogenic in infants than was the currently used whole-cell vaccine. These results were comparable to those reported for a similar study done earlier with 60 infants aged 18–24 months old [80]. However, it has also been reported recently that there is considerable controversy over estimates of the efficacy of an acellular vaccine that has been tested in Japan and most recently in Sweden [77]. More comprehensive studies, time, and hopefully improved methodologies should provide clearer trends for this type of vaccine in the future.

### A. Synthetic Peptides

Pertussis toxin is composed of five different subunits, S1 through S5, with molecular weights ranging from 26,000 (S1) to 11,000 (S5) [81]. The toxin is structured in an A-B model with A (S1) being the ribosylating enzymatic moiety, while B (S2–S5) contains the target-cell binding activity [82]. Subunit S1 has been reported to be immunodominant [81,83]. Several polypeptide sequences that contain epitopes of the S1 subunit have been identified [83] by synthetizing peptides (15–22 amino acids) based upon DNA sequencing of the pertussis toxin genes [84,85] and sequencing the amino-terminal end of the toxin [86]. Five of these synthetic peptides react with hyperimmune sera from mice and humans immunized with either a whole-cell vaccine or with pertussis toxoid [87]. These results and further characterization of the epitopes involved could lead to the basis for developing a synthetic peptide pertussis vaccine that could be free of many of the problems and concerns presented above.

### B. Recombinant DNA Methodologies

Pertussis toxin has been shown to be the major protective antigen in aerosol and intracerebral challenge models [88]. Monoclonal antibodies against the S1 subunit have been shown to be protective in animal models [89]. The structural genes for all five subunits of pertussis toxin have been cloned and sequenced [84,85,90]. Expression has been achieved in *E. coli* of subunit S1 under control of the *lac* promoter and the translation signals of the *lac*Z gene [81]. Of three different constructs made, a truncated S1 subunit contains the predicted immunodominant region between amino acid residues 109 and 187 and lacks the enzymatically active site that confers toxicity on the A subunit (complete S1). If this subunit proves to contain the neutralizing epitope, it could be useful in developing a genetically engineered pertussis vaccine. Several problems that need to be overcome have been outlined with regard to low-yield expression of pertussis toxin components in the *E. coli* system [81]. Some improvement in expression levels utilizing pT7-7 (a T7 promotor plasmid) recently has been reported [91].

## IV. TETANUS TOXIN VACCINES

Tetanus has been known as a disease since the earliest days of medicine. The disease, the organism *(Clostridium tetani)* that produces the protein neurotoxin (tetanus toxin or tetanospasmin) that causes tetanus, and the history of tetanus vaccines have been reviewed elsewhere [92,93]. Surprisingly little is understood about the molecular mode of action of the toxin, but active (vaccine) and passive (antitoxin) immune therapy are essentially the only effective agents available to control the disease.

Existing tetanus vaccine is frequently administered as a component of the DTP vaccine described earlier, but it can be administered alone or combined with one or more of several other bacterial or viral vaccines [93]. It is prepared from tetanus toxin collected from culture filtrates. Tetanus toxin is a simple, highly toxic protein with a molecular weight close to 150,000 [94]. Prior to diffusion out of the bacterial cells it exists as a single polypeptide chain (''intracellular'' toxin). When the toxin is released to the medium through cellular lysis it is cleaved by endogenous proteases to form a nicked (''extracellular'') toxin with two chains held together by one or more disulfide bridges [95]. Nicked toxin, when reduced, separates into fragment A (light chain, $M_r$ 50,000) and fragment BC (heavy chain, $M_r$ 100,000) [96]. It is possible to further cleave the BC fragment with papain into subunits B and C [97].

It has been known for more than 50 years that prolonged formaldehyde treatment of the toxin yields a completely inactive and highly immunogenic product [94]. Today human tetanus vaccines usually consist of semipurified formalinized toxoids adsorbed most commonly on aluminum hydroxide, but also on aluminum phosphate [93].

There are technical difficulties in assessing the value of protective levels of tetanus toxin antibodies in humans. Clinical assessments of the vaccines are done indirectly and most conclusions are generally derived from the dramatic drop observed in tetanus cases in actively immunized populations [93]. The route of injection is also a source of variation in immune response and side effects. The deep intramuscular route is most commonly used and causes fewer side reactions than subcutaneous or intradermal injection. Attempts to develop an aerosol (respiratory route) vaccine [98,99] and an oral vaccine [100] have been reported. Recommended vaccine injection schedules also vary, although it seems that vaccine-induced immunity should last 10 years and even longer in older children and adults [93].

Some authors have expressed concern regarding booster vaccination of any individual who is at immediate risk of developing tetanus [101]. Development of a protective level of circulating antibodies in response to a booster shot can take as long as 4 days, while the disease can develop in as little as 24 hours, which means that passive immune therapy may be indicated in such cases. Mouse-human-human monoclonal antibodies against tetanus toxoid [102] have been developed that perhaps could some day obviate problems associated with injection of hyperimmune globulins.

Tetanus vaccine can cause untoward reactions ranging from mild to severe local or systemic reactions [103]. These include immediate, delayed-type, and Arthus-like reactions and, very rarely, neurological complications. Most of these reactions, however, have been correlated with preexisting high levels of circulating toxin antibodies [103] and possibly tetanus toxoid hypersensitivity [93,104]. Many reactions may be due to impurities in the toxoid preparations. The incidence of reactions to highly purified toxoid tends to be lower than that to partly purified material [93]. Recent reports have raised some questions about possible effects of the toxoid on chromosomes in sperm cells [105] and stimulation of rheumatoid factor (RF) [106], but the extent of these effects has not been studied sufficiently to allow firm conclusions at this time.

Because a relatively safe and effective immunogen has been available for quite some time in the form of the current tetanus toxoid vaccine, new approaches to vaccine development have not been pursued as actively as for other vaccines, like pertussis. Since 1980 the following developments toward an improved tetanus vaccine have been reported.

### A. Conventional Toxoid Vaccines

Glutaraldehyde has been used to inactivate the toxin [107] in only 10 min at room temperature. This toxoid appears to be as immunogenic in the liquid form as conventional adsorbed vaccine and has been used successfully to immunize large numbers of infants and children without noticeable adverse reactions [93].

### B. Subunit Vaccines

While it has been demonstrated that tetanus toxin subunits can be used successfully to immunize animals against tetanus [108], subunit tetanus vaccines present the problem of impractical purification schemes necessary to isolate the required components. A procedure using a column of immobilized gangliosides that may allow isolation of subchains with high yield for a subunit vaccine has been reported [93].

### C. Recombinant DNA Methodologies

DNA-encoding portions of fragments C and B of tetanus toxin have been cloned and expressed in *E. coli* [96]. Nontoxic protein fragments from both B and C and C alone have also been expressed in *E. coli;* these fragments have been shown to induce protective neutralizing antibodies against tetanus toxin challenge [109]. As more work is done to identify and locate protective epitopes on the toxin molecule [110,111], the possibilities will increase of developing a genetically engineered or synthetic peptide vaccine that would be free of the side effects seen when conventional vaccine is used.

## V. CHOLERA TOXIN VACCINES

Because it is perennially a major cause of illness and death throughout the world, cholera has been studied extensively. The disease is caused by *Vibrio cholerae,* which produces a protein enterotoxin ($M_r$ 84,000) called cholera toxin, or choleragen. The toxin contains two subunits. One is the exterior subunit, called choleragenoid, which actually is five (or six [112]) identical noncovalently associated moieties, each called a B subunit (i.e., there are five B subunits of $M_r$ 11,500 each). Subunit B is the immunodominant portion of the holotoxin and is responsible for its binding to host-cell membrane receptors containing the oligosaccharide of ganglioside Gm1 [18,113]. The other subunit is the interior subunit A, a single polypeptide chain ($M_r$ 28,000) that is further distinguished into fragments A1 and A2. Of these, A1 peptide ($M_r$ 21,000) is the biologically active portion of the holotoxin. A general overview of cholera disease and toxin, as well as a list of related reviews, has recently been published [114].

Parenteral cholera vaccines consisting of whole killed cells have been in use for nearly 100 years, but the efficacy of these vaccines, especially in young children (the most susceptible group), is not very good [115], and protection is generally not long lasting [116,117]. None of these vaccines is in general use at this time.

Cholera itself is a highly effective immunizing process (second bouts of cholera are rare) [18], therefore it should be possible to duplicate this immunity artificially and without unacceptable side effects [118]. Although cholera is noninvasive, the immune response to disease involves circulating IgM and IgG antibodies as well as secretory IgA antibodies against the LPS somatic antigen, the enterotoxin, colonization factors, flagella, membrane proteins, and other lesser-studied antigens [114]. It appears that only antibodies present in the gut lumen can protect against cholera, and it is believed that secretory IgA antibodies produced locally in the gut are assisted by the serum IgM "spill-over" in protecting against the disease [119].

The human host, presented with all the products of the cholera vibrio at the local level, undergoes maximal stimulation of secretory antibody. It appears at this point that immunity against cholera would best be conferred by a living attenuated strain of *V. cholerae* that would colonize the gut and stimulate immunity while not causing disease or untoward reactions [118]. A summary of recent approaches to development of a cholera vaccine based upon cholera toxin is presented below.

Vaccines against cholera based on toxoid alone have not proved to date to be very effective [116,117]. Formaldehyde-lysine–inactivated cholera toxin (toxoid) has proved to be antigenic but not immunogenic in humans [120]. Glutaraldehyde toxoid produces significant protection for up to 6 months, after which protection rates drop dramatically [121].

### A. Subunit Vaccines

Choleragenoid (composed of five B subunits) has been studied as a subunit vaccine against cholera. It has been found [114] to be immunogenic in rabbits, rats, mice, and dogs [122], though it is less effective than crude culture filtrates, purified cholera toxin, or procholeragenoid [123]. However, in heavy endemic areas peroral doses of choleragenoid have been reported to induce a secretory IgA antibody response [124].

Procholeragenoid is the stable, relatively nontoxic aggregate formed by controlled heating of purified choleragen (further heating results in the formation of choleragenoid) [116]. Procholeragenoid's residual toxicity can be reduced by formaldehyde treatment. In this form it appears to be at least as effective as cholera toxin in inducing antitoxic antibodies [125].

Contemporary research is focusing upon combining subunit B of cholera toxin with killed whole-cell vaccines that are reported to be without side effects [126]. An oral cholera vaccine (B + WCV) consisting of subunit B and whole killed cholera vibrios [127,128] has been developed. This vaccine has been found to stimulate mucosal IgA antibodies to cholera toxin and bacterial LPS in a manner similar to that seen in persons convalescing from cholera disease [127]. North American volunteers given this vaccine were significantly protected against clinically severe diarrhea after challenge with virulent *V. cholerae* 01 [129]. Another study has been reported in which this type of vaccine was studied over a 5 year period in Sweden (a cholera-free area) [126]. It was found that the immune response (IgA and IgG) to the B-subunit component is associated with long-lasting memory comparable to that evoked by natural immunization (via cholera disease) that was not dependent upon being sustained by inapparent, low-dose exposures to *V. cholerae,* as can be anticipated in endemic areas. Significant antitoxin levels were observed even 5 years after the initial vaccination with B + WCV.

A related study in Bangladesh (endemic for cholera) has demonstrated that B subunit + whole cell (BS-WC) vaccine affords greater protection than whole cell (WC) alone during the first 8 months after vaccination [130]. In persons over 5 years of age, high-grade sustained protection was observed with both WC and BS-WC vaccines, while protection was transient in younger recipients. At the same time, another recent report states that no significant difference between WC and BS-WC vaccination in an endemic cholera area was found with regard to reduction of diarrhea treatment or mortality, though there was a reduction in severe diarrhea by BS-WC over WC vaccine [131]. It has also been reported that immunization with BS-WC can result in a dramatic (67%) reduction in heat-labile enterotoxin-enterotoxigenic *E. coli* (LT-ETEC) diarrhea in an endemic area [132]. Previous exposure to *E. coli* diarrhea also can prepare the intestine for an anamnestic IgA antitoxin response to BS-WC vaccine [133]. Because cholera toxin is structurally, functionally, and immunologically related to enterotoxin from *E. coli,* salmonella, and several other enteric pathogens [134], research toward development of vaccines based upon common portions of these toxins may lead to a broad-spectrum vaccine that would simultaneously protect against multiple members of this enteric group.

### B. Synthetic Peptide Vaccines

Nine different synthetic peptides, which account in total for about 80% of the B subunit of cholera toxin, have been synthesized by solid-phase methodology [134,135]. These have been designated by the letters CTP (cholera toxin peptide) and a number, such as CTP3, CTP3a, CTP6, etc. The immunogenicity of these peptides has been studied by coupling them with tetanus toxoid as a carrier and injecting them into rabbits. One peptide, designated CTP3, was particularly effective in inducing antibodies that neutralize both cholera toxin and *E. coli* human (H-LT) and porcine (P-LT) heat-labile toxin [136]. For some peptides, an increase in length (number of amino acids) increased the immune response, but with CTP3 an increase in peptide length reduced its effectiveness [137].

It has been reported that these synthetic peptides can successfully prime the immune system to develop a significant immune response to a subsequent challenge with otherwise subimmunizing doses of cholera toxin or *E. coli* H-LT and P-LT [138]. This phenomenon is important to vaccine development for endemic areas where the population is constantly exposed to low levels of infectious organisms. These low levels are insufficient to induce an immune response under normal circumstances. However, if the immune system were to be primed, as by injection of carrier-coupled synthetic peptides, the low levels of exposure may repeatedly induce a protective secondary immune response [134].

### C. Recombinant DNA Methodologies

Recombinant DNA methodologies in cholera have primarily been used to develop attenuated vaccines [139,140]. A pathogenic strain of *V. cholerae* has been attenuated by deletion of DNA sequences encoding the A subunit of cholera toxin. A restriction endonuclease fragment encoding the A1 but not the A2 or B sequences was deleted in vitro from cloned cholera toxin genes. The mutation was then reinserted into *V. cholerae* strain Ogawa 395 [134]. This genetically engineered strain affords significant protection but also causes significant diarrheal problems in recipients [141]. However, a similarly constructed strain with a deleted A subunit may produce a more limited diarrheal response in vaccinees [142]. Genetically engineered strains deleted of the ability to produce any portions of the cholera holotoxin induce a diarrheal response in vaccine volunteers that is similar to the response induced by strains that produce only nontoxic portions of the cholera toxin [143]. While diarrhea is milder than that induced by cholera disease, it is apparent that the virulence of these organisms depends upon more than just cholera toxin production. It has been proposed that cholera-toxin–negative strains of *V. cholerae* may cause mild diarrheas as a consequence of the production of other secretogenic substances by the engineered strains [143], including hemolysin/cytotoxins, a Shiga-like toxin [144], and perhaps a previously unrecognized "new" variety of protein toxin similar to cholera toxin [145]. It is also possible that colonization of the gut, necessary for a protective immune response, in and of itself causes a diarrheal response that, if true, makes a single-dose nonreactogenic live oral cholera vaccine very difficult if not impossible to achieve [143]. All of these problems theoretically could be obviated by expressing immunogenic portions of the cholera toxin in a genetically engineered nonpathogenic normal resident of the intestine such as *E. coli*. A recent report describes the transposon metagenesis of *E. coli* strain N100 and the isolation of "cell surface leaky" mutants that do not trap expressed proteins internally [146]. When transformed with plasmids containing cloned cholera toxin genes, the *E. coli* N100:Tn*5* mutant synthesized and excreted the toxin. Such a strain has the potential to be developed into a live oral vaccine that would not carry with it some of the problems outlined above.

Recently, plasmids containing a synthetic gene coding for region 50 to 54 in the amino acid sequence of cholera toxin (corresponding to the synthetic peptide CTP 3 described above) were inserted in phase into the gene coding for *E. coli* β-galactosidase [147]. Immunization with the fusion protein produced did not elicit a significant titer of cholera toxin antibodies. But, when followed by a booster injection of a subimmunizing amount of cholera toxin or *E. coli* LT, a substantial level of anti-cholera-toxin neutralizing antibodies developed [134]. As work continues along these lines hopefully a combination of these methodologies will eventually lead to an effective and long-lasting vaccine, especially in young children who seem to be the most susceptible group in endemic areas.

## VI. *ESCHERICHIA COLI* TOXIN VACCINES

Diarrhea due to *Escherichia coli* is common in most, if not all, developing countries [18]. Enterotoxigenic *E. coli* (ETEC) elaborate one or more enterotoxins (coded for by extrachromosomal plasmid DNA) that cause typical disease symptoms. The heat-labile toxin (LT) is a protein structurally, functionally, and immunologically similar to cholera toxin (see above). The LT produced by human isolates (LTh) is distinct from LT produced by porcine isolates (LTp).

Heat-stable toxin (ST) on the other hand, is the name for a heterogeneous group of small proteins ranging in molecular weight from 1,000 or 2,000 to 10,000 [148]. A somewhat inconsistent nomenclature has been applied to the heat-stable toxins in the literature. One can encounter $ST_I$ and $ST_{II}$ [149], $ST_a$ (or $ST_A$) and $ST_b$ (or $ST_B$) [148], and $ST_h$ (or ST-H) and $ST_p$ (or ST-P)

[148,150]. Apparently $ST_a$, $ST_A$, and $ST_I$ are the same toxin, just as $ST_b$, $ST_B$, and $ST_{II}$ are different designations for a second heat-stable toxin from ETEC. $ST_h$ (18 amino acids) and $ST_p$ (19 amino acids) are human (h)- and porcine (p)-derived forms of $ST_a$ ($ST_A$, $ST_I$), respectively. The $ST_a$ forms cause gut fluid secretion via stimulation of guanylate cyclase activity in intestinal epithelial cells.

Usually both heat-labile and heat-stable toxin classes are associated with ETEC in humans, although either alone can cause disease [151,152]. Increases in serum and secretory IgA anti-LT antibody levels are seen in persons who have been infected with ETEC that produce LT, while no anti-ST binding or neutralizing antitoxin is produced in persons infected with ETEC that produce only ST [153,154]. *E. coli* LT and ST, ETEC disease, and vaccine developments have recently been reviewed in depth [148,154–156].

Because ETEC and *V. cholerae* are similar in some important respects and the major toxins produced by each (LT and CT) are very closely related, some of the problems and approaches to vaccine development are interrelated as well. One apparent difference is *E. coli*'s dependence upon fimbriae (pili) for attachment and colonization of the gut surface in order to cause disease (no adhesin has been identified with a clear role in *V. cholerae* attachment). While it has been demonstrated that isolated fimbriae alone can cause fluid accumulation in the gut of experimental animals, fimbriae are not usually considered collectively to be a toxin as such. Their importance to vaccine development is nevertheless central in that the most effective vaccines will probably stimulate both antitoxic and antibacterial surface antigen immunity. Vaccines based upon colonization factor antigens (CFA = fimbriae = pili) such as CFA/I, CFA/II, and type 1 somatic pili have been reviewed elsewhere [155,156].

### A. Subunit Vaccines

ST toxins are small peptides, therefore they also are poor immunogens [157]. Studies have been reported in which ST has been conjugated (with consequent attenuation of ST toxicity) to a carrier such as porcine IgG [158], LT [159], or the B subunit of LT [160,161] (also see VI.B and VI.C below). Such conjugates have proved to be immunogenic in animal models. Other reasonable candidates to which the ST could be coupled are procoligenoid, procholeragenoid, and CFA fimbriae.

Because of the similarities between *E. coli* LT and cholera toxin, immunity induced by CT or LT is cross-protective against challenge by the complementary organism. However, formaldehyde and glutaraldehyde LT toxoids have been studied as immunogens without much success [156]. When heated, LT holotoxin forms a high-molecular-weight toxoid (procoligenoid) that may be a useful component in a vaccine. Like the B subunit of CT, the B subunit of LT is immunodominant and can induce protective antibodies, though the holotoxin has been reported to be a superior antigen to the B subunit [153,156]. Vaccines for piglets, already commercially available, are based upon LT B-subunit and adhesion antigens. They normally are administered parenterally to the sow, who in turn protects her suckling offspring via colostrum antibodies [149,162]. The development of methodologies for successful large-scale production of B subunit would encourage its application to human vaccine development.

### B. Synthetic Peptides

A nontoxic, completely synthetic peptide vaccine that consists of the 18-amino-acid ST and a 26-amino-acid epitope of the LT B subunit has been produced [163]. When given orally to human volunteers, neutralizing antitoxin titers to both toxin components were raised fourfold in serum samples and sevenfold in jejunal aspirates over preimmunization control titers. No adverse reactions were noted in any of the 13 recipients of the vaccine.

The synthetic peptides CTP3 and CTP1 (described in section V), which are based upon portions of the B subunit of cholera toxin, also have been found to induce antibodies that cross-neutralize ETEC LT and in fact cross-react with LT from more different strains than do antibodies against whole cholera toxin [134,164].

### C. Recombinant DNA Methodologies

The genes for both LT and ST are found on three groups (based on size) of plasmids, designated ENT, and have been characterized and studied extensively [155]. Given that the genetics of *E. coli* have certainly been more thoroughly studied than those of any other organism, the possibilities for genetic manipulation to create attenuated strains for vaccine development are numerous. Aside from the details already presented in section V, the recent reports described below present possibilities for application of recombinant DNA methodologies to *E. coli* vaccine development.

In one report, critical amino acids in the ST toxin molecule were substituted by oligonucleotide-directed site-specific mutagenesis [150]. It was found that specific substitutions in the amino acid at position 11 could dramatically reduce toxicity of the ST while not hindering reactivity with monoclonal and polyclonal antibodies against native ST. In another report, DNA that codes for ST was fused with DNA for the LT B subunit [165] and a hybrid molecule was translated (see VI.A above). This information could be used in constructing mutants that produce ST toxoid or innately coupled ST and LT subunits that potentially would be immunogenic but not toxic.

Just as a "new" cholera toxin has been described (see above) [145], so has a "new" *E. coli* LT (designated LT-II) been described [166,167]. While the LT-II appears to have a similar polypeptide unit structure (though perhaps different primary structure) and mode of action as LT, it is not neutralized by antisera against LT. And the 0.8 kilobase LT-II DNA probe studied by the investigators does not hybridize with the cloned structural genes for LT. These newly discovered toxins may explain, in part, why strains of *V. cholerae* and *E. coli* attenuated for known toxic factors continue to cause diarrhea in vaccinees.

### D. Passive Immunization

An interesting study was recently reported [168] in which dairy cows were immunized with cholera toxin, *E. coli* LT, and glutaraldehyde-inactivated whole *E. coli* cells. Lyophilized milk immunoglobulins were prepared from the colostrum of the immunized cows, suspended in buffered water, and given orally to volunteers three times daily. After 3 days of this treatment all volunteers were protected from challenge with virulent ETEC. This type of passive prophylaxis could prove very effective for travelers to areas endemic for enteropathogenic *E. coli*.

## VII. PSEUDOMONAS TOXIN VACCINES

*Pseudomonas aeruginosa* is an opportunistic pathogen that only rarely causes serious illness in normal healthy individuals [169]. However, burn victims, those suffering from cystic fibrosis or granulocytopenia, and general-admission hospital patients frequently become infected with pseudomonas. Mortality rates for some infections can surpass 75%. Disease due to *P. aeruginosa* and vaccine research have been reviewed elsewhere [169,170].

While no less than eight extracellular and cellular virulence factors produced by *P. aeruginosa* have been identified [171], the pathogenesis of pseudomonas disease is not well understood. One of these factors, exotoxin A, is more toxic than any of the others (at least 20,000 times more toxic than *P. aeruginosa* endotoxin, for example) [172,173]. In its native form exotoxin A ($M_r$ 71,500) is a proenzyme consisting of an enzymatically active fragment A ($M_r$ 26,000) and an enzymatically inactive fragment B ($M_r$ 45,000), which is the binding fragment [174,175]. The toxin (an ADP ribosyl transferase similar in activity to diphtheria toxin) acts to inhibit protein synthesis by interfering with polypeptide translocation on messenger RNA [176].

### A. Subunit Vaccines

Circulating antibodies against toxin A have been positively correlated with survival in bacteremic patients [177,178] and play a significant role in protecting against pseudomonas infections [175]. Although exotoxin A–deficient mutants can be far less virulent in certain animal models than wild-type strains [179,180], absence of the capability to produce toxin A does not usually alter a strain's ability to initiate an infection [181].

Formalin and formalin-lysine inactivation of toxin A for vaccine development have met with only limited success [182], as has glutaraldehyde toxoid adsorbed to aluminum phosphate [169]. A formalinized toxoid with MDP as adjuvant has been reported to afford moderate protection against *P. aeruginosa* challenge in a burned mouse model [183].

Investigators have isolated a mutant strain of *P. aeruginosa* that produces an immunologically cross-reactive material (CRM), designated CRM 66, that is nontoxic because of a mutation within the structural gene for toxin A that adversely affects its enzymatic activity [184]. CRM 66 is antigenically indistinguishable from native toxin A and could have valuable potential for vaccine development. However, it would first be necessary to develop methodologies for large-scale production of the CRM, which have not been reported to date.

Vaccines whose dominant antigen is LPS (Pseudogen: Parke-Davis, Warner-Lambert; PEV: Wellcome Biotechnology) have proved effective in protecting burn patients from pseudomonas infection [185,186], though they can cause significant adverse reactions upon administration [187]. A new vaccine has been developed that incorporates a nontoxic portion of LPS (O-polysaccharide; O-PS) conjugated to toxin A through reductive amination [188]. The resultant conjugate has been shown to be nontoxic and nonpyrogenic but is able to evoke an immune response that recognizes critical determinants of LPS and toxin A. In a study with humans, volunteers were immunized at days 0 and 28 by subcutaneous injection of a conjugate ($M_r$ $2 \times 10^6$) containing approximately 70% toxin A and 30% O-PS [187]. No systematic reactions, but infrequent (20%) mild reactions such as tenderness at the site of injection, were noted. Ninety percent of vaccinees responded with a fourfold or higher rise in titer of IgG antibodies against both LPS and toxin A. Further analysis of the immune response to this vaccine has shown that the anti-LPS IgG response was predominantly within the IgG1 and IgG2 subclasses while the antitoxin A response was within the IgG1 subclass [189]. Maximal titers to both antigens were seen at 42 days after immunization. Antitoxin A neutralizing and opsonic antibodies remained elevated over the 14 month period of the study. A follow-up study reported recently indicates that after 15 months both anti-LPS and antitoxin A titers had markedly declined [190]. However, a booster shot of the same vaccine at 15 months evoked a vigorous antitoxin A IgG response with 100% of vaccinees having a fourfold or greater rise in neutralizing antibody compared to preimmunization levels. In contrast, the booster had no effect on anti-LPS IgG antibodies. After 24 months only antitoxin A titers were significantly higher than preimmunization levels. Clearly there is room for improvement in this vaccine, and the need is evident for application of recombinant DNA techniques such as described above for other organisms to enable the simultaneous expression of nontoxic but immunogenic components that contribute to the virulence and immunogenicity in pseudomonas infections.

## VIII. OTHER TOXIN VACCINES

### A. Anthrax Toxin

Human anthrax cases in the United States are rare enough (1 case reported since 1984) to warrant news status when one occurs [191]. However, interest in anthrax vaccine continues for several reasons: 1) an outbreak, however rare, often mandates that others who may have been exposed to the contamination source be immunized to prevent them from contracting the disease; 2) vaccination is important in underdeveloped countries where the incidence of anthrax is higher than in the West; 3) vaccination is desirable for handlers of potentially contaminated animals or animal products (as well as obvious veterinary applications); and 4) the theoretical possibility unfortunately persists that *Bacillus anthracis* could some day be applied to military strategies (germ warfare) [192,193]. *B. anthracis,* its complex toxin composed of edema factor (EF), lethal factor (LF), and protective antigen (PA), and anthrax disease have recently been reviewed elsewhere [194].

Studies have indicated that purified PA [195], LF, and EF [196,197] could serve as protective antigens in a vaccine. Protective antigen (PA) is the major component of the commercial anthrax vaccines currently produced in the United States and Britain [198–200]. Work continues [201] to further characterize this antigen in terms of epitopes and structure–function relationships that could enable improved vaccine formulation. Studies on the genetics of *B. anthracis* indicate that toxin production [202] and encapsulation [203] appear to be plasmid mediated. Restriction fragments of *B. anthracis* DNA have been cloned into an *E. coli* host vector system that produces functional PA [203]. As these studies continue it should be possible to develop a strain that produces immunogenic antigen for incorporation into a safe, efficacious vaccine.

### B. *Staphylococcus aureus* Toxin

The role of toxic shock syndrome toxin-1 (TSST-1) in the pathogenesis of toxic shock syndrome (often associated with the widespread use of tampons by menstruating women) is considered substantial [204]. Greater than 90% of menstrually associated isolates of *S. aureus* produce the toxin [205]. In an experimental rabbit model, immunization with purified TSST-1 protected against toxic shock syndrome symptoms and mortality [206]. The *tst* (TSST-

1) gene has been cloned and incorporated into a plasmid in *S. aureus* strain RN4710 [207]. This work is a beginning that could lead to a vaccine that would offer immune protection from toxic shock syndrome in high-risk individuals.

### C. Group A Streptococci (Protein M)

While not considered to be a toxin per se, protein M is a major virulence factor of group A streptococci that is being used in the development of vaccines against streptococcal disease. The history of vaccines against streptococcus has been reviewed [208], as has the work done to develop a vaccine based on synthetic peptides [209] and conjugates of capsular polysaccharides to toxins such as tetanus and diphtheria toxin [210]. A recent study reported success in covalently linking a synthetic peptide based upon protein M to cholera toxin B subunit as a mucosal adjuvant [211]. It is interesting to add that other recent reports describe a similar utilization of cholera toxin [212] and *E. coli* LT [213] as adjuvants to stimulate mucosal immunity to unrelated protein antigens.

## REFERENCES

1. Sun M: The vexing problems of vaccine compensation. Science 227:1012, 1985.
2. Alouf JE: From "diphtheritic" poison to molecular toxinology. ASM News 53:547, 1987.
3. Dorner F, Drews J: "Pharmacology of Bacterial Toxins." New York: Pergamon Press, 1986.
4. Alouf JE, Fehrenbach FJ, Freer JH, Jeljaszewicz J: "Bacterial Protein Toxins." New York: Academic Press, 1984.
5. Harris JB: "Natural Toxins: Animal, Plant, and Microbial." Oxford: Clarendon Press, 1986.
6. Gill DM: Bacterial toxins: Description. In Laskin AI, Lechevalier HA (eds): "CRC Handbook of Microbiology, Vol. VIII, Toxins and Enzymes." Boca Raton, FL: CRC Press, 1987, p. 3.
7. McDonel JL, Dorner F, Drews J: The role of toxins in bacterial pathogenesis. In Dorner F, Drews J (eds): "Pharmacology of Bacterial Toxins." New York: Pergamon Press, 1986, p 1.
8. Germanier R: "Bacterial Vaccines." Orlando, FL: Academic Press, 1984, p xi.
9. White WG, Barnes GM, Barker E, Gall D, Knight P, Griffith AN, Morris-Owen RM, Smith JWG: Reactions to tetanus toxoid. J Hyg (Camb) 71:283, 1973.
10. Schwick HG, Helging TB, Zwisler O: Bacterial split vaccines. In Mizrahi A, Hertman I, Klingberg MA, Kohn A (eds): "New Developments With Human and Veterinary Vaccines." New York: Alan R. Liss, 1980, p. 143.
11. Northrup RS, Chisari FV: Response of monkeys to immunization with cholera toxoid. J Infect Dis 125:471, 1972.
12. Shinnick TM, Sutcliffe JG, Green N, Lerner RA: Synthetic peptide immunogens as vaccines. Annu Rev Microbiol 37:425, 1983.

13. Fridkin M, Buchta R: Chemistry of immunogenic peptide synthesis. In Arnon R (ed): "Synthetic Vaccines." Boca Raton, FL: CRC Press, 1987, Vol. I, p 7.
14. Meienhofer J, Gross E (eds): "The Peptides." New York: Academic Press, 1979–1985, Vols. 1–7.
15. Bodanszky M, Bodanszky A: "The Practice of Peptide Synthesis." Berlin: Springer-Verlag, 1984.
16. Barany G, Merrifield RB: Solid-phase peptide synthesis. In Gross E, Meienhofer J (eds): "The Peptides." New York: Academic Press, 1979, Vol. 2, p 1.
17. Steward MW, Howard CR: Synthetic peptides: A next generation of vaccines? Immunol Today 8:51, 1987.
18. Dorner F, McDonel JL: Bacterial toxin vaccines. Vaccine 3:94, 1985.
19. Chedid L: Use of adjuvants for synthetic vaccines. In Arnon R (ed): "Synthetic Vaccines." Boca Raton, FL: CRC Press, 1987, Vol. I, p. 93.
20. Audibert F, Jolivet M, Chedid L, Arnon R, Sela M: Successful immunization with a totally synthetic diphtheria vaccine. Proc Natl Acad Sci USA 79:5042, 1982.
21. Ada GL: Discussion—Peptide-based vaccines. In Bell R, Torrigiani G (eds): "Progress Towards Better Vaccines." Oxford: Oxford University Press, WHO, 1986, p. 217.
22. Sela M: The choice of carrier. In Arnon R (ed): "Synthetic Vaccines." Boca Raton, FL: CRC Press, 1987, Vol. I, p 83.
23. Morein B, Simons K: Subunit vaccines against enveloped viruses: Virosomes, micelles and other protein complexes. Vaccine 3:83, 1985.
24. Morein B: Membrane proteins as micelles or iscoms in vaccination experiments. In Bell R, Torrigiani G (eds): "Progress Towards Better Vaccines." Oxford: Oxford University Press, WHO, 1986, p 119.
25. Morein B, Sundquist B, Hoglund S, Dalsgaard K, Osterhaus A: Iscom, a novel structure for antigenic presentation of membrane proteins from enveloped viruses. Nature 308: 457, 1984.
26. Arnon R (ed): "Synthetic Vaccines." Boca Raton, FL: CRC Press, 1987, Vols. I, II.
27. Devany MA: New approaches to animal vaccines utilizing genetic engineering. Crit Rev Microbiol 15:269, 1988.
28. Locht C, Marchitto KS, Keith JM: Expression of recombinant pertussis toxin S1 subunit: An important protective antigen in whooping cough vaccines. In Chanock RM, Lerner RA, Brown R, Ginsberg H (eds): "Vaccines 87—Modern Approaches to New Vaccines." Cold Spring Harbor, NY: Cold Spring Harbor Laboratory, 1987, p 273.
29. Reiser J, Germanier R: Expression in *Escherichia coli* of a *Bordetella pertussis* toxin-coding region. In Brown F, Chanock RM, Lerner RA (eds): "Vaccines 86—New Approaches to Immunization." Cold Spring Harbor, NY: Cold Spring Harbor Laboratory, 1986, p 235.
30. Jilg W, Schmidt M, Zoulek G, Lorbeer B, Wilkse B, Deinhardt F: Clinical evaluation of a recombinant hepatitis B vaccine. Lancet ii:1174, 1984.
31. Panicalli D, Paoletti E: Construction of pox viruses as cloning vectors: Inserting of the thymidine kinase gene from herpes simplex virus into the DNA of infectious vaccinia virus. Proc Natl Acad Sci USA 79:4927, 1982.
32. Paoletti E, Weinberg RL, Davis SW, Davis M: Genetically engineered poxviruses: A novel approach to the construction of live vaccines. Vaccine 2:204, 1984.
33. Sacks DL: Anti-idiotypic vaccines. In Arnon R (ed): "Synthetic Vaccines." Boca Raton, FL: CRC Press, 1987, Vol. I, p 105.
34. Lambert PH: Anti-idiotypic vaccines. In Bell R, Torrigiani G (eds): "Progress Towards Better Vaccines." Oxford: Oxford University Press, WHO, 1986, p 89.

35. McNamara MK, Ward RE, Kohler H: Monoclonal idiotype vaccine against *Streptococcus pneumonia* infection. Science 226:1325, 1984.
36. Stein K, Soderstrom T: Neonatal administration of idiotype or anti-idiotype primes for protection against *E. coli* K13 infection in mice. J Exp Med 160:1001, 1984.
37. Simonsen O, Kjeldsen K, Vendborg HA, Heron I: Revaccination of adults against diphtheria. I: Responses and reactions to different doses of diphtheria toxoid in 30–70-year-old persons with low serum antitoxin levels. Acta Pathol Microbiol Immunol Scand 94:213, 1986.
38. Mortimer J, Melville-Smith, Sheffield F: Diphtheria vaccine for adults. Lancet 22:1182, 1986.
39. Simonsen O, Klaerke M, Klaerke A, Bloch AV, Hansen N, Hald N, Haw C, Heron I: Revaccination of adults against diphtheria. II: Combined diphtheria and tetanus revaccination with different doses of diphtheria toxoid 20 years after primary vaccination. Acta Pathol Microbiol Scand [C] 94:219, 1986.
40. Bjorkholm D, Grandstrom M, Wahl M, Hedstrom C-E, Hagberg L: Adverse reactions and immunogenicity in adults to regular and increased dosage of diphtheria vaccine. Eur J Clin Microbiol 6:637, 1987.
41. Anonymous: Vaccine injury. In: ''Vaccine Supply and Innovation.'' Washington, DC: National Academy Press, 1985, p 65.
42. Baraff LF, Shields WD, Beckwith L, Strome G, Marcy SM, Cherry JD, Manclark CR: Infants and children with convulsions and hyporesponsive episodes following diphtheria-tetanus-pertussis immunization: Follow-up evaluation. Pediatrics 789, 1988.
43. Pappenheimer AM Jr: Diphtheria. In Germanier R (ed): ''Bacterial Vaccines.'' Orlando, FL: Academic Press, 1984, p 1.
44. Smith WP, Tai PC, Murphy JR, Davis BD: Precursor in cotranslational secretion of diphtheria toxin. J Bacteriol 141:184, 1980.
45. Murphy JR: Diphtheria toxin, cross-reacting nontoxic proteins and the development of new biologicals. In Weinstein L, Fields BN (eds): ''Seminars of Infectious Diseases, Vol. 4: Bacterial Vaccines.'' New York: Thieme-Stratton, 1982, p 81.
46. Triebel F, Autran B, DeRoquefeuil S, Falmagne P, Debre P: Immune response to diphtheria toxin and to different cyanogen bromine fragments: Evidence for different B and T cell reactivities. Eur J Immunol 16:47, 1986.
47. Audibert F, Jolivet M, Chedid L, Arnon R, Sela M: Successful immunization with a totally synthetic diphtheria vaccine. Proc Natl Acad Sci USA 79:5042, 1982.
48. Pappenheimer AM Jr: Diphtheria toxin. Annu Rev Biochem 45:69, 1977.
49. Colier RJ: Diphtheria toxin: Mode of action and structure. Bacteriol Rev 39:54, 1975.
50. Audibert F: Utilization of diphtheria peptides for the study of adjuvants and of optimized presentation of protective epitopes in synthetic vaccines. In Arnon R (ed): ''Synthetic Vaccines.'' Boca Raton, FL: CRC Press, 1987, Vol. II, p 1.
51. Jolivet ME, Francoise M, Audibert H, Tartar AL, Schlesinger DH, Wirtz R, Chedid LA: Induction of biologically active antibodies by a polyvalent synthetic vaccine constructed without carrier. Infect Immun 55:1498, 1987.
52. Uchida T, Gimm DM, Pappenheimer AM Jr: Mutation in the structural gene for diphtheria toxin carried by temperate phage β. Nature 233:8, 1971.
53. Vernon WB: The production, purification, and properties of a mutant diphtherial toxin. Thesis, Harvard College, Cambridge, MA, 1976.
54. Laird W, Groman NB: Isolation and characterization of *tox* mutants of corynebacteriophage *beta*. J Virol 19:220, 1976.
55. Holmes RK: Characterization and genetic mapping on nontoxinogenic *(tox)* mutants of corynebacteriophage *beta*. J Virol 19:195, 1976.

56. Pappenheimer AM Jr, Uchida T, Harper AA: An immunological study of the diphtheria toxin molecule. Immunochemistry 9:891, 1972.
57. Porro M, Saletti M, Nencioni L, Tagliaferri L, Marsili I: Immunogenic correlation between cross-reacting material (CRM 197) produced by a mutant of *C. diphtheriae* and diphtheria toxoid. J Infect Dis 142:716, 1980.
58. Rappouli R, Michel JL, Murphy FR: Integration of corynebacteriophages $\beta^{tox+}$, $\alpha^{tox+}$ and $\lambda^{tox-}$ into two attachment sites on the *C. diphtheriae* chromosome. J Bacteriol 153: 1210, 1983.
59. Bishal WR, Rapprioli R, Murphy JR: High-level expression of a proteolytically sensitive diphtheria toxin fragment in *Escherichia coli.* J Bacteriol 169:5140, 1987.
60. Yakubovich NC, Adanovskii AG, Yanulaitis AA, Ynakovskii NK, Dababov VG: Construction of diphtheria toxin gene derivatives and their expression in *Escherichia coli* cells. Mol Genet Mikrobiol Virusol O(12):11, 1987.
61. Barvieri JT, Collier RJ: Expression of a mutant, full-length form of diphtheria toxin in *Escherichia coli.* Infect Immun 55:1647, 1987.
62. Yakubovich NC, Novikova IS, Bychkova MA, Mazurova IK, Unakovskii NK: Cloning the structural gene for diphtheria toxin from *Corynebacterium diphtheriae* PW8 and expression of its deleted derivative in *Escherichia coli.* Mol Genet Mikrobiol Virusol O(11):13, 1987.
63. Phalipon A, Kaczorek M: Genetically engineered diphtheria toxin fusion proteins carrying the hepatitis B surface antigen. Gene 55:255, 1987.
64. Hinman AR, Koplan JP: Pertussis and pertussis vaccine. Reanalysis of benefits, risks, and costs. JAMA 251:3109, 1984.
65. Koplan JP, Schoenbaum SC, Weinstein MC, Fraser DW: Pertussis vaccine—Analysis of benefits, risks and costs. N Engl J Med 301:906, 1979.
66. Fox JL: Contemplating a new pertussis vaccine. ASM News 53:135, 1987.
67. Miller DL, Alderslade R, Ross EM: Whooping cough and whooping cough vaccine: The risks and benefits debate. Epidemiol Rev 4:1, 1982.
68. Cody CL, Baraff LF, Cherry JD, Marcy SM, Manclark CR: Nature and rates of adverse reactions associated with DTP and DT immunizations in infants and children. Pediatrics 68:650, 1981.
69. Manclark CR, Cowell JL: Pertussis. In Germanier R (ed): ''Bacterial Vaccines.'' Orlando, FL: Academic Press, 1984, p 69.
70. Kanai K: Japan's experience in pertussis epidemiology and vaccination in the past 30 years. J Med Sci Biol 33:107, 1980.
71. Wardlaw AC, Parton R: *Bordetella pertussis* toxins. In Dorner F, Drews J (eds): ''Pharmacology of Bacterial Toxins.'' Oxford: Pergamon Press, 1986, p. 327.
72. Toumanen E, Weiss A: Characterization of two adhesins of *Bordetella pertussis* for human ciliated respiratory-epithelial cells. J Infect Dis 152:118, 1985.
73. Toumanen EI, Hendley JO: Adherence of *Bordetella pertussis* to human respiratory epithelial cells. J Infect Dis 148:125, 1983.
74. Sato Y, Kimura M, Fukumi H: Development of a pertussis component vaccine in Japan. Lancet i:122, 1984.
75. Aoyama T, Murase Y, Kato T, Iwata T: Efficacy of an acellular pertussis vaccine. Jpn J Pediatr 107:180, 1985.
76. Noble GR, Bernier RH, Esber EC, Hardegree MC, Hinman AR, Klein D, Saah AJ: Acellular and whole-cell pertussis vaccine in Japan. Report of a visit by US scientists. JAMA 257:1351, 1987.
77. Anonymous: Swedish trial leaves pertussis vaccine issue unsettled. ASM News 54:164, 1988.

78. Fine PEM, Clarkson JA: Reflections on the efficacy of pertussis vaccines. Rev Infect Dis 9:866, 1987.
79. Anderson EL, Belshe RB, Bartram J: Differences in reactogenicity and antigenicity of acellular and standard pertussis vaccines combined with diphtheria and tetanus in infants. J Infect Dis 157:731, 1988.
80. Lewis K, Cherry JD, Holroyd HJ, Baker LR, Dudenhoeffer FE, Robinson RG: A double-blind study comparing an acellular pertussis-component DTP vaccine with a whole-cell pertussis-component DTP vaccine in 18-month-old children. Am J Dis Child 140:872, 1986.
81. Locht C, Marchitto KS, Keith JM: Expression of recombinant pertussis toxin S1 subunit: An important protective antigen in whooping cough vaccines. In Chanock RM, Lerner R, Brown F, Ginsberg H (eds): "Vaccines 87—Modern Approaches to New Vaccines." Cold Spring Harbor, NY: Cold Spring Harbor Laboratory, 1987, p 273.
82. Tamura M, Nogimori K, Murai S, Yajima M, Ito K, Katada UM, Ishii S: Subunit structure of islet-activating protein, pertussis toxin, in conformity with the A-B model. Biochemistry 21:5516, 1982.
83. Askelof P, Rodmalm K, Abens J, Unden A, Bartfai T: Use of synthetic peptides to map antigenic sites of *Bordetella pertussis* toxin subunit S1. J Infect Dis 157:738, 1988.
84. Locht C, Keith JM: Pertussis toxin gene: Nucleotide sequence and genetic organization. Science 232:1258, 1986.
85. Nicosia A, Perugini M, Franzini C, Casagli MC, Borri MG, Antoni G, Almoni M, Neri P, Ratti G, Rappuoli R: Cloning and sequencing of the pertussis toxin genes: Operon structure and gene duplication. Proc Natl Acad Sci USA 83:4631, 1986.
86. Engstrom O, Rodmalm K, Jornvall H, Lundquist G, Kalman M, Simoncsits A, Bartfai T, Lofdahl S, Askelof P: Characterization of the N-terminal structure of pertussis toxin subunit S1 and hybridization of oligodeoxyribonucleotide probes with a *Bordetella pertussis* DNA fragment. FEMS Microbiol Lett 36:219, 1986.
87. Hedenskog S, Granstrom M, Olin P, Tiru M, Sato Y: A clinical trial of a monocomponent pertussis toxin vaccine. Am J Dis Child 141:844, 1987.
88. Sato H, Sato Y: *Bordetella pertussis* infection in mice: Correlation of specific antibodies against two antigens, pertussis toxin, and filamentous hemagglutinin with mouse protectivity in an intracerebral or aerosol challenge system. Infect Immun 46:415, 1984.
89. Sato H, Ito A, Chiba J, Sato Y: Monoclonal antibody against pertussis toxin: Effect on toxin activity and pertussis infections. Infect Immun 46:422, 1984.
90. Locht C, Barstad PA, Coligan JE, Mayer L, Munoz JJ, Smith S, Keith JM: Molecular cloning of pertussis toxin genes. Nucleic Acids Res 14:3251, 1986.
91. Brown DR, Lobet Y, Keith JM: Expression of the S1 subunit of pertussis toxin in Escherichia coli, using various vectors (abstract B-45). In: "Abstracts of the Annual Meeting of the American Society for Microbiology." Washington, DC: American Society for Microbiology, 1988, p 37.
92. van Heynigen S: Tetanus toxin. In Dorner F, Drews J (eds): "Pharmacology of Bacterial Toxins." Oxford: Pergamon Press, 1986, p 549.
93. Bizzini B: Tetanus. In Germanier R (ed): "Bacterial Vaccines." Orlando, FL: Academic Press, 1984, p 37.
94. Holmes MJ, Ryan WL: Amino acid analysis and molecular weight determination of tetanus toxin. Infect Immun 3:133, 1971.
95. Matsuda M, Yoneda M: Isolation and purification of two antigenically active, "complementary" polypeptide fragments of tetanus neurotoxin. Infect Immun 12:1147, 1975.
96. Fairweather NF, Lyness VA, Pickard DJ, Allen G, Thomson RO: Cloning, nucleotide

sequencing, and expression of tetanus toxin fragment C in *Escherichia coli*. J Bacteriol 165:21, 1986.
97. Helting TB, Zwisler O: Structure of tetanus toxin. I. Breakdown of the toxin molecule and discrimination between polypeptide fragments. J Biol Chem 252:187, 1977.
98. Fontanges R, Fournier JM, Monteil H, Darmenton P, Garrard A: L'immunisation par aerosols. VII. Vaccination des singes abouins par l'anatoxine tetanique lyophilisee et par le melange a. tetanique a. diphtherique. Ann Inst Pasteur (Paris) 120:831, 1971.
99. Wigley FM, Wood SH, Waldman RH: Aerosol immunization of humans with tetanus toxoid. J Immunol 103:1096, 1969.
100. Baljer G: Oral immunization against tetanus. Dev Biol Stand 33:63, 1976.
101. Simonsen O, Klaerke M, Jensen JEB, Kjeldsen K, Hau C, Heron L: Revaccination against tetanus 17 to 20 years after primary vaccination: Kinetics of antibody response. J Trauma 27:1358, 1987.
102. Kazuaki K, Iwamoto K, Shintani Y, Akiyama S-I: Effective production of a human monoclonal antibody against tetanus toxoid by selection of high productivity clones of a heterohybridoma. J Immunol Methods 109:9, 1988.
103. Relyveld EH, Henocq E, Bizzini B: Studies on untoward reactions to diphtheria and tetanus toxoids. Dev Biol Stand 43:33, 1979.
104. Bizzini B, Turpin A, Henocq E, Maro B: Tetanus toxin. In Rosenberg P (ed): "Toxins: Animal, Plant and Microbial." Oxford: Pergamon Press, 1978, p 971.
105. Das RK, Noyak R: Spermatocyte chromosome alterations in mice inoculated with antirabies vaccine and tetanus toxoid separately and in combination. Mutat Res 204:277, 1988.
106. Levinson A, Tar L: *In vitro* IgM rheumatoid factor production induced by tetanus toxoid. J Allergy Clin Immunol 81:730, 1988.
107. Relyveld EH, Labusquiere R, Gateff C, La Bourthe F, Ravisse P, Lemarinier G, Chambon L: Antitetanus vaccination and neonatal protection in developing countries. Prog Immunobiol Stand 5:517, 1972.
108. Helting TB, Nau HH: Analysis of the immune response to papain digestion products of tetanus toxin. Acta Pathol Microbiol Scand [C] 92:59, 1984.
109. Fairweather NF, Lyness VA, Maskell DJ: Immunization of mice against tetanus with fragments of tetanus toxin synthesized in *Escherichia coli*. Infect Immun 55:2541, 1987.
110. Kenimer JG, Habig WH, Hardegree MC: Monoclonal antibodies as probes of tetanus toxin structure and function. Infect Immun 42:942, 1983.
111. Volk WA, Bizzini B, Snyder RM, Bernhard E, Wagner RR: Neutralization of tetanus toxin by distinct monoclonal antibodies binding to multiple epitopes on the toxin molecule. Infect Immun 45:604, 1984.
112. Ludwig DS, Ribi HO, Schoolnik GK, Kornberg RD: Two-dimensional crystals of cholera toxin B subunit-receptor complexes: Projected structure at 17A resolution. Proc Natl Acad Sci USA 83:8586, 1986.
113. Fernandes PB, Welsh KM, Bayer ME: Characterization of membrane-bound nicotinamide adenine dinucleotide glycohydrolase of *Vibrio cholerae*. J Biol Chem 254:9254, 1979.
114. Finkelstein RA, Dorner F: Cholera enterotoxin (Choleragen). In Dorner F, Frews J (eds): "Pharmacology of Bacterial Toxins." Oxford: Pergamon Press, 1986, p 161.
115. Mosley WH: Vaccines and somatic antigens. The role of immunity in cholera. A review of epidemiological and serological studies. Tex Rep Biol Med 27:227, 1969.
116. Finkelstein RA: Cholera. In Germanier R (ed): "Bacterial Vaccines." New York: Academic Press, 1984, p 107.

117. Levine MM, Kaper JB, Black RE, Clements ML: New knowledge on the pathogenesis of bacterial enteric infections as applied to vaccine development. Microbiol Rev 47:510, 1983.
118. Anonymous: The prospects for immunizing against *Vibrio cholerae*. In: "New Vaccine Development, Vol. II, Diseases of Importance in Developing Countries." Washington, DC: National Academy Press, 1986, p 376.
119. Sack D, Islam A, Holmgren J, Svennerholm AM: Development of methods for determining the intestinal immune response to *V. cholerae* in humans. In: "15th Joint Conference on Cholera. The US–Japan Cooperative Medical Science Program, Cholera Panel." Washington, DC: NIH Publication No. 80-2003, 1980, p 423.
120. Ohtomo N: Safety and potency tests of cholera toxoid lot II in animals and volunteers. In Fukumi H, Zinnaka Y (eds): "Proceedings of the 12th Joint Conference US–Japan Cooperative Medical Science Program Symposium on Cholera, Sapporo." Tokyo: National Institute of Health, 1977, p 286.
121. Curlin G, Levine R, Axix KMA, Rahman ASMM, Verwey WF: Field trial of cholera toxoid. In: "Proceedings of the 11th Joint Conference on Cholera, US–Japan Cooperative Medical Science Program, New Orleans." Washington, DC: National Institutes of Health, 1976, p 314.
122. Pierce NF, Cray WC Jr, Sacci JB Jr, Craig JP, Germanier R, Furer E: Procholeragenoid: A safe and effective antigen for oral immunization against experimental cholera. Infect Immun 40:112, 1983.
123. Holmgren J, Lonnroth I, Svennerholm L: Fixation and inactivation of cholera toxin by Gml ganglioside. Scand J Infect Dis 5:77, 1973.
124. Svennerholm AM, Holmgren J: Synergistic protective effect in rabbits of immunization with *Vibrio cholerae* lipopolysaccharide and toxin/toxoid. Infect Immun 13:735, 1976.
125. Germanier R, Furer E, Varallyay S, Inderbitzen TM: Preparation of a purified antigenic cholera toxoid. Infect Immun 13:1692, 1976.
126. Jertborn M, Svennerholm AM, Holmgren J: Five-year immunologic memory in Swedish volunteers after oral cholera vaccination. J Infect Dis 157:374, 1988.
127. Svennerholm AM, Jertborn M, Gothefors L, Karim AMMM, Sack DA, Holmgren J: Mucosal antitoxic and antibacterial immunity after cholera disease and after immunization with a combined B subunit-whole cell vaccine. J Infect Dis 149:884, 1984.
128. Clemens JD, Sack DA, Harris JR, Chakraborty J, Khan MR, Stanton BF, Kay BA, Khan MU, Yunus MD: Field trial of oral cholera vaccines in Bangladesh. Lancet ii:124, 1986.
129. Black RE, Levine MM, Clements ML, Young CR, Svennerholm AM, Holmgren J: Protective efficacy in humans of killed whole-Vibrio oral cholera vaccine with and without the B subunit of cholera toxin. Infect Immun 55:1116, 1987.
130. Clemens JD, Harris JR, Sack DA, Chakraborty J, Ahmed F, Stanton FB, Khan MU, Kay BA, Huda N: Field trial of oral cholera vaccines in Bangladesh: Results of one year of follow-up. J Infect Dis 158:60, 1988.
131. Clemens JD, Sack DL, Harris JR, Chakraborty J, Khan MR, Stanton BF, Ali M, Ahmen F, Yunus MD: Impact of B subunit killed whole-cell and killed whole-cell only oral vaccines against cholera upon treated diarrheal illness and mortality in an area endemic for cholera. Lancet i:1375, 1988.
132. Clemens JD, Sack DA, Harris JR, Chakraborty PK, Stanton NB, Huda N, Kahn MU, Kay BA: Cross-protection by B subunit-whole cell cholera vaccine against diarrhea associated with heat-labile toxin-producing enterotoxigenic *Escherichia coli:* Results of a large-scale field trial. J Infect Dis 158:372, 1988.
133. Holmgren J, Svennerholm AM, Gothefors L, Jertborn M, Stoll B: Enterotoxigenic *Escherichia coli* diarrhea in an endemic area prepares the intestine for an anamnestic

immunoglobulin A antitoxin response to oral cholera B subunit vaccination. Infect Immun 56:230, 1988.
134. Jacob CO: Synthetic approaches for vaccine development against the cholera/coli family of toxins. In Arnon R (ed): "Synthetic Vaccines." Boca Raton, FL: CRC Press, 1987, Vol. II, p 15.
135. Jacob CO, Sela M, Arnon R: Antibodies against synthetic peptides of the B subunit of cholera toxin: Cross reaction and neutralization of the toxin. Proc Natl Acad Sci USA 80:7611, 1983.
136. Jacob CO, Arnon R, Finkelstein RA: Immunity towards heat-labile enterotoxins of porcine and human *Escherichia coli*. Infect Immun 52:562, 1986.
137. Jacob CO, Harari I, Arnon R, Sela M: Antibodies to cholera toxin enteric peptides of increasing size and their reactivity with related toxins. Vaccines 4:95, 1986.
138. Jacob CO, Grossfeld S, Sela M, Arnon R: Priming immune response to cholera toxin induced by synthetic peptides. Eur J Immunol 16:1057, 1986.
139. Meklanos JJ, Swarts DJ, Pearson DM, Harford N, Groyne G, DeWilde M: Cholera toxin genes: Nucleotide sequence, deletion analysis and vaccine development. Nature 306:551, 1983.
140. Kaper JB, Lockman H, Baldini MM, Levine MM: Recombinant nontoxinogenic *Vibrio cholerae* strains as attenuated cholera vaccine candidates. Nature 308:655, 1984.
141. Kaper JB, Levine MM, Lockman HA, Baldini MM, Black RE, Clements ML, Morris JG: Development and testing of a recombinant live oral cholera vaccine. In Lerner RA, Chanock RM, Brown F (eds): "Vaccines 85: Molecular and Chemical Basis of Resistance to Parasitic, Bacterial, and Viral Diseases." Cold Spring Harbor, NY: Cold Spring Harbor Laboratory, 1985, p 107.
142. Kaper JB, Levine MM, Herrington DA, Michalski J, Glaen J, Hall R, Ketley J: Recent advances in developing a safe and effective live oral attenuated *Vibrio cholerae* vaccine. Abstracts of the U.S.–Japan Cooperative Medical Science Program, 22nd Joint Conference on Cholera, Toyama, Japan, 1986, p 26.
143. Levine MM, Kaper JB, Herrington D, Losonsky G, Morris JG, Clements L, Black R, Tall B, Hall R: Volunteer studies of deletion mutants of *Vibrio cholerae* 01 prepared by recombinant techniques. Infect Immun 56:161, 1988.
144. O'Brien AD, Chen ME, Holmes RK, Kaper JB, Levine MM: Environmental and human isolates of *Vibrio cholerae* and *Vibrio parahaemolyticus* produce a *Shigella dysenteriae* 1 (Shiga-like) cytotoxin. Lancet i:77, 1984.
145. Sanyal SC: Live oral cholera vaccines: A new toxin produced by candidate strains. In Harris JB (ed): "Natural Toxins: Animal, plant, and microbial." Oxford: Clarendon Press, 1986, p 304.
146. Dasgupta U, Guhathakurta I, Das J: Excretion of cholera toxin from *Escherichia coli:* A potential oral vaccine for cholera. Biochem Biophys Res Commun 153:967, 1988.
147. Jacob CO, Leitner M, Zamir A, Salomon D, Arnon R: Priming immunization against cholera toxin and *E. coli* heat-labile toxin by cholera toxin short peptide β-galactosidase hybrid synthetized in *E. coli*. EMBO J 4:3339, 1985.
148. Greenberg RN, Guerrant RL: *E. coli* heat-stable enterotoxin. In Dorner F, Drews J (eds): "Pharmacology of Bacterial Toxins." Oxford: Pergamon Press, 1986, p 115.
149. Dougan G: The role of enterotoxin components in future vaccines against enterotoxigenic *Escherichia coli*-induced diarrhoea and cholera. In Harris JB (ed): "Natural Toxins: Animal, Plant, and Microbial." Oxford: Clarendon Press, 1986, p 321.
150. Okamoto K, Okamoto K, Yukitake J, Miyama A: Reduction of enterotoxic activity of *Escherichia coli* heat-stable enterotoxin by substitution for an asparagine residue. Infect Immun 56:2144, 1988.

151. Wood LV, Ferguson LE, Hogan P, Thurman D, Morgan DR, DuPont HL, Ericsson CD: Incidence of bacterial enteropathogens in food from Mexico. Appl Environ Microbiol 46:328, 1983.
152. Wood LV, Wolfe WH, Ruiz-Palacios G, Foshee WS, Corman LI, McCleskey F, Wright JA, DuPont HL: An outbreak of gastroenteritis due to a heat-labile enterotoxin (LT) producing strains of *Escherichia coli.* Infect Immun 41:931, 1983.
153. Anonymous: The prospects for immunizing against *Escherichia coli.* In: ''New Vaccine Development, Vol. II, Diseases of Importance in Developing Countries.'' Washington, DC: National Academy Press, 1986, p 178.
154. Levine ML, Kaper JB, Black RE, Clements JL: New knowledge on pathogenesis of bacterial enteric infections as applied to vaccine development. Microbiol Rev 74:510, 1983.
155. Robertson DC, McDonel JL, Dorner F: *E. coli* heat-labile enterotoxin. In Dorner F, Drews J (eds): ''Pharmacology of Bacterial Toxins.'' Oxford: Pergamon Press, 1986, p 77.
156. Levine MM: *Escherichia coli* infections. In Germanier R (ed): ''Bacterial Vaccines.'' Orlando, FL: Academic Press, 1984, p 187.
157. Dougan G, Morrissey P: Molecular analysis of the virulence determinants of enterotoxigenic *Escherichia coli* isolated from domestic animals: Applications for vaccine development. Vet Microbiol 10:241, 1985.
158. Klipstein FA, Engert RF, Clements JD: Protection in rats immunized with *Escherichia coli* heat-stable enterotoxin. Infect Immun 34:637, 1981.
159. Klipstein FA, Engert RF, Clements FJ: Development of a vaccine of cross-linked ST-LT that protects against *Escherichia coli* heat-labile and heat-stable enterotoxins. Infect Immun 37:550, 1982.
160. Klipstein FA, Engert RF, Houghton RA: Protection in rabbits immunized with a vaccine of *Escherichia coli* heat-stable toxin cross-linked to the heat-labile toxin B subunit. Infect Immun 40:888, 1983.
161. Klipstein FA, Engert RF, Clements JD, Houghten RA: Vaccine for enterotoxigenic *Escherichia coli* based on synthetic heat-stable toxin cross-linked to the B subunit of heat-labile toxin. J Infect Dis 147:318, 1983.
162. Dorner F, Mayer P, Leskova R: Immunity to Escherichia in piglets: The role of colostral antibodies directed against heat-labile enterotoxin in experimental neonatal diarrhea. Zentralbl Veterinarmed 27:207, 1980.
163. Klipstein FA, Engert RF, Houghton RA: Immunisation of volunteers with a synthetic peptide vaccine for enterotoxigenic *Escherichia coli.* Lancet i:471, 1986.
164. Jacob CO, Pines M, Arnon R: Neutralization of heat-labile toxin of *E. coli* by antibodies to synthetic peptides derived from the B subunit of cholera toxin. EMBO J 3:2889, 1984.
165. Guzman-Verduxco L-M, Kupersztoch YM: Fusion of *Escherichia coli* heat-stable enterotoxin and heat-labile enterotoxin B subunit. J Bacteriol 169:5201, 1987.
166. Pickett CL, Twiddy EM, Belisle BW, Holmes RK: Cloning of genes that encode a new heat-labile enterotoxin of *Escherichia coli.* J Bacteriol 165:348, 1986.
167. Pickett CL, Weinstein DL, Holmes RK: Genetics of type IIa heat-labile enterotoxin of *Escherichia coli:* Operon fusions, nucleotide sequence, and hybridization studies. J Bacteriol 169:5180, 1987.
168. Tacket CO, Losonsky G, Link H, Hoang Y, Guesry P, Hilpert H, Levine MM: Protection by milk immunoglobulin concentrate against oral challenge with enterotoxigenic *Escherichia coli.* N Engl J Med 318:1240, 1988.
169. Cryz SJ Jr: *Pseudomonas aeruginosa* infections. In Germanier R (ed): ''Bacterial Vaccines.'' Orlando, FL: Academic Press, 1984, p 317.

170. Bodey GP, Bolivar R, Fainstein V, Jadeja L: Infections caused by *Pseudomonas aeruginosa.* Rev Infect Dis 5:279, 1983.
171. Peterson PK: "Host Defense Against *Pseudomonas aeruginosa.* International Symposium." Bern: Hans Huber, 1980, p 103.
172. Liu PV: Extracellular toxins of *Pseudomonas aeruginosa.* J Infect Dis 130(Suppl):S94, 1974.
173. Liu PV, Yoshii S, Hsieh H: Exotoxins of *Pseudomonas aeruginosa* II. Concentration, purification, and characterization of exotoxin A. J Infect Dis 128:514, 1973.
174. Vasil ML, Kabat D, Iglewski BH: Structure-activity relationship of an exotoxin of *Pseudomonas aeruginosa.* Infect Immun 16:353, 1974.
175. Pollack M: Pseudomonas toxin and toxoid. In Weinstein L, Fields BN (eds): "Seminars in Infectious Diseases, Vol. 4: Bacterial Vaccines." New York: Thieme-Stratton, 1982, p 64.
176. Iglewski BH, Kabat D: NAD-dependent inhibition of protein synthesis by *Pseudomonas aeruginosa* toxin. Proc Natl Acad Sci USA 72:2284, 1975.
177. Pollack M, Young LS: Protective activity of antibodies to exotoxin A and lipopolysaccharide at the onset of *Pseudomonas aeruginosa* septicemia in man. J Clin Invest 63:276, 1979.
178. Cross AS, Sadoff JC, Iglewski BH, Sokol PA: Evidence for the role of toxin A in the pathogenesis of infection with *Pseudomonas aeruginosa.* J Infect Dis 142:538, 1980.
179. Ohman DE, Burns RP, Iglewski BH: Corneal infections in mice with toxin A and elastase mutants of *Pseudomonas aeruginosa.* J Infect Dis 142:547, 1980.
180. Woods DE, Cryz SJ Jr, Freiedman RL, Iglewski BH: Contribution of toxin A and elastase to virulence of *Pseudomonas aeruginosa* in chronic lung infections of rats. Infect Immun 36:1223, 1982.
181. Blackwood LL, Stone RM, Iglewski BH, Pennington JE: Evaluation of *Pseudomonas aeruginosa* exotoxin A and elastase as virulence factors in acute lung infection. Infect Immun 39:198, 1983.
182. Cryz SJ Jr, Furer E, Germanier R: Protection against *Pseudomonas aeruginosa* infection in a murine burn wound sepsis model by passive transfer of antitoxin A, antielastase, and antilipopolysaccharide. Infect Immun 39:1072, 1983.
183. Pavlovskis OR, Edman DC, Leppla SH, Wretlind B, Lewis LR, Martin KE: Protection against experimental *Pseudomonas aeruginosa* infection in mice by active immunization with exotoxin A toxoids. Infect Immun 32:681, 1981.
184. Cryz SJ Jr, Pavlovskis OR, Iglewski BH: Chemical and genetic approaches to making *Pseudomonas aeruginosa* toxin A toxoid. In Weinstein L, Fields BN (eds): "Seminars in Infectious Diseases, Vol. 4: Bacterial Vaccines." New York: Thieme-Stratton, 1982, p 70.
185. Alexander JW, Fisher MW: Immunization against Pseudomonas infection after thermal injury. J Infect Dis 130(Suppl):S152, 1974.
186. Miler JM, Spilsbury JF, Jones RJ, Roe EA, Lowbury EJL: A new polyvalent Pseudomonas vaccine. J Med Microbiol 10:19, 1977.
187. Cryz SJ Jr, Furer E, Sadoff JC, Germanier R, Pastan I, Willingham MC, FirzGerald DJP: Use of *Pseudomonas aeruginosa* toxin A in the construction of conjugate vaccines and immunotoxins. Rev Infect Dis 9(Suppl):S644, 1987.
188. Cryz SJ Jr, Furer E, Sadoff JC, Germanier R: *Pseudomonas aeruginosa* immunotype 5 polysaccharide-toxin A conjugate vaccine. Infect Immun 52:161, 1986.
189. Cryz SJ Jr, Sadoff JC, Ohman D, Furer E: Characterization of the human immune response to a *Pseudomonas aeruginosa* O-polysaccharide-toxin A conjugate vaccine. J Lab Clin Med 111:701, 1988.

190. Cryz SJ Jr, Sadoff JC, Furer E: Immunization with a *Pseudomonas aeruginosa* immunotype 5 O-polysaccharide-toxin A conjugate vaccine: Effect of a booster dose on antibody levels in humans. Infect Immun 56:1829, 1988.
191. Anonymous: Rare U.S. case of human anthrax. ASM News 54:471, 1988.
192. Coggins CH: Weapons of mass destruction. Milit Rev 42:43, 1963.
193. Rich V: Incident at military village no. 19. Nature 284:294, 1980.
194. Stephen J: Anthrax toxin. In Dorner F, Drews R (eds): "Pharmacology of Bacterial Toxins." Oxford: Pergamon Press, 1986, p 381.
195. Fish DC, Mahlandt BG, Dobbs JP, Lincoln RE: Purification and properties of *in vitro*-produced anthrax toxin components. J Bacteriol 95:907, 1968.
196. Mahlandt BG, Klein F, Lincoln RE, Haines BW, Jones WI Jr, Friedman RH: Immunological studies of anthrax. IV. Evaluation of the immunogenicity of three components of anthrax toxin. J Immunol 96:727, 1966.
197. Stanley JL, Smith H: The three factors of anthrax toxin: Their immunogenicity and lack of demonstrable enzymic activity. J Gen Microbiol 31:329, 1963.
198. Hambleton P, Carman JA, Melling J: Anthrax: The disease in relation to vaccines. Vaccine 2:125, 1984.
199. Little SF, Knudson GB: Comparative efficacy of *Bacillus anthracis* live spore vaccine and protective antigen vaccine against anthrax in the guinea pig. Infect Immun 52:509, 1986.
200. Turnbul PCB, Broster MG, Carman JA, Manchee RJ, Melling J: Development of antibodies to protective antigen and lethal factor components of anthrax toxin in humans and guinea pigs and their relevance to protective immunity. Infect Immun 52:356, 1986.
201. Little SF, Leppla SH, Cora E: Production and characterization of monoclonal antibodies to the protective antigen component of *Bacillus anthracis* toxin. Infect Immun 56: 1807, 1988.
202. Mikesell P, Ivins BE, Ristroph JD, Dreier TM: Evidence for plasmid-mediated toxin production in *Bacillus anthracis*. Infect Immun 39:371, 1983.
203. Uchida I, Sekizaki R, Hashimoto K, Terakado N: Association of the encapsulation of *Bacillus anthracis* with a 60 megadalton plasmid. J Gen Microbiol 131:363, 1985.
204. Best GK, Scott DR, Kling M, Thompson MR, Adinolfe L, Bonventre PF: Protection of rabbits in an infection model of toxic shock syndrome (TSS) by a TSS toxin-1-specific monoclonal antibody. Infect Immun 56:998, 1988.
205. Bonventre PF, Thompson MR, Adinolfi LE, Gillis ZA, Parsonnet J: Production of staphylococcal enterotoxin F and pyrogenic exotoxin C by *Staphylococcus aureus* isolates from toxic shock syndrome-associated sources. Infect Immun 54:1023, 1983.
206. Scott DF, Kling JM, Best GK: Immunological protection of rabbits infected with *Staphylococcus aureus* isolates from patients with toxic shock syndrome. Infect Immun 53:441, 1986.
207. deAzavedo JCS, Foster TJ, Hartigan PJ, Arbuthnott JP, O'Reilly M, Kreiswirth BN, Novick RP: Expression of the cloned toxic shock syndrome toxin 1 gene *(tst) in vivo* with a rabbit uterine model. Infect Immun 50:304, 1985.
208. Austrian R: Pneumococcal infections. In Germanier R (ed): "Bacterial Vaccines." New York: Academic Press, 1984, p 257.
209. Dale JB, Beachey EH: Synthetic streptococcal vaccines. In Arnon R (ed): "Synthetic Vaccines." Boca Raton, FL: CRC Press, 1987, Vol. I, p 131.
210. Anonymous: The prospects for immunizing against *Streptococcus pneumoniae*. In: "New Vaccine Development, Vol. II, Diseases of Importance in Developing Countries." Washington, DC: National Academy Press, 1986, p 357.
211. Bessen D, Fischetti VA: Influence of intranasal immunization with synthetic peptides

corresponding to conserved epitopes of M protein on mucosal colonization by group A streptococci. Infect Immun 56:2666, 1988.

212. Liang X, Lamm ME, Nedrud JG: Oral administration of cholera toxin-sendai virus conjugate potentiates gut and respiratory immunity against sendai virus. J Immunol 141:1495, 1988.

213. Clements JD, Hartzog HM, Lyon FL: Adjuvant activity of *Escherichia coli* heat-labile enterotoxin and effect on the induction of oral tolerance in mice to unrelated protein antigens. Vaccine 6:269, 1988.

Bacterial Vaccines, pages 35–55


# Tetanus Vaccine: Present Status

**Maria Wirz, Giuliano Gentili, and Clelia Collotti**

Laboratorio di Immunologia, Istituto Superiore di Sanità, 00161 Rome, Italy



## I. INTRODUCTION

Tetanus is a severe disease characterized by convulsive tonic contractions of voluntary muscles. It has been known since ancient times; indeed, description of its symptoms can be found in Hippocrates' books. Its infectious nature, however, was not recognized until 1884, first by Carle and Rattone [1], followed by Nicolaier [2]. Carle and Rattone transmitted tetanus to rabbits by injecting them with pus from a lesion of a human case; soon after, Nicolaier succeeded in reproducing the disease by injecting laboratory animals with small quantities of soil and observed a bacillus in the pus taken from the lesions produced at the injection site.

The causative agent *Clostridium tetani* was isolated in pure culture in 1889 by Kitasato, who demonstrated its causal role and proved the limited invasiveness of the microorganisms, assuming for the first time that a soluble venom (toxin) produced during bacterial growth was responsible for clinical tetanus [3]. The following year Faber discovered tetanus toxin [4], a potent neurotoxin that acts mainly on the central nervous system [5–7] and is the only cause of the clinical picture of tetanus.

Tetanus shows some epidemiological characteristics that make it unique: it is accidental in origin, not contagious; its incidence is mainly related to poor socioeconomic and hygienic conditions, which explains its high occurrence in developing countries; the illness itself does not immunize, because the toxin is so powerful that it acts on the nervous system at doses too small to be sufficient for immunization.

The annual world case fatality rate is very high and difficult to quantitate; however, WHO estimates the approximate number of deaths from tetanus each year as over one million, with more than 90% of deaths occurring in developing countries [8].

The only effective weapon to fight this disease is mass vaccination, along with improved standards of living and hygiene. In fact, during the last 30 to 40 years, tetanus incidence has declined dramatically in industrialized countries, whereas the disease, especially neonatal tetanus [9], still remains a serious health problem in developing countries.

One of the objectives of the WHO program "Health for all by the year 2000" is the elimination of tetanus through vaccination [10]. In particular, the Expanded Programme on Immunization (EPI) emphasizes immunizing women of childbearing age (especially pregnant women) and ensuring hygienic delivery and postdelivery care with the purpose of controlling neonatal tetanus [8].

## II. *CLOSTRIDIUM TETANI* PRESENCE IN NATURE

*Clostridium tetani* is a gram-positive, spore-forming, strict anaerobic microorganism. Its wide distribution is accounted for by the great resistance of its spores. *C. tetani* is mainly found in soil, in animal and human feces, and in house dust. It is not defined as ubiquitous organism because its presence is dependent on the deposition of contaminated fecal material, but the penetration of dust everywhere explains why the spreading of tetanus spores is so extensive.

The presence of the bacilli is somewhat associated with nature of the soil: *C. tetani* growth is favored by alkaline calcareous clay, whereas volcanic andgranitic soils, which are practically devoid of calcium carbonate, have an acid pH adverse to bacterial growth [11].

An accurate study on the isolation of *C. tetani* from different soil specimens has shown the presence of bacillus in 35% of the samples examined, with a higher frequency among those taken from the shores of ponds and rivers and from fields than from other places (i.e., school and hospital grounds, home yard, farmyard, etc.) [12]. Moreover, in 28% of cases, the minimal quantity of soil required for isolation was less than 1 mg. This explains how wound contamination with just a very small amount of infected soil can result in disease in nonimmune subjects.

Portals of entry into the body for *C. tetani* are numerous [13–15]. The most common and best known is contamination of a traumatic or surgical wound or of the severed umbilical cord as a result of contact with infected soil; however, tetanus can also result from traditional practices such as circumcision, excision, tattooing, piercing of ear lobes and alae nasi, and ritual incisions, when they are made without fundamental sanitary measures.

## III. TETANUS TOXIN AND TOXOID

*C. tetani* produces two proteic exotoxins—tetanolysin, an oxygen-labile hemolysin that does not play a role in the pathogenesis of tetanus, and tetanospasmin, which is responsible for the whole clinical picture of the disease.

Tetanospasmin is synthesized as a single polypeptide chain with a molecular weight ranging from 140 to 160 kDa. The toxin is discharged into the culture medium at the time of cell autolysis (extracellular or filtrate toxin) [16], though it may be extracted from the cells before it has diffused in the medium (intracellular or extracted toxin).

The proteases present in the culture medium cleave the toxin in two polypeptide chains linked by disulfide bonds (''nicked'' toxin). Reduction of nicked molecule results in a heavy chain of about 100 kDa, which is involved in the binding to cell receptors (gangliosides), and a light chain of about 50 kDa responsible for the toxic activity [17]. Toxin may be measured on the grounds of its specific action or its immunogenicity.

When injected into laboratory animals, tetanus toxin causes paralysis or death, according to the dose. Minimum paralyzing dose (MPD) and minimum lethal dose (MLD) are defined as the smallest amounts of toxin that, when injected by a given route, cause paralysis or death of the treated animals, respectively, within a stated period of time. Usually, the doses that cause paralysis or death of 50% of the treated animals ($PD_{50}$ or $LD_{50}$) are used, as their evaluation is more accurate.

The assessment of immunogenicity requires the utilization of the WHO International Standard for Tetanus Antitoxin (or another antitoxin of known potency, calibrated in IU). The following values may be defined [18]: 1) Lf, the amount of toxin (or toxoid) that when reacted with 1 IU (standard for flocculation test) of antitoxin gives a flocculation in the shortest period of time; 2) Lp and L+, the smallest amounts of toxin that, when mixed with 1 IU of antitoxin and injected into laboratory animals by a specified route, cause paralysis or death of the treated animals within a stated period of time.

The toxin values defined by means of the antiserum are called "indirect" values, in contrast to those "direct" values obtained simply by injecting the toxin into susceptible laboratory animals.

Tetanus neurotoxin, along with toxins of *Corynebacterium diphtheriae, Clostridium botulinum,* and *Shigella dysenteriae,* is one of the most powerful poisons known. Highly purified toxin has a specific activity of $3–20 \times 10^7$ MLD (mouse) per milligram protein N. Tetanus neurotoxin has been obtained as a pure crystalline protein by Pillemer et al. [19].

Although several serological types of *C. tetani* have been recognized, only one antigenic type of toxin is responsible for tetanus. This has allowed the development of an effective monotypic toxoid for vaccination.

As noted above, pathogenesis of tetanus is related to the action of the toxin on the central nervous system. In the natural course of events, material contaminated with tetanus spores penetrates into the body through open wounds. When favorable conditions of anaerobiosis occur (as a consequence of vascular damage, bacterial contamination, tissue necrosis, etc.), spores germinate; however, bacterial growth is limited to the portal of entry, while toxin locally produced spreads around, reaching the central nervous system by a retrograde motion along the peripheral motor nerves. Its main action is on the motor neurons of the anterior horns of the spinal cord.

Among animals, horses are particularly susceptible to tetanus, whereas birds are almost resistant to the natural disease. Among laboratory animals, guinea pigs and mice are the most susceptible and have been extensively used in experimental work on tetanus toxin.

It has been shown that several reagents (iodine, diazonium salts, formaldehyde, etc.) are able to denature toxin in such a way that it loses its toxicity while retaining its immunogenicity (toxoid).

The reactive most widely used for toxin inactivation is formaldehyde. Formaldehyde treatment results in the formation of both intramolecular and intermolecular bonds that promote toxin polymerization [20], with consequent moderate reduction of flocculating activity and immunogenicity as compared to untreated toxin [21].

Tetanus vaccines now in use are predominantly made from formalinized toxin and may be defined as preparations of formalin toxoid, plain or adsorbed on a mineral carrier.

## IV. BRIEF HISTORY OF TETANUS VACCINATION

As a result of the identification of the neurotoxin as the unique cause of clinical tetanus, the attention of investigators was quickly focused on the possibility of controlling the disease.

In 1890 Von Behring and Kitasato [22] were able to produce an antitoxin by immunizing rabbits with attenuated toxin. Animal antitoxic sera opened the door to serotherapy. Passive protection represented the first success in tetanus control.

Various attempts at producing inactivated tetanus toxin suitable for active immunization failed until Ramon showed, in 1923, that the combined treatment of diphtheria toxin with heat and formaldehyde resulted in a biologically inactive but highly immunogenic preparation, able to stimulate the production of neutralizing antibodies in animals [23]. The same detoxification procedure was applied to tetanus toxin, and in 1926 Ramon and Zoeller successfully immunized human volunteers with tetanus toxoid [24].

The first demonstration of the effectiveness of tetanus vaccination was provided by data relative to the vaccinees during World War II. Practically all soldiers of the British and American forces had been immunized against tetanus. Thirty-five tetanus cases were recorded among British soldiers (against 2,385 during World War I) and 12 among American soldiers [25]. By contrast, a large number of cases occurred in the German and Japanese forces, whose soldiers were not regularly vaccinated.

At present, tetanus vaccination, whether mandatory or not, is largely applied in industrialized countries. In Italy, for example, this vaccination has been compulsory since 1963 for certain job categories "at risk" and since 1968 for all newborns, in combination with diphtheria vaccination. In different countries, several schedules of vaccine administration have proved to be adequate.

## V. PREPARATION AND CONTROL OF TETANUS VACCINE

The general principle for preparation of tetanus vaccine is to subject the neurotoxin to a detoxification procedure by chemical or physical means such that the toxin retains its immunogenicity; moreover, chosen methods must exclude reversion from toxoid to toxin.

Each manufacturer has to submit production protocols for approval to the proper national authorities. After the initial approval, any possible modification in the production process also is submitted to the same authorities for approval before its implementation.

Whatever the production method, tetanus vaccine must fulfill the requirements of the national pharmacopoeia, where it exists, or of the International Pharmacopoeia.

WHO has prepared and distributed a manual for the preparation and control of tetanus vaccine that provides careful indications on procedures to be adopted. The method described in the manual does not necessarily require the most expensive and modern equipment, but it is known to yield toxoids of good quality. The manual's purpose is to help countries not yet active in toxoid preparation to start their own production [26].

## A. Equipment

The production equipment need not be particularly sophisticated. In addition to basic instrumentation of a good microbiological laboratory (sterilizing devices, vertical laminar flow systems, thermostatic rooms, balances, microscopes, etc.), stainless-steel fermentors for bacterial growth, a freeze-drying apparatus, and automatic equipment for vial filling and sealing are required. An animal house is also necessary. Exact information on instrumentation, materials, and suppliers can be found in the WHO manual [26].

It is essential that production equipment be considered as a series of units in a flow line, each unit complementing the other so that vessels for media, growth, and holding are of the same or similar capacity; centrifuges, filters, and ultrafilters should be capable of processing the output from a previous stage without undue delay and consequent risk of contamination.

## B. Production of Tetanus Vaccine

**1. Strains and media.** Tetanus vaccine production is based on a seed lot system. The choice of *C. tetani* strain plays a very important role, as toxin yield essentially depends on the strain and its adaptation to the selected medium. Usually, the highly toxigenic Harvard strain is used; it originates from the New York Department of Health and can be obtained through the WHO Secretariat.

The media for bacterial growth should not contain components that may be present in the final product and cause toxic or allergic reactions in man. Techniques suitable for testing for the presence of allergenic substances in media and vaccines have been described by Cox and Hardegree [27]. In addition, neither mammalian protein nor human blood group substances should be present in the final vaccine [28].

The medium most commonly used for an optimal toxin production in combination with the Harvard strain is a modification of that originally described by Mueller and Miller. Its main constituent is a tryptic digest of casein that is used as the nitrogen source. *C. tetani* strains able to grow on Latham semisynthetic medium have also been selected. Information on these media can be found in the WHO manual [26].

**2. Production of toxin.** Industrial production of tetanus toxin is carried out in large stainless-steel fermentors under controlled conditions of temperature and stirring [26,29,30].

The inoculum consists of 24 h working seed cultures that should be checked for purity [26] to ensure that the characteristics of the original strain have been retained.

After an incubation of 5–7 days at 35°C, the toxin is spontaneously discharged into culture medium at the time of cell autolysis.

The average yield of toxin under optimal conditions should range between 60 and 80 Lf/ml of culture medium.

**3. Inactivation of toxin.** Nowadays the production of tetanus toxoid is based on the procedure developed by Ramon. Detoxification is performed by addition of formaldehyde to give a final concentration of 0.5% (v/v). The pH is adjusted to 7.6 with sterile sodium carbonate or bicarbonate and the mixture is incubated at 37°C for 4 weeks [26].

At the end of the incubation period the following control tests should be carried out: specific toxicity, sterility, determination of antigen content, and total combining power [26,28]. In the case of residual toxicity, an additional quantity of 0.1% (v/v) formaldehyde is added for 2 more weeks.

Inactivation can be carried out before or after the toxin purification from culture medium. Modern ultrafilters used for this purpose offer increased security for production staff. It is preferable to operate on partially purified toxin for two reasons: 1) inactivation of crude culture filtrates results in a highly variable immunogenicity of vaccines [31] and 2) incorporation of foreign proteins present in the culture medium into the vaccine is avoided, thus removing a possible cause of untoward reactions [11,32].

Inactivation after partial purification produces toxoids with a specific activity of 1,000–2,000 Lf/mg protein N. Toxins with higher degrees of purification are not used for vaccine production as yields of extreme purification are low and economically disadvantageous [32].

Although vaccines prepared by formaldehyde treatment have proved to be safe and effective, sometimes untoward reactions can occur [33]. For this reason other methods of toxin inactivation have been studied.

A significant achievement was obtained by substituting glutaraldehyde for formaldehyde [34,35]. It has been demonstrated that a quantity of tetanus toxin equal to 400 Lf/ml is completely and irreversibly inactivated by treatment with glutaraldehyde 0.0263 M for 10 min at 37°C [36]. This resulted in a toxoid (named Polan-T) that is a good immunogen and does not need adsorption onto mineral adjuvants [35]. The same procedure has been applied to preparation of combined vaccines.

Field trials in elderly people with Polan-T and in babies with Polan-DTP demonstrated excellent tolerance of these preparations because no adverse reactions were observed. Circulating antibody titers after DTP immunization were comparable to those induced by immunization with two different preparations adsorbed onto calcium phosphate or aluminum hydroxide [37]. Nevertheless, glutaraldehyde-inactivated toxoids are not yet widely used, at least in Europe.

**4. Purification of toxoid.** Purification of crude tetanus toxoid is obtained by ultrafiltration followed by precipitation with ammonium sulfate [26]. Of particular importance is the ammonium sulfate concentration, which is why a precipitation curve by a pilot experiment is performed [26].

In brief, by plotting the ammonium sulfate concentration on the abscissa and the toxoid content (Lf) on the ordinate, a characteristic S-shaped curve is obtained, in which lower asymptote indicates the salt concentration optimal for the precipitation of impurities, while upper asymptote indicates the lowest concentration at which toxoid precipitates in maximal yield.

Results of pilot experiments are then applied to purification of toxoid, which is subjected to a first precipitation at low ammonium sulfate concentration (12–15%) to remove impurities. The supernatant is then subjected to a second precipitation at a higher concentration (about 25%). The precipitate is dissolved in the smallest possible volume and subjected to a prolonged dialysis (8 days) against distilled water until all the ammonium sulfate has been removed. Thimerosal is then added as a preservative at a final concentration of 0.01%.

Bulk toxoid has an average purity varying between 1,000 and 1,200 Lf/mg protein N and should be subjected to a series of control tests—sterility, specific toxicity, potency, determination of antigen content, purity, Thimerosal and free formaldehyde contents, irreversibility—before further processing [28].

**5. Plain tetanus vaccine.** Plain tetanus vaccine is prepared by diluting the bulk toxoid with buffered saline to the concentration established by the health authorities of a given country.

**6. Adsorbed tetanus vaccine.** To enhance their immunogenicity, vaccines are adsorbed on mineral carriers (adjuvants), either aluminum hydroxide or aluminum phosphate or calcium phosphate. The adsorption may be achieved by adding the toxoid to the aluminum gels [26].

According to the requirements of WHO [28] and the European Pharmacopoeia [38], the final concentration of aluminum shall not exceed 1.25 mg per single human dose.

Aluminum compounds are the most widely used adjuvants for preparation of adsorbed vaccines. Adverse reactions due to the presence of aluminum have occasionally been reported. Reactions have been either local or systemic, varying in intensity from mild to severe, with the most common being local erythema, swelling, pain, and tenderness [33,39,40].

Some studies [41–45] have shown that aluminum compounds enhance the production of homocytotropic antibodies in animal species and of IgE antibodies in humans after injection of adsorbed diphtheria and tetanus vaccines. Moreover, a few reports relate the occurrence of anaphylactic reactions to the administration of adsorbed vaccines [46–48].

In order to minimize the incidence of adverse reactions, tetanus toxoid has been adsorbed onto calcium phosphate. Vaccines adsorbed onto calcium phosphate offer a number of advantages: the adjuvant is a natural constituent of the human body and therefore is well tolerated, and the entrapment of antigen in the three-dimensional network of precipitate results in a delayed release [32]. In addition, nonadsorbed impurities can be removed merely by washing the precipitate; this results in very stable preparations and in a decrease of untoward reactions [49]. Nevertheless, these vaccines are not largely used outside of France, where they were developed.

**7. Filling into final containers and storage.** The process of filling containers with vaccine is an especially critical step in production, in that a final product possibly contaminated during this phase can neither be recovered nor be terminally sterilized. Therefore, filling should be carried out with completely automated equipment, so as to ensure strict aseptic conditions throughout the operation, thus minimizing risks of contamination [26].

The filling process should be checked periodically by filling a suitable number of vials with nutrient medium and incubating them for several days. If more than 1% of the vials are shown to be contaminated, all of the equipment should be controlled and sterilized [26].

It is also important to remember that the quality of glass containers and rubber closures can affect the keeping quality of the final product. It is necessary, therefore, to use neutral borosilicate containers and butyl rubber stoppers that have been checked for compatibility with the vaccine and its preservative.

Each vial in each final lot should be inspected visually and those showing abnormalities should be discarded [28]. The vaccine should then be stored away from light at a temperature of 2–8°C [28,50].

The expiration date is calculated from the beginning of the potency test in animals. In the prescribed conditions of storage, the expiration date should be within 3 years of manufacture [28,50]. The expiration date is related to the stability of the product.

Each new production method should be checked for vaccine stability by potency tests [28]. Control tests on the final product should be carried out according to the WHO requirements [28].

**TABLE I. Controls Required for Tetanus Vaccine***

| In process | On the final product |
|---|---|
| On the toxin | |
| Culture purity | Identity |
| Determination of MLD | Specific toxicity |
| Determination of MTV[a] | Sterility |
| Determination of L+/10 | Innocuity |
| | Potency test |
| | pH |
| On the toxoid | Adjuvant content |
| Specific toxicity | Free formaldehyde content |
| Sterility | |
| Determination of Lf/mg protein N | Thimerosal content |
| Total combining power | |
| Test for reversal | |

*Details on the tests can be found in the WHO manual [26].
[a]Maximum toxin value.

## C. Control of Tetanus Vaccine

In order to ensure the quality of tetanus vaccine, a series of control tests are carried out during production and on the final product [28,38]; these prescribed controls are listed in Table 1.

One of the most significant tests is the potency assay, whose purpose is to ensure that tetanus toxoid is biologically effective and sufficient to induce full protection in vaccinees. For many years potency of tetanus vaccines has been assessed in animals (guinea pigs) previously immunized with the vaccine being tested by measuring the antitoxin level in their sera or by challenging their resistance to a lethal dose of the toxin. These simple and relatively inexpensive methods have had an excellent record of ensuring vaccine efficacy and are still used in some countries, e.g., the United States [32].

In the past, the antigen content of tetanus toxoid was expressed in Lf units. In 1977, the European Pharmacopoeia made it compulsory to measure the activity of tetanus toxoids in IU and at the same time made available a European Standard. Toxoid adsorbed on aluminum hydroxide calibrated in IU against the International Standard. Almost at the same time, this method was adopted by WHO [28].

Potency is evaluated by comparing the dose of vaccine being tested that protects the animals (guinea pigs or mice) from the lethal or paralyzing effect of a stated amount of toxin given subcutaneously with the dose of the standard vaccine necessary to give the same protection ($ED_{50}$).

For each assay, three groups of animals (16 per group) are immunized with three twofold dilutions of the vaccine under test and three groups are given three analogous dilutions of the standard. After 28 days, animals are given a paralytic or lethal challenge with 50 $PD_{50}$ or $LD_{50}$ of tetanus toxin per each animal. Dilutions of both preparations must be such that the intermediate dose protects about 50% of the animals. Results of the assay are then statistically analyzed for determination of vaccine potency [51,52].

The potency assay has been criticized [32,53] for many reasons.

1. Potency tests require many animals (over 100 per each assay), and thus are very expensive; this creates difficulties in developing countries and concern among animal welfare groups in Western Europe [54].

2. Wide variations in the potency of preparations are observed, depending on the species of animal selected [55,56].

3. Nonparallelism of dose-response relationships due to dissimilarities between standard and test vaccines and difficulties in estimating the $ED_{50}$ have been recorded. Moreover, the unitage of the new (2nd) International Standard on the basis of results obtained in mice has encountered serious difficulties [57].

4. The method could be unreliable because vaccines used for immunization of the animals are diluted in saline and the adjuvant effect is no longer present. It has been shown in mice that the antibody response is very low or even absent when they are immunized with vaccines diluted in saline. On the contrary, titers are high when dilutions are made in the adjuvant [32].

5. Potency of vaccines is not correlated with protection in humans. In fact, high levels of tetanus antitoxin were found after vaccination of children with tetanus toxoids of low potency as expressed in IU. The response in humans depends not only on antigen content, but also on the immunization schedule [58].

There is not a simple relationship between the antigenic content and the immunogenic potency of tetanus vaccines [59]. Any estimate of potency based on antigen content alone, which does not take proper account of other factors (such as degree of detoxification, state of aggregation, properties of other antigenic components, composition, and allomorphic condition of the adjuvant), will not distinguish between good and poor vaccines. The need for potency tests, therefore, remains inescapable. However, the number of animals required for potency tests could be reduced if quantitative in vitro measures of the immune response were substituted for the present paralytic or lethal challenge method.

It has been suggested [53] that the potency test be replaced with a test based on three vaccinations of a few animals with a long follow-up. Such a test could show whether the vaccine being examined is able to induce a high level of long-lasting immunity.

It has also been proposed [60] that when consistency in production and control have been proven, a test requiring fewer animals be used provided that the assays are statistically valid. In this respect it is worth mentioning that WHO requirements [28] do not indicate the number of animals for each vaccine dilution. The statement is made that sufficient animals should be used to achieve a 95% confidence interval smaller than 50–200%.

According to some authors, potency of tetanus vaccine can be ensured by using a number of animals that is considerably less than is required at present. The data presented in one study show that the frequency with which tetanus assays provide valid assurance of potency is not decreased when the number of animals per group is reduced to five [54].

As an alternative method, it has been proposed to carry out the potency test using a single dose of both the vaccine to be examined and the standard [60,61]. At any rate, the debate on biological standardization of tetanus vaccines is still open.

## VI. AVAILABLE VACCINES

Tetanus toxoid can be used as plain or adsorbed vaccine. Usually the adsorbed preparation is preferred because it induces a higher and long-lasting immunity. On the other hand, the fluid vaccine may be used without problems in individuals who are reactogenic to adjuvants.

Toxoid adsorbed onto aluminum hydroxide is available with an antigenic content varying from 5 to 30 Lf per single human dose. The primary immunization is achieved with two or three doses followed by a booster [11].

Toxoid adsorbed onto aluminum phosphate is available in many countries; the antigenic content of a human dose varies from 5 to 20 Lf. It is used in primary immunization, with two or three doses followed by a booster injection.

Tetanus vaccines adsorbed onto calcium phosphate are produced in France by the Pasteur Institute; they contain 30 Lf/dose. The immunization schedule requires one or two inoculations, followed by a third injection 1 year later [11].

Both plain and adsorbed tetanus vaccines are available in Italy, with different contents for pediatric or adult use. In the case of adults, vaccines with a lower antigen content are used. A combined adsorbed tetanus-diphtheria (double) or tetanus-diphtheria-pertussis (triple) vaccine is used for primary immunization of children.

As only tetanus and diphtheria vaccinations are compulsory in Italy, triple vaccine is left to the discretion of parents and physicians, who often prefer to leave out pertussis vaccination because of the possible adverse reactions.

In most countries, tetanus toxoid is available associated with other vaccines either as a double vaccine: tetanus-diphtheria, tetanus-polio, tetanus-flu; triple vaccine: diphtheria-tetanus-pertussis, diphtheria-tetanus-typhoid; quadruple vaccine: diphtheria-tetanus-pertussis-polio; or quintuple vaccine: diphtheria-tetanus-polio-pertussis-measles.

Field trials carried out in Sweden [62] showed high circulating antibody levels when primary tetanus immunization was performed with a fluid triple vaccine (DTP), followed by a booster with aluminum hydroxide–adsorbed vaccine.

The advantage of combined vaccines no doubt is to give good protection with a minimal number of injections. This is particularly important for developing countries, in which, on the one hand, admittance to the health services is difficult for logistic reasons, and, on the other hand, diseases as tetanus, diphtheria, poliomyelitis, measles, and hepatitis are causes of high mortality.

Several studies have shown the good tolerance and effectiveness of combined vaccines. A vaccination trial carried out in Mali among 320 children 3 to 24 months old [63] has shown that quadruple vaccine (DTP-polio) was well tolerated and highly immunogenic. Laboratory results showed serologic response to all antigens was excellent and only two doses of vaccine were needed to give a good protection; this experience suggested that DTP-polio could allow the immunization schedule of infants in developing countries to be simplified.

The above results are in complete agreement with those of a serologic follow-up performed to evaluate the validity of an immunization schedule providing for the use of a combined DTP-polio vaccine given in two doses, followed by a booster 6 months later [64].

Immunization of Senegalese children from 3 to 24 months old with a quadruple DTP-polio vaccine associated with hepatitis B vaccine has been performed according to a simplified schedule of three doses given at 6 month intervals [65]. This study has shown that 1) the immune response to the two vaccines injected simultaneously was similar to the immune response observed after administration of each single vaccine; 2) adverse reactions were not reported; and 3) with such a vaccine containing 30 Lf of tetanus toxoid, two vaccinations at 6 month intervals were sufficient to induce a satisfactory immune response, both as seroconversion (98.6%) and antibody titers.

A particular application of calcium phosphate–adsorbed tetanus vaccine has been its use as a solvent for freeze-dried meningococcal [66] and rabies vaccines.

Following experimental studies on mice [67], a clinical trial on the effectiveness of tetanus-rabies vaccination, applied in pre- and postexposure schemes, was undertaken [68]. This study has shown that immunized subjects developed satisfactory antibody titers to both components; adverse reactions were not reported. This observation is important in that exposure to rabies virus is often associated with risk of tetanus.

Because of difficulties in vaccine storage and distribution in particularly poor regions, schemes of reduced immunization with two inoculations of vaccines formulated to give satisfactory levels of protection have been proposed [64,65]. In a controlled field trial in infants in the Philippines, using a concentrated adsorbed DTP vaccine, a two-dose schedule with an interval of 6 months between injections was evaluated [69]. The serologic response to the three components of the vaccine was satisfactory, whereas the side effects in the concentrated vaccine group did not differ from those observed in a control DTP group. On the basis of the results of this trial, the Immunization Committee of the Philippines has decided to adopt the above-outlined two-dose scheme for immunization with the concentrated DTP vaccine.

Many strategies have been attempted with the purpose of ensuring a proper mass vaccination. In a rural area of India, for example, a "cluster approach" has been adopted. A mobile health team visited villages once a month for vaccination of children, after an active campaign to make people aware of such problems. This approach resulted in a 95% coverage of children with measles vaccine, 50% with three doses of DTP, and 79% with two doses of tetanus toxoid [70]. Other strategies have attempted oral or nasal immunization with tetanus toxoid at high doses. It has been demonstrated that oral vaccination is not able to stimulate an immune response [71], while better results have been obtained with nasal immunization.

Tetanus vaccines presently available are safe and effective, although untoward reactions sometimes occur stimulating research for production of totally harmless vaccines.

In addition to the inactivation and adsorption methods already described, a monomeric purified 7 S tetanus toxoid [72] and subunit vaccines [73] have been prepared. Their efficacy has not been sufficiently studied.

A tetanus toxoid purified by affinity chromatography, prepared to minimize the side effects attributable to contaminants, was compared with conventionally purified material in a double-blind trial in 205 volunteers in Scotland [74]. There was no difference in immunogenicity or in side reactions between the two vaccines.

## VII. IMPORTANCE AND RESULTS OF VACCINATION

Tetanus vaccination induces a high, long-lasting immunity. Its widespread application in the European region has enormously reduced tetanus incidence. During the period between 1951 and 1970, 3,000 fatal cases per year were reported; over the decade 1971–1980, the number of fatal cases fell to less than 1,000 per year [75]. In countries such as Denmark [76] and Sweden [77], the high immunization coverage of the population has practically eradicated tetanus (0–2 cases/year).

In most countries, cases still occurring are predominantly limited to adult women who were never vaccinated, elderly people, or people with an inadequate vaccination [75]. In France, 107 cases of tetanus were reported in 1987, with a mortality rate of 25%; 68 cases were studied in detail. There were three times as many women as men; 87.1% of the patients were over 60 years of age. Of 64 individuals for whom information about vaccination was available, 50 (78.1%) had never been vaccinated and only 4 of the 14 immunized patients appeared to have received a booster within the last 10 years [78].

In Italy, tetanus morbidity has decreased from 3,503 cases in 1956–1960 to 1,116 in the period 1976–1980. Trend analysis for tetanus age-specific incidence rates indicates that until 1965 the highest incidence was observed in the under-one-year and the over-60-years age groups, whereas after the introduction of mandatory vaccination for newborns, the highest incidence was recorded among elderly people [79]. Presently, the annual number of reported tetanus cases is about 100.

In the United States, the annual incidence of tetanus for the 1985–1986 period was 0.03/100,000 inhabitants, compared with 0.39/100,000 in 1947. Seventy-one percent of the 147 cases reported in the biennium 1985–1986 occurred among people over 50 years of age [80].

On the whole, the above data show that in the industrialized countries, vaccination has virtually eliminated tetanus as a problem of infancy. In contrast, tetanus, especially neonatal tetanus, is still a serious health problem in most developing countries [81]. WHO estimates that 800,000 newborns die from tetanus every year [8], with variable rates in different countries [81]. In recent years [82], mortality rates varying from 67 per 1,000 livebirths in rural areas of India to 3 per 1,000 in urban Egypt have been recorded.

A cluster survey study conducted in three different districts in Kenya has shown death rates from neonatal tetanus of 6, 11, and 16 per 1,000 livebirths. From this study it has been determined that between 8,000 and 12,000 children die from tetanus in Kenya every year [83].

Immunization of pregnant women is a very effective means to protect newborns from tetanus as a consequence of both a transplacental passage of maternal IgG antibodies [84] and a transplacental immunization of the fetus, so that offspring have antitetanus IgM at birth [85].

During the 8th International Conference on Tetanus (1987), an immunization schedule for women of childbearing age was proposed, according to which the first dose of tetanus toxoid should be given as early as possible during pregnancy, followed by a second dose after 4 weeks and a third dose 6–12 months from the second one or during a subsequent pregnancy [8].

The effectiveness of tetanus vaccination is influenced by the route of injection and the immunization schedule. Usually, adsorbed vaccines are given by deep intramuscular injection so as to reduce the risk of formation of a granuloma [11].

The second important factor for the success of active immunization is the schedule selected for vaccination. Schedules are rather variable, but usually a primary immunization with two doses suitably spaced (6–8 weeks apart) is recommended, with a third dose being given 6–12 months later. Booster injections should be given every 10 years or, preferably, when antibody level is decreasing. Indeed, the presence of a high level of circulating tetanus antitoxin could result in the development of an Arthus-like reaction after a vaccine dose [11].

Infants should receive immunization as early in life as possible in order to minimize the risk of contracting the disease. Infants born to immune women receive maternal antibodies that protect them during the first few weeks of life. Although these antibodies may interfere with the newborn's immune response, it has been shown that the first dose of DTP (or DT) vaccine can be given effectively at 4 weeks of life, even if final titers may sometimes be lower than after immunization started at 3–6 months of life [86].

The knowledge of the immunity status of a population against tetanus fulfills various practical needs; for example, it allows the verification of the effectiveness of the immunization schemes and the evaluation of the persistence of immunity. Moreover, it may provide useful indications for a correct prophylaxis [87]. One method for the evaluation of the effectiveness of vaccination, other than the survey of tetanus clinical incidence, is the determination of the level of circulating antitoxin following immunization.

The classical method for the detection of tetanus antibodies is the in vivo seroneutralization test (SNT), which measures exclusively the antibodies able to neutralize tetanus toxin, the only ones implied in the protection. SNT is carried out in mice or guinea pigs [88], and the serum to be examined is assayed in comparison with the International Standard for Tetanus Antitoxin, thus allowing the antitoxic content of the unknown serum to be expressed in IU/milliliter. Presently, a level of 0.01 IU/ml is accepted as protective in humans [89].

The SNT is rather difficult to perform; besides being expensive and time consuming, it requires animal facilities. Therefore it is not suitable for seroepidemiological studies requiring rapid, simple, and possibly inexpensive assays. Several in vitro tests for the evaluation of tetanus antibodies have been developed, among which passive hemagglutination [90–92], radioimmunoassay (RIA) [93,94], and immunoenzymatic assays (ELISA) [93,95–97] appear to be the most useful.

ELISA has been applied to the evaluation of tetanus antibodies both in commercial preparations of specific immunoglobulin [98] and in human sera [99]; its good correlation with SNT has been demonstrated.

The ELISA assay has been used in a seroepidemiological study recently carried out in Italy [100]. About 5,000 young men (recruits) were tested for their antitetanus immunity. About 33% were not protected, and a relation between absence of protection and region of residence, sociocultural level, and urban or rural residence was established. On the other hand, 11% of the immune subjects were found to be hyperimmunized, showing an antitoxic titer ≥5 IU/ml [101]. Apart from hyperimmunization, another risk, possible even in countries where tetanus vaccination is mandatory, is the escape from vaccination. There are several reasons explaining this phenomenon. One that should be corrected is inequality in the health-care delivery system in a given country.

## REFERENCES


1. Carle A, Rattone G: Studio sperimentale sull'eziologia del tetano. G Accad Med 32:174, 1884.
2. Nicolaier A: Uber infectiosen Tetanus. Dtsch Med Wochenschr 10:842, 1884.
3. Kitasato SB: Uber den Tetanusbacillus. Z Hyg Infektionskr 7:225, 1889.
4. Faber K: Die Pathogenese des Tetanus. Berl Klin Wochenschr 27:717, 1890.
5. Mellanby J, Green J: How does tetanus toxin act? Neuroscience 6:281, 1981.
6. Wellhoner HH: Tetanus neurotoxin. Rev Physiol Biochem Pharmacol 93:1, 1982.
7. Fedinec AA: Current studies on pathogenesis of tetanus. In Nisticó G, Mastroeni P, Pitzurra M (eds): "Seventh International Conference on Tetanus." Rome: Gangemi, 1985, p 61.
8. Expanded Programme on Immunization: Tetanus control. Wkly Epidemiol Rec 50:380, 1987.
9. Lambo T: The social aspects of tetanus. In Nisticó G, Mastroeni P, Pitzurra M (eds): "Seventh International Conference on Tetanus." Rome: Gangemi, 1985, p 345.
10. Expanded Programme on Immunization. Wkly Epidemiol Rec 56:9, 1981.
11. Bizzini B: Tetanus. In Germanier R (ed): "Bacterial Vaccines." New York: Pergamon Press, 1984, p 37.
12. Ebisawa I, Kurata M: A quantitative study of *C. tetani* in the earth. In Nisticó G, Mastroeni P, Pitzurra M (eds): "Seventh International Conference on Tetanus." Rome: Gangemi, 1985, p 7.
13. Bourgeade A, Kadio A: Inoculations traditionelles et inoculations iatrogénique du tétanos en Afrique tropicale. In: "Proceedings of the 4th International Conference on Tetanus." Lyon: Collection Fondation Mérieux, 1975, p 77.
14. Vakil BJ, Dalal NJ: Ways of inoculation. Portal of entry of tetanus and risk evaluation of a wound. In: "Proceedings of the 4th International Conference on Tetanus." Lyon: Collection Fondation Mérieux, 1975, p 23.

15. Rey M, Guillaumont P, Majnoni d'Intignano B: Benefits of immunization versus risk factors in tetanus. Dev Biol Stand 43:15, 1979.
16. Bizzini B: The chemistry of tetanus toxin as a basis for understanding its immunological and biological activities. In Nisticó G, Mastroeni P, Pitzurra M (eds): ''Seventh International Conference on Tetanus.'' Rome: Gangemi, 1985, p 11.
17. Helting TB, Parschat S, Engelhardt H: Structure of tetanus toxin. Demonstration and separation of a specific enzyme converting intracellular tetanus toxin to the extracellular form. J Biol Chem 254:10728, 1979.
18. European Pharmacopoeia: ''General Notices.'' I edition, 1975, Vol. III, p 12.
19. Pillemer L, Wittler RG, Burrel JL, Grossberg DB: The immunochemistry of toxins and toxoids—IV. The crystallization and characterization of tetanal toxin. J Exp Med 88:205, 1948.
20. Bizzini B, Raynaud M: La détoxication des toxines protéiques par le formol: Mécanismes supposés et nouveaux développements. Biochimie 56:297, 1974.
21. Raynaud M, Bizzini B, Turpin A: Hétérogénéité de l'anatoxine tétanique. Ann Inst Pasteur 120:801, 1971.
22. Von Behring EA, Kitasato SB: Ueber das Zustandekommen der Diphtherie-Immunitat und der Tetanus-Immunitat bei Thieren. Dtsch Med Wochenschr 16:1113, 1890.
23. Ramon G: Sur le pouvoir floculant et sur les propriétés immunisantes d'une toxine diphtérique rendue atoxique (anatoxine). C R Acad Sci 177:1338, 1923.
24. Ramon G, Zoeller C: De la valeur antigène de l'anatoxine tétanique chez l'homme. C R Acad Sci 182:245, 1926.
25. Boyd J: Symposium on tetanus. Tetanus in two world wars. Proc R Soc Med 52:109, 1959.
26. World Health Organization Expert Committee: ''Manual for the Production and Control of Vaccines.'' Geneva: WHO, 1977.
27. Cox CB, Hardegree MC: Diphtheria and tetanus toxoids: Results of assays for the presence of bovine antigens. J Biol Stand 4:81, 1976.
28. World Health Organization Expert Committee on Biological Standardization: Requirements for tetanus toxoid. WHO Tech Rep Ser 638:81, 1979.
29. Bizzini B, Turpin A, Raynaud M: Production et purification de la toxine tétanique. Ann Inst Pasteur 116:686, 1969.
30. Cameron J: Some problems—and their solution—in the production of DTP (diphtheria, tetanus, pertussis) vaccine. Dev Biol Stand 41:45, 1978.
31. Bizzini B: Vaccini antitetanici. Boll Ist Sieroter Milan 62:550, 1983.
32. Relyveld EH: Immunological, prophylactic and standardization aspects in tetanus. In Nisticó G, Mastroeni P, Pitzurra M (eds): ''Seventh International Conference on Tetanus.'' Rome: Gangemi, 1985, p 215.
33. Relyveld EH, Henocq E, Bizzini B: Studies on untoward reactions to diphtheria and tetanus toxoids. Dev Biol Stand 43:33, 1979.
34. Relyveld EH: Préparation de vaccins antitoxiques et antimicrobiens à l'aide de glutaraldéhyde. C R Acad Sci 277:613, 1973.
35. Relyveld EH: Detoxification of microbial toxins with glutaraldehyde and their use in the preparation of vaccines. In Rosenberg P (ed): ''Toxins: Animal, Plant and Microbial.'' Oxford: Pergamon Press, 1978, p 1049.
36. Relyveld EH, Ben-Efraim S: Preparation of vaccines by the action of glutaraldehyde on toxins, bacteria, viruses, allergens, and cells. In Langone JJ, Van Vunakis H (eds): ''Methods in Enzymology.'' New York: Academic Press, 1983, p 24.
37. Guérin N, Fillastre C: Vaccin D-T-C Polan. In: ''Proceedings of the 6th International Conference on Tetanus.'' Lyon: Collection Fondation Mérieux, 1982, p 477.

38. European Pharmacopoeia: Vaccinum Tetanicum Adsorbatum. I edition, 1977, vol. III (suppl), p 184.
39. Collier LH, Polakoff S, Mortimer J: Reactions and antibody responses to reinforcing doses of adsorbed and plain tetanus vaccines. Lancet i:1364, 1979.
40. Collier LH: Reactions after plain and adsorbed tetanus vaccines. Lancet i:148, 1980 (letter).
41. Vassilev TL: Aluminium phosphate but not calcium phosphate stimulates the specific IgE response in guinea pigs to tetanus toxoid. Allergy 33:155, 1978.
42. Vijay HM, Lavergne G, Huang H, Bernstein IL: Preferential synthesis of IgE reaginic antibodies in rats immunized with alum-adsorbed antigens. Int Arch Allergy Appl Immunol 59:227, 1979.
43. Pauwels R, Bazin H, Platteau B, Van Der Straeten M: The influence of different adjuvants on the production of IgD and IgE antibodies. Ann Immunol (Inst Pasteur) 130C:49, 1979.
44. Nagel JE, White C, Lin MS, Fireman P: IgE synthesis in man. II: Comparison of tetanus and diphtheria IgE antibody in allergic and nonallergic children. J Allergy Clin Immunol 63:308, 1979.
45. Matuhasi T, Ikegami H: Elevation of levels of IgE antibody to tetanus toxin in individuals vaccinated with diphtheria-pertussis-tetanus vaccine. J Infect Dis 146:290, 1982.
46. Staak M, Wirth E: Zur problematik anaphylaktischer reaktionen nach aktiver Tetanus-Immunisierung. Dtsch Med Wochenschr 98:110, 1973.
47. Levine L, Edsall G: Tetanus toxoid: What determines reaction proneness? J Infect Dis 144:376, 1981.
48. Jacobs RL, Lowe RS, Lamer BQ: Adverse reactions to tetanus toxoid. JAMA 247:40, 1982.
49. Relyveld EH: Preparation and use of calcium phosphate adsorbed vaccines. Dev Biol Stand 65:131, 1986.
50. Farmacopea Ufficiale della Repubblica Italiana: Vaccini per uso umano, IX edition. 1985, Vol. II, p. 1733.
51. Reed LJ, Muench H: A simple method of estimating fifty per cent endpoints. Am J Hyg 27:493, 1938.
52. Worcester J, Wilson EB: A table determining $LD_{50}$ or the fifty per cent end-point. Proc Natl Acad Sci USA 29:207, 1943.
53. Lyng J: Potency assay of diphtheria and tetanus toxoid: Some theoretical and practical considerations. Dev Biol Stand 64:47, 1986.
54. Knight PA, Roberts PAG: Studies on the minimal number of animals required to achieve assurance of satisfactory potency in diphtheria and tetanus vaccines. Dev Biol Stand 65:245, 1986.
55. Hardegree MC, Pittman M, Maloney CJ: Influence of mouse strain on the assayed potency (unitage) of tetanus toxoids. Appl Microbiol 24:120, 1972.
56. Hardegree MC, Fornwald RE, Farber J, London WT, Parks F, Kessler MJ, Rastogi SC: Titrations of tetanus toxoids in international units: Relationship to antitoxin responses in rhesus monkeys. In: "Proceedings of the 6th International Conference on Tetanus." Lyon: Collection Fondation Mérieux, 1982, p 409.
57. Lyng J, Nyerger G: The second international standard for tetanus toxoid adsorbed. J Biol Stand 12:121, 1984.
58. Huet M: La standardisation des vaccins tetaniques. In: "Proceedings of the 6th International Conference on Tetanus." Lyon: Collection Fondation Mérieux, 1982, p 425.
59. Knight PA: Are potency tests for tetanus vaccines really necessary? Dev Biol Stand 64:39, 1986.

60. Kreeftenberg JG: Report of an informal meeting about alternative methods for the potency control of the diphtheria and tetanus components in vaccines. Dev Biol Stand 65:261, 1986.
61. Stainer DW, Jakus CM, Sparkes JD: Reduction in animal usage for potency testing of diphtheria and tetanus toxoids. Dev Biol Stand 65:241, 1986.
62. Ullberg-Ollson K, Wickbom B, Eriksson E: Immunization against tetanus with aluminium free versus aluminium containing triple vaccine. Dev Biol Stand 43:39, 1979.
63. Drucker J, Soula G, Diallo O, Fabre P: Evaluation of a new combined inactivated DPT-polio vaccine. Dev Biol Stand 65:145, 1986.
64. Swartz TA, Roumiantzeff M, Peyron L, Stopler T, Drucker J, Epstein I, Leitner L, Goldblum N: Use of a combined DTP-polio vaccine in a reduced schedule. Dev Biol Stand 65:159, 1986.
65. Coursaget P, Yvonnet B, Relyveld EH, Barres JL, Diop-Mar I, Chiron JP: Simultaneous administration of diphtheria-tetanus-pertussis-polio vaccine and hepatitis B vaccine in a simplified immunization programme. Dev Biol Stand 65:169, 1986.
66. Relyveld EH, Huet M, Causse GY: Studies on the effectiveness of the association of adsorbed tetanus toxoid with meningococcal polysaccharide vaccines. In: ''Proceedings of the 6th International Conference on Tetanus.'' Lyon: Collection Fondation Mérieux, 1982, p 179.
67. Relyveld EH, Rollin PE, Sureau P: Combined tetanus rabies vaccination. Experimental studies in mice. In Nisticó G, Mastroeni P, Pitzurra M (eds): ''Seventh International Conference on Tetanus.'' Rome: Gangemi, 1985, p 413.
68. Lery L, Rotivel Y, Trabaud MA, Parvaz P, Mouterde S, Relyveld EH: Combined tetanus-rabies vaccination. Dev Biol Stand 65:209, 1986.
69. Mangay-Angara A, Fulgencio L, Casabal G, Gudani L, Sumpaico J, Ocampo A, Nagel J, Hagenaars AM, van Hemert PA, Cohen H: A two-dose schedule for immunization of infants using a more concentrated DTP-vaccine. Dev Biol Stand 41:15, 1978.
70. Jajoo UN, Chhabra S, Gupta OP, Jain AP: Annual cluster (pulse) immunization experience in villages near Sevagram, India. J Trop Med Hyg 88:277, 1985.
71. Mayr A, Baljer G, Baljer E, Sailer J: Controlled field trials with oral and nasal immunization against tetanus in humans. Dev Biol Stand 41:31, 1978.
72. Bizzini B, Turpin A, Henocq E, Maro B: Tetanus immunization with a fluid pure monomeric toxoid administered by intradermal route. In Roseberg P (ed): ''Toxins: Animal, Plant and Microbial.'' Oxford: Pergamon Press, 1978, p 971.
73. Schwick HG, Helting TB, Zwisler O: Bacterial split vaccines. Prog Clin Biol Res 47:143, 1980.
74. Leen CLS, Barclay GR, McClelland DBL, Shepherd WM, Langford DT: Double-blind comparative trial of standard (commercial) and antibody-affinity-purified tetanus toxoid vaccines. J Infect 14:119, 1987.
75. Bytchenko BD, Dittmann S: Elimination of diseases from Europe through use of vaccines. Dev Biol Stand 65:3, 1986.
76. Simonsen O, Bloch AV, Heron I: Epidemiology of tetanus in Denmark 1920–1982.
77. Christenson B, Bottiger M: Epidemiology and immunity to tetanus in Sweden. Scand J Infect Dis 19:429, 1987.
78. Expanded Programme on Immunization: Tetanus. Wkly Epidemiol Rec 63:197, 1988.
79. Rosmini F, Wirz M, Gentili G, Collotti C, Rossino Ricci M, Franco E, Terzi I, Pasquini P: Year of birth, sex and residence as ''determinants'' of tetanus incidence and immunity in Italy. Eur J Epidemiol 3:377, 1987.
80. Anonymous: Tetanus—United States 1985–1986. MMWR 36:477, 1987.
81. Galazka A, Cook R: Neonatal tetanus today and tomorrow. In Nisticó G, Mastroeni P,

Pitzurra M (eds): "Seventh International Conference on Tetanus." Rome: Gangemi, 1985, p 349.
82. Expanded Programme on Immunization: Prevention of neonatal tetanus. Wkly Epidemiol Rec 57:137, 1982.
83. Melgaard B, Mutie DM, Kimani G: A cluster survey of mortality due to neonatal tetanus in Kenya. Int J Epidemiol 17:174, 1988.
84. Rethy L, Rethy AL: Active anti-tetanus immunisation of females to control neonatal tetanus. Lancet i:616, 1986.
85. Gill TJ, Papaevangelou G, Repetti CF, Cortese AL, Nikolopoulos D, Argyroundis E: Vaccination of the fetus by immunization of the mother: Transplancental immunization of the human fetus to tetanus. In Nisticó G, Mastroeni P, Pitzurra M (eds): "Seventh International Conference on Tetanus." Rome: Gangemi, 1985, p 509.
86. Halsey N, Galazka A: The efficacy of DPT and oral poliomyelitis immunization schedules initiated from birth to 12 weeks of age. Bull WHO 63:1151, 1985.
87. Evans AS: The need for serologic evaluation of immunization programs. Am J Epidemiol 112:725, 1980.
88. European Pharmacopoeia, 2nd edition. Sainte-Ruffine: Maisonneuve, 1981, 1985.
89. Levine L, McComb JA, Dwyer RC, Latham WC: Active-passive tetanus immunization. N Engl J Med 274:186, 1966.
90. Bytchenko BD, Grin E, Matveev KI, Cvjetanovic B, Vorst A, Stroganova M: The level of tetanus immunity in the population of villages in Bosnia and Herzegovina. Bull WHO 45:431, 1971.
91. Newell KW, Leblanc DR, Edsall G, Levine L, Christensen H, Montuori MH, Ramirez N: The serological assessment of a tetanus toxoid field-trial. Bull WHO 45:773, 1971.
92. Rey M, Fontanges R, Robert D, Tayot JL, Relyveld EH, Vincent-Falquet JC, Durand B: Le test d'hémagglutination passive dans l'evaluation de l'immunite' antitetanique. Dev Biol Stand 41:55, 1978.
93. Layton GT: A micro-enzyme-linked immunosorbent assay (ELISA) and radioimmunosorbent technique (RIST) for the detection of immunity to clinical tetanus. Med Lab Sci 37:323, 1980.
94. Habermann E, Horvath E, Schaeg W: A radioimmunoassay for tetanus antibodies using protein A-containing *Staphylococcus aureus*. Med Microbiol Immunol 163:261, 1977.
95. Stiffler-Rosenberg G, Fey H: Measurement of human tetanus antitoxin using an enzyme-linked immunosorbent assay. Schweiz Med Wochenschr 107:1101, 1977.
96. Melville-Smith ME, Seagroatt VA, Watkins JT: A comparison of enzyme-linked immunosorbent assay (ELISA) with the toxin neutralization test in mice as a method for the estimation of tetanus antitoxin in human sera. J Biol Stand 11:137, 1983.
97. Gentili G, Wirz M, Collotti C: *Clostridium tetani* toxin antibodies. In Bergmeyer HU (ed): "Methods of Enzymatic Analysis," 3rd edition. Weinheim: Verlag Chemie, 1986, Vol. 11, p 44.
98. Gentili G, Pini C, Collotti C: The determination of the potency of human tetanus immunoglobulins by an enzyme-linked immunosorbent assay. J Biol Stand 12:167, 1984.
99. Gentili G, Pini C, Collotti C: The use of an immunoenzymatic assay for the estimation of tetanus antitoxin in human sera: a comparison with seroneutralization and indirect haemagglutination. J Biol Stand 13:53, 1985.
100. Rosmini F, Gentili G, Wirz M, Collotti C, Paná A, Terzi I, Conti S, Pasquini P: Immunity to tetanus among italians born between 1956 and 1963. Eur J Epidemiol 3:302, 1987.
101. Wirz M, Gentili G, Rosmini F, Collotti C, Paná A, Terzi I, Conti S, Pasquini P: Prevalence of hyperimmunization against tetanus in a national sample of 18–26 year old immune subjects in Italy. Vaccine 5:211, 1987.

**Bacterial Vaccines, pages 57–86**


# Polysaccharides and Membrane Vaccines

**Jan T. Poolman**

Department of Bacterial Vaccine Development, National Institute of Public Health and Environmental Protection, 3720 BA Bilthoven, The Netherlands

## I. INTRODUCTION

Bacterial vaccines in current use are killed or attenuated whole-cell vaccines such as *Bordetella pertussis* and bacillus Calmette-Guérin (BCG) and toxoid, chemically inactivated toxin, vaccines such as diphtheria and tetanus vaccines. It can be expected that capsular polysaccharides chemically conjugated to proteins, i.e., conjugate vaccines, will be the third type of bacterial vaccines in general use. These three types of vaccines can be produced in high yields by existing production facilities.

The development of polysaccharide conjugate vaccines is the first breakthrough after the introduction of diphtheria-tetanus-pertussis vaccines. The first such vaccine to be applied is the *Haemophilus influenzae* type b (Hib) conjugate vaccine, designed to prevent by parenteral immunization systemic Hib diseases such as meningitis. If the introduction of this vaccine is successful, more polysaccharide conjugate vaccines will be introduced to prevent diseases caused by other pathogenic bacteria, e.g., pneumococci, meningococci, and *Pseudomonas aeruginosa,* among others.

Further vaccine development is being carried out to use other bacterial surface components such as lipopolysaccharides, outer-membrane proteins, flagellae, and pili or fimbriae. Lipopolysacharides (LPS) with long-chain O-repeating saccharide units are comparable to capsular polysaccharides with regard to accessibility on the bacterial surface, and the induction of systemic immunity can be expected to be comparable. The production of vaccines consisting of O-repeating units, however, is more complicated as compared to capsular polysaccharides because of yield and the necessity to remove toxic lipid A. In some of these cases chemical synthesis of the needed saccharides will be more practical. Lipopolysaccharides without long-chain repeating units, sometimes called lipooligosaccharides, and outer-membrane proteins of gram-negative bacteria can be targets of bactericidal defense mechanisms because of the membrane-disruptive properties of complement. These outer-membrane–type vaccines depend upon a different defense mechanism than the capsular polysaccharide/O-polysaccharide–based vaccines because of the difference in location on the bacterial surface. Other surface appendices, such as pili or fimbriae, flagellae, and streptococcal M proteins, are also being studied for vaccine development. Immune mechanisms involved in the prevention of bacterial diseases in these cases are phagocytosis and prevention of colonization and motility.

Often these outer-membrane components and surface appendices cannot be isolated in high yields from bacterial cultures. This implicates that improvement of yield by recombinant DNA methodology will often be necessary, either in homologous or heterologous systems. This methodology has other vaccine production advantages such as the preparation of genetically constructed toxoids and the construction of multiepitope products in the case of type specificity of immunity. Another approach with great potential is chemical synthesis in order to prepare synthetic peptide or saccharide vaccines.

This chapter does not intend to be complete in its examples, but rather describes general principles, although each infectious disease will require its own approach. Such general principles can, however, be described allowing general strategies for preventive measures. To summarize, the following types of vaccines are in general use or under various stages of development: 1) killed or attenuated whole-cell vaccines, 2) toxoid vaccines, 3) polysaccharide-protein conjugate vaccines, 4) outer-membrane–based vaccines, and 5) vaccines based on protein surface appendices. In this chapter the third, fourth, and fifth types of bacterial vaccines will be discussed.

## II. HISTORY OF POLYSACCHARIDE VACCINES

The early work on polysaccharides as antigens and immunogens was focused on the capsular polysaccharides of *Streptococcus pneumoniae,* the pneumococcus. Pneumococcal pneumonia was a common disease causing tremendous morbidity and mortality in the preantibiotic era. In 1923, Heidelberger and Avery demonstrated that polysaccharides were the type-specific antigens of pneumococci [1]. This culminated in the demonstration in 1945 that purified pneumococcal polysaccharides as parenteral vaccines are able to prevent pneumococcal pneumonia [2]. Since pneumococci can be divided into over 80 capsular polysaccharides, multivalent polysaccharide vaccines were developed that contain the polysaccharides from the most prevalent serotypes causing disease; a 23-valent vaccine is currently available. Two important facts have prevented a general use of such pneumococcal polysaccharide vaccines: 1) the enormous success of the introduction of highly efficacious antibiotics such as penicillin drastically reduced the pneumococcal disease load and 2) plain polysaccharide vaccines appeared to have serious drawbacks because of their T-cell independency: no immunological memory is formed and the immune response starts late in the ontogeny. The practical consequences of these immunological phenomena are that polysaccharide vaccines are poorly immunogenic in young children, a group at great risk for pneumococcal diseases, and that other risk groups with different kinds of immunodeficiencies also tend to respond poorly to the polysaccharide vaccines; varying protection rates have been found among such risk groups. In people who mount a reasonable response, this immunity disappears rather quickly over time.

In the late 1920s, Avery and Goebel [3] demonstrated that pneumococcal polysaccharides reveal enhanced immunogenicity and booster responses when the polysaccharides were chemically conjugated to proteins. The protein ensures that T-helper cells are involved in the activation of B lymphocytes, causing a different immune response that includes the formation of memory, an earlier response in the ontogeny, and a maturation of affinity.

In addition to the pneumococcal polysaccharides, other polysaccharide vaccines, e.g., meningococcal and Hib polysaccharides, were developed in the 1970s. It was shown in a field trial that these polysaccharide vaccines are protective, albeit above a certain age group [4]. The age dependency of protection has prohibited these plain polysaccharide vaccines from being introduced into general vaccine programs. The meningococcal group A polysaccharide vaccine is used on a large scale every once and a while when an epidemic of group A meningococcal meningitis occurs, usually in Africa and Asia. The duration of protection of such vaccination programs has been demonstrated to be relatively short because of limited formation of immunological memory.

The principle of chemical coupling of protein as demonstrated with the pneumococcal polysaccharide was also applied to the Hib and meningococcal polysaccharides with good laboratory results [5,6]. Recently, in Finland, it was shown that such polysaccharide-protein conjugate vaccines protect at an early age against invasive Hib diseases [7]. This success will lead to conjugate vaccines that can be used in general vaccine programs and that will make the plain polysaccharide vaccines obsolete, although plain polysaccharides can boost a primary response being induced by a conjugate.

The dominant defense mechanism against the pneumococcus is the process of phagocytosis that takes place by way of antibodies against the capsular polysaccharides, with complement being involved in the opsonization process. It is likely therefore that such mechanisms are also sufficient and dominant in the defense against other encapsulated bacteria such as Hib and meningococci. The direct bactericidal process by the deposition of antibodies on the polysaccharide layer has also been noted to play an important role in the defense against gram-negative bacteria. However, it is difficult to understand how bacterium–antibody interactions at a considerable distance from the outer membrane can catalyze such reactions.

In addition to bacteria being covered with capsular polysaccharides, having only limited direct association with the bacterial body by way of fatty acids, the principle of polysaccharide or conjugate vaccines can also be applied to bacteria covered with LPS with long-chain repeating saccharide units, e.g., *Pseudomonas aeruginosa, Salmonella typhimurium.*

One important phenomenon needs to be mentioned before shifting to the polysaccharide conjugate vaccines: not every capsular polysaccharide serotype causes a similar disease load. For instance, serotype b among *Haemophilus influenzae* causes most of the invasive disease caused by this species.

Similar observations have been made regarding group A, B, and C meningococci and the dominant pneumococcal serotypes. Thus, these capsular polysaccharides do more than just form a physical barrier in order to frustrate the defense mechanisms. An important activity of certain polysaccharides, depending upon the chemical structure, is the interaction with the complement system in order to prohibit complement, particularly the alternative pathway, from becoming activated by structures such as peptidoglycan and LPS.

The precise chemical criteria by which polysaccharides have such biological effects are unknown, except that it is known that sialic acid can have complement-binding activity. Sialic acid is the building block of group B and C meningococcal polysaccharides and others. But why can meningococcal group A polysaccharide, mannosaminephosphate, do the job, whereas group X meningococcal polysaccharide, glucosaminephosphate, cannot. We don't know. The same holds true for haemophilus type b polysaccharide, riboseribitolphosphate, and type a polysaccharide, glucoseribitolphosphate [8]. The precise chemical structures that are able to interact with complement still need to be elucidated.

### A. T-Cell Independency of Polysaccharides and Ways to Circumvent the Immunological Consequences

Athymic mice still give normal antibody responses to large polymeric molecules with repeating epitopes [9]. Polysaccharides are one example of these molecules, with great practical consequences regarding vaccine application. Different kinds of such T-cell independency exist: most bacterial polysaccharides are TI-2 [10] antigens, not able to cause an immune response in early postnatal life and devoid of polyclonal activating activity. Lipopolysaccharides are able to cause an early immune response and are polyclonal activators called TI-1 antigens. The characteristics of a T-cell–dependent immune response are not found in a polysaccharide-specific response: no memory is demonstrable, so no booster reactions occur, and the antibody response is of short duration, no antibody affinity maturation is found. The specific binding of polysaccharides to T cells has never really been demonstrated. The structural demands of T-cell epitopes, i.e., association with T-cell receptor and major histocompatibility complex (MHC), are not fully understood. Some saccharide epitopes may fulfill these characteristics, but have not yet been defined simply because many more peptide epitopes have been studied as compared to saccharide epitopes. Recently it was shown that pneumococcal polysaccharides activate antigen specific T-helper and T-suppressor cells. At immunogenic antigen doses for a primary IgM response, Ts activity masks Th activity (Milligan and Braley-Mullen, Cell Immunol 119: 222, 1989).

Polysaccharides often demonstrate poor immunogenicity per se, and TI-2 polysaccharides reveal delayed ontogeny of immune responses. The necessary B cells, however, are present early in life. In order to be immunogenic, large-size polymers that often stick together are needed, most likely because of the presence of fatty acid tails. Different antibody specificities involving different epitopes or different sizes of the same saccharide structures can be recognized [11]. The antibody response consists primarily of IgM antibodies, but other isotypes depend on T-cell involvement are made. Non-IgM formation does not take place in athymic mice [12]. Athymic mice also reveal a slower decline of the antibody response as compared to euthymic mice. The isotype switch activity and suppressor activity most likely are due to anti-idiotypic regulation [13,14].

The first approach described to improve the immunogenicity of polysaccharides in order to prepare applicable vaccines was chemical conjugation to proteins, which transforms the saccharide epitopes into T-cell–dependent antigens. Other approaches that have been studied in order to be able to prepare better polysaccharide vaccines include 1) the use of anti-idiotypic antibodies as mirrors of the primary antigen [15], 2) the use of peptides that mimic carbohydrate antigens [16], and 3) saccharide synthesis in vitro by chemical or enzymatic methods [17]. Basically, these alternative approaches are intended to achieve the same results: to present antigenic B-cell epitopes representing the involved carbohydrate in a T-cell–dependent manner by way of inclusion of T-cell epitopes. This also is the essential criterium for saccharide-protein conjugate vaccines. Anti-idiotypic carbohydrate vaccines may not necessarily contain all possible B-cell epitopes present on the carbohydrate being imitated, and the T-cell epitopes are also limited, leading to MHC restrictiveness of the antibody response to the T-cell sites. If peptides on antibodies can mimic carbohydrate antigens, although saccharides on antibodies clearly can play a role too, it is a more direct approach to chemically synthesize the peptides and to conjugate them to a carrier containing the necessary T-cell sites. Peptide synthesis using nonnatural amino acids may allow the possible synthesis of surrogate carbohydrate antigens, but still it is limited. The imitation of carbohydrate antigens will be different for each antigen and may require chemical structures that we would not expect or predict. Examples of unpredicted antigenic cross-reactivity have been described [18]. Chemical or enzymatic synthesis of saccharide antigens will become increasingly important if carbohydrate antigens that cannot easily be purified from natural sources are going to be used. Examples of this are carbohydrate antigens from parasites or bacteria such as mycobacteria [19,20] and carbohydrate antigens that are part of molecular structures that cannot be accepted as vaccine components, such as the small carbohydrate structures present in bacterial LPS without long-chain polysaccharides. Even in the case of long-chain LPS, saccharide synthesis can become an alternative source of the carbohydrate antigen.

A very special case is the group B polysaccharide of *Neisseria meningitidis,* a homopolymer of α(2-8)sialic acid that is poorly immunogenic probably because of tolerance due to cross-reactivity with host biomolecules containing α(2-8)sialic acid. This poor immunogenicity prohibits a broadly protective meningococcal vaccine from becoming reality. However, the B polysaccharide contains a number of epitopes, some of which may not be cross-reactive with host tissue [18,21]. Some of these epitopes can be mimicked with unrelated chemical structures [18]. This implicates a challenge in order to separate the ''good'' and ''bad'' epitopes by way of mimotoping or by using anti-idiotypic antibodies.

The carbohydrate antigen, however obtained, then needs to be presented to the immune system in the direct presence of T-cell epitopes in such a way that MHC restrictiveness does not play a role. The easiest way to achieve this is the conjugation of the carbohydrate antigen to a carrier protein containing multiple T-cell epitopes, such as tetanus toxoid. As more carbohydrate vaccines are developed, more carrier proteins are needed, because the same carrier protein cannot be used too often because of hyperimmunization effects. T-cell epitopes that do not reveal MHC restrictiveness need to be studied in molecular terms to better understand the MHC–T-cell receptor association [22].

TI-1 antigens have an intermediate position between TI-2 and TD antigens. One important difference between TI-1 and TI-2 antigens is the immunogenicity of TI-1 antigens early in life. The polyclonal activating activity limits the applicability of TI-1 antigens. Studies using conjugation of carbohydrate antigens to various fatty acids in different configurations are needed [23].

Currently, the directly applicable approach for carbohydrate antigens to become vaccines is the conjugation to carrier proteins; this approach is described in the next section.

## III. POLYSACCHARIDE-PROTEIN CONJUGATE VACCINES

As mentioned above, the principle of the beneficial immunological effects of chemical conjugation of protein to polysaccharide was already discovered in the late 1920s by Avery and Goebel [3]. Although polysaccharide-protein conjugate vaccines have been developed to be used against a number of bacterial diseases, this chapter focuses on conjugate vaccines using the capsular polysaccharide from *Haemophilus influenzae* type b (Hib). The reason for this limitation comes from the great efforts that are currently being made in the development of Hib conjugate vaccines for human use. More is currently known about the production, control, serology, etc., of Hib conjugate vaccines as compared to other conjugate vaccines. Consistency of production, potency, toxicity, stability of products, reproducibility of results in humans, and a critical mass of results in infants have allowed the evaluation of the Hib conjugate vaccines to be more thorough than that of other conjugate vaccines.

Different Hib conjugate vaccines have reached the stage of testing in infants and the effectiveness of one such vaccine has been shown in a placebo-controlled trial in infants in Finland. Registration of Hib conjugate vaccines has been realized or is expected soon. Because of conflicting results in some studies in the United States and because the vaccine that was effective in Finland showed no protection in infants in Alaska, registration of Hib conjugate vaccines for infants awaits further results from clinical trials that are under way. The current age limit for the application of Hib conjugate vaccines is 18 months, which is the age limit being approved for use of the plain Hib polysaccharide vaccine. This age limit comes from the results of a trial in the 1970s in Finland with the plain polysaccharide vaccine. At the age of 24 months, 90% protection was detected with the plain polysaccharide vaccine, which correlated with an antibody level of 1 μg/ml antipolysaccharide detected 3 weeks after vaccination by radioimmunoassay. At the age of 18 months, some protection could be detected. These results are now being debated in the United States because of postmarketing results obtained in case-control studies [24]. Conjugate vaccines induce higher antibody responses at lower ages and protection appears to occur at lower antibody levels [7]. These results fit into the T-cell–dependent immune response against the saccharide epitopes. Since immunological memory is playing a part in the immunity in case of conjugate vaccines, a correlation of measurable antibodies with protection is more complicated when compared to the situation of plain polysaccharide vaccines.

Also, because we are dealing with only a few epitopes being present in the riboseribitolphosphate structure of the Hib polysaccharide, genetic restrictiveness of the immune response may play a role, and certain risk groups may not react adequately to these epitopes, even when present in conjugate form. This may cause great variation in efficacy of such vaccines depending upon the population structure of different geographic areas. Fortunately, known bad responders to the plain Hib polysaccharide vaccines, e.g., IgG2-deficient patients, do respond very well to Hib conjugate vaccines [25]. These results implicate that known risk groups for Hib diseases are likely to be protected by the use of Hib conjugate vaccines. The restricted immune response against the plain polysaccharide vaccine can also be demonstrated against conjugate vaccines: the IgG subclass distribution is comparable [26].

IgG2-deficient patients, however, produce mainly IgG1 antibodies. IgG1 and IgG2 antipolysaccharide antibodies have comparable antibacterial activity [27]. The booster response induced by conjugate vaccines is mainly due to IgG antibodies, and the affinity can be expected to increase [28].

Four different Hib conjugate vaccines have been developed to a stage of testing in infants and are being produced industrially:

1. PRP-T: native polysaccharide activated with CNBr, derivatized with adipic dihydrazide, and conjugated to tetanus toxoid by condensation with carbodiimide [28].
2. PRP-D: heated, sized, and activated polysaccharide conjugated to diphtheria toxoid, activated and derivatized with adipic dihydrazide [29].
3. HbOC: polysaccharide activated to dialdehydes and fragmented into oligosaccharides by periodate treatment, sizing and conjugation to a mutant diphtheria toxoid, CRM197, by reductive amination [30].
4. PRP-OMP: native polysaccharide activated to contain an alpha halopropionyl group and conjugated to a mixture of meningococcal outer-membrane proteins, activated to contain a thiol group, resulting into a thioether linkage [31].

It is clear that these conjugate vaccines differ in a number of essential criteria such as molecular weight, length of saccharide molecules, the carrier protein (T-helper epitopes included), the linkage, and the substitution degree of protein with saccharide. These differences will make it difficult to relate potency to molecular architecture of conjugates. Essential information to relate potency to structure of each individual conjugate vaccines is protected because of industrial involvement. The saccharide and protein constituents are applied within the normal vaccine dose ranges: 10–20 μg/dose saccharide and 10–50 μg/dose protein.

A number of criteria are important in conjugate vaccines:

1. Length of saccharide chain. Short saccharide chains are highly immunogenic when conjugated to a carrier in correct configuration. Short chains will intrinsically have little T-independent properties, but TI properties of high-molecular-weight polysaccharides can be limited sufficiently whenever a correct substitution degree is realized.
2. Molecular weight. A lower limit of molecular weight must be realized in order for conjugate vaccines to be immunogenic without the need for the addition of the new generations of adjuvants. This lower limit is not very well defined, nor is the higher molecular weight limit. The latter depends on production possibilities such as filtrability.
3. Stability of the linkage.

4. Architecture of conjugate molecules, i.e., substitution degree, number of linkages, protein:saccharide ratio.

5. Creation of neoepitopes due to linking.

6. Effect of conjugation on epitopes. The saccharide epitopes are limited and simple antigenic and immunogenic tests will give the information needed. The T-helper carrier epitopes, however, are ill defined and therefore the effects of conjugation are difficult to estimate. It will be helpful if this can be tested by in vitro T-helper potency tests of the carrier, in addition to in vivo saccharide antibody formation tests. The effects of conjugation on the B-cell epitopes of the carrier protein can best be tested by using a collection of critical monoclonal antibodies.

The minimal length of the saccharide chain that is able to induce, in conjugate formulation, an immune response comparable to potent conjugates, considering the whole repertoire of epitopes involved and the characteristics of antibodies such as affinity and clonal heterogeneity, is not known. Synthetic trimeric riboseribitolphosphate in conjugate form does mount an appreciable antibody response [32] The combination of results obtained with the four conjugate vaccines tested in infants demonstrate the following: 1) after three doses PRP-D induces an anti-PRP antibody response of 0.2–1.0 μg/ml. There appear to be differences between lots that can influence the protective efficacy, depending on the socioeconomic and genetic background of the population being immunized [33–35]. 2) After two immunizations HbOC induces about 5 μg/ml of antipolysaccharide antibodies [33]. 3) PRP-OMP and PRP-T induce similar amounts of antibodies as compared to HbOC, but induce a significant antibody response after just the first dose. The last three vaccines can be expected to protect against Hib diseases after the second dose, knowing the results obtained with PRP-D in Finnish field trials that showed that 0.42 μg/ml anti-PRP correlates with protection. Ongoing field trials with HbOC, PRP-T, and PRP-OMP are addressing this question.

Aside from the direct issue of anti-PRP antibody formation and protection, the issue of possible interference of Hib conjugate vaccines and existing DTP, polio, and MMR vaccines is important. On the basis of experiments in infant rhesus monkeys, it can be concluded that concurrent injection of tetanus toxoid along with PRP-T enhances the Hib polysaccharide antibody response, existing anti-TT titers did not influence the polysaccharide antibody response, and reinjection of the carrier protein after conjugate immunization did not differ from schedules without the conjugate [28]. The issue of overimmunization with the carrier protein will become more important as more conjugate vaccines are introduced for infant immunization programs. Eventually, the carrier protein itself can best be a vaccine component in order to avoid possible hyperimmunization effects. Homologous carrier proteins can in some cases be preferable. In the case of *Haemophilus influenzae,* outer-membrane proteins may be interesting carriers, in addition to being a vaccine. However, protection obtained with bacterial outer-membrane proteins has yet been proved in a field trial. This awaits further investigation, and the first such experimental vaccines to be tested are the meningococcal outer-membrane proteins. Separate immunizations of Hib conjugate vaccines and DTP given at the same time do not seem to influence each other. For ease of administration, the combination of DTP or DTP-polio (inactivated) with Hib conjugate vaccines (and future infant vaccines such as meningococcal vaccines) is preferable. Careful investigations are needed to see if this combination causes interference problems. There is a precedent for this: the addition of aluminum gel to PRP-T drastically reduces the immunogenicity of the conjugate [36]. When more new vaccines become available, such as acellular pertussis and meningococcal vaccines, interference studies will become more important.

## IV. LIPOPOLYSACCHARIDE VACCINES

Lipopolysaccharides are a major constituent of the gram-negative outer membrane, being located on the outside of this peculiar asymmetric membrane. The lipid portion of this amphipathic molecule is composed of a disaccharide, N-acetylglucosamine, that is linked through ester and amide linkages with a number of fatty acids such as B-hydroxymyristate. This molecule, lipid A, or sometimes different ones such as lipid X, is different from host, such as human, constituents, and a number of biological phenomena are triggered by lipid A. The molecular recognition systems that cause these biological phenomena are unknown. These biological phenomena include antigen-independent mitogenicity for B lymphocytes, triggering of lymphokine production by macrophages, etc. One of the serious consequences of these biological activities is the occurrence of so-called gram-negative shock, a disease syndrome induced by bacterial LPS that is released in the bloodstream. Gram-negative septicemia is still an infectious disease problem because of the effects of LPS or endotoxin. Antibiotic treatment alone is not fully satisfactory and may even enhance the release of free LPS from the bacteria. Because of this, supplementary treatments such as anti-TNF (tumor necrosis factor) and anti-LPS treatments are being investigated. Immune therapy may turn out to be beneficial in addition to antibiotic treatment of gram-negative septicemia.

The lipid A molecule is linked on the hydrophilic, outer surface side with a number of rather specific sugars, such as heptose and ketodesoxyoctonate (KDO), phosphoethenolamine groups are often linked to these saccharides.

This region of the LPS molecule is often called the core region, and although different among various bacteria, it appears to have biological activities such as binding to macrophages and complement, again not being defined on the molecular level.

From here, a major separation occurs among gram-negative bacteria. Some bacteria synthesize long-chain saccharide repeating units starting from the core region. Gut-associated gram-negative bacteria express such long saccharide chains, also called O-chains, and this expression can be considered essential for survival in the gut, probably in order to protect the outer membrane from being attacked by bile acids and the like. Examples of such bacteria are *Escherichia coli, Salmonella* sp., *Vibrio cholerae,* and *Shigella* sp. *Pseudomonas aeruginosa* also expresses O-chains when isolated from burn-wound patients, but isolated strains from pneumonia patients produce an alginate capsule and do not express O-chains. A number of pathogenic bacteria do not express O-chains: *Neisseria meningitidis, Neisseria gonorrhoeae, Haemophilus influenzae, Bordetella pertussis,* and *Chlamydia trachomatis.* These bacteria colonize the nasopharynx and genital tracts, and these environments probably require different properties for bacteria to survive. They appear to synthesize additions to the basic heptose-KDO core structure that allow them to interact with and survive within the host. Considering immune prophylactic measures, the difference between O-chain–producing bacteria and non-O-chain–producing bacteria is critical and will be discussed separately.

### A. O-Chain–Producing Bacteria

O-chain–producing bacteria resemble encapsulated bacteria inasmuch that the polysaccharide structure is similarly located. This polysaccharide layer is directly accessible to the immune system and is a target for antibacterial defense mechanisms. The polysaccharide layer covers most other surface components, although O-chain–producing bacteria probably express more dense polysaccharide layers as compared to encapsulated bacteria. Antimembrane immunity in case of encapsulated bacteria can therefore also be effective, depending upon the density and kind of polysaccharide.

In the case of O-chain–producing bacteria, antimembrane immunity is less likely to be an important factor in the antibacterial defense mechanisms. O-chain–dependent immune prophylactic measures resemble those dependent upon capsular polysaccharides: the O-chain polysaccharides are good vaccines whenever presented in a T-cell–dependent formulation by conjugation to a protein carrier. The production of such vaccines is complicated by the need to detoxify the isolated LPS, i.e., to remove the lipid A before conjugation to protein. These procedures have been successfully used in order to prepare saccharide-protein conjugate vaccines to protect against diseases caused by *Pseudomonas aeruginosa* and *Salmonella typhimurium* [37,38]. These types of vaccines are complicated by the serological heterogeneity of LPS within species, implying the need for multivalent polysaccharide-based vaccines, a situation comparable to the heterogeneity of pneumococcal capsular polysaccharides.

Another complicating phenomenon is the need for local immunity in some cases. Whenever most of the disease load caused by a particular organism is due to systemic or locally invasive infection, parenteral immunization with saccharide conjugates can give protection. However, whenever most of the disease effects are due to a strictly mucosal infection, e.g., cholera, parenteral immunization may not be very beneficial. Oral immunization with attenuated strains in such cases appears to be a better option. Such vaccines can also be produced at lower cost, which will make application possible in areas that need it most [39].

### B. Non-O-Chain–Producing Bacteria

The core region of the LPS of non-O-chain–producing bacteria is extended with additional nonrepeating saccharide and phosphoethanolamine groups, which makes these structures different from the core region of R-LPS mutants from O-chain–producing bacteria. Such structures appear to cause limited serological diversity within non-O-chain–producing bacterial species. Gram-negative bacteria with short-chain LPS, sometimes called lipooligosaccharides or LOS, are susceptible to attack by antibodies to outer-membrane proteins, but antibodies to the LPS are also able to kill these bacteria [40]. Good examples of such bacterial species are the pathogenic neisseria, gonococci, and meningococci. Antibodies against neisserial LPS are present in normal human sera and are able to kill these bacteria by direct bactericidal mechanisms [41]. Antibody-independent complement activation also occurs through the alternative pathway. Some strains are not susceptible to these bactericidal mechanisms because of altered LPS structures, but can be killed with LPS-serotype–specific antibodies. Also, gonococci appear to be able to use the biosynthetic mechanisms present in serum, and as a consequence become serum resistant to the bactericidal activity of normal human serum, probably involving the alternative complement pathway [42]. In order to kill such bacteria, the presence is needed of LPS-serotype–specific antibodies. The whole array of epitopes in such short-chain LPS is presently unknown; the potential protective epitopes therefore need further characterization. Essential to this is a chemical characterization of these complicated molecules, especially the position of the phosphoethanolamine groups [43]. Some meningococcal serotype-specific LPS epitopes are understood in terms of structure, others are not. Meningococci synthesize a neolactotetraose that is also present on some human cells, implying a second mimicry mechanism used by the meningococcus in addition to the a(2-8)sialic acid of the group B capsular polysaccharide. Such phenomenon may very well be present among other pathogenic non-O-chain–producing bacteria.

The characteristics of short saccharide chains, including structures cross-reacting with host components, will make vaccine production very complicated, also realizing that lipid A needs to be eliminated. The yields will be very poor whenever the right mutant strains without cross-reacting epitopes can be made or isolated. It is likely that in this area chemical synthesis of the minimal structures needed for induction of bactericidal antibodies can be an approach that will allow production of short-chain LPS-based vaccine [44].

## V. OUTER-MEMBRANE PROTEIN VACCINES

Outer-membrane protein (OMP) vaccines are the next in the row of vaccines after whole-cell, toxoid, and polysaccharide (conjugate) vaccines. However, demonstration of protective efficacy of OMP vaccines has up until now not been proven in a field trial. Currently, the first such experimental OMP vaccines are being tested in Chile, Cuba, and Norway in order to demonstrate protection against meningococcal diseases. Because of this, the OMP vaccines discussed here will mainly be meningococcal OMP vaccines since no other OMP vaccines has been studied to such a level.

In general, OMP vaccines can be expected to have more chances for success against diseases caused by non-O-chain–producing as opposed to O-chain–producing gram-negative bacteria. The dense O-chains appear to mask the OMP antigens, although some effect of anti-OMP immunity has been published on experimental pseudomonas infection. Comparison of monoclonal antibodies against pseudomonas LPS and OMP revealed a much greater protective capacity of anti-LPS monoclonals [45]. In the case of meningococcal OMPs, a strong protective immunity by passive immunization with monoclonal antibodies, comparable to anticapsular immunity, was shown to occur using an experimental infant rat meningitis model [46].

Similar findings showing protection against Hib meningitis by passive immunization with monoclonal antibodies have been published [47]. *Bordetella pertussis* is another gram-negative bacteria with short-chain LPS against which OMP vaccines may be applicable. These findings have stimulated the developed of experimental OMP vaccines, particularly in the case of meningococcal OMPs, because we lack an immunogenic polysaccharide vaccine in the case of group B meningococci.

Haemophilus OMP vaccines directed against nontypeable *Haemophilus influenzae* strains causing upper respiratory tract infectious diseases is another area of great interest [48]. Because the incidence of otitis media is a lot higher than meningitis, haemophilus OMP vaccines can be tested in the field

**TABLE I. Influence of Growth Media on Bactericidal Activity of Monoclonal Antibodies (Test Strain 2996 [B:2b:P1.2])**

| | Media | | |
|---|---|---|---|
| Monoclonal antibody | TSB | MHB | MHB + glucose 0.25% (w/v) |
| MN2C3B(2b) | 1:4,000 | <1:10 | 1:4,000 |
| MN4C5E(P1.2) | 1:8,000 | 1:4,000 | 1:8,000 |
| MN8D6A (LOS) | 1:8,000 | 1:2,000 | 1:8,000 |

Reprinted from Poolman et al. [56] with permission of Kluwer Academic Publishers.

a lot easier than meningococcal OMP vaccines. Haemophilus OMPs reveal serotype heterogeneity, as do meningococcal, gonococcal, and chlamydial OMPs.

*Bordetella pertussis* OMPs are extremely homogeneous and as such are interesting vaccine candidates. Bordetella probably deals with the immune system expressing immunoregulatory toxins, so OMP heterogeneity is not needed for species survival. Antitoxin immunity is needed, but anti-OMP immunity can then give additional protection.

### A. Meningococcal OMP Vaccines: Protein Characteristics

Meningococci express four classes of major OMPs: class 1, class 2/3, class 4, and class 5. Class 1 OMP appears to be a cation-specific porin (Tommassen et al., unpublished results). Class 2/3 OMPs (mutually exclusive) have been demonstrated to be an anion-specific porin [49]. The function of class 4 OMP is unknown. Class 5 OMP is likely to be an adhesion-promoting protein analogous to gonococcal protein II [50]. Class 5 OMP appears to be highly variable among meningococci: higher inter- and intrastrain variability have been demonstrated [51]. The human immune response appears to be directed toward the variable portion of these proteins [52]. The class 4 OMP may induce antibodies that block the bactericidal activity of antibodies against other bacterial surface components, analogous to the findings with gonococcal protein III [53]. Meningococcal class 1 and 2/3 OMPs form the basis for a serosubtyping system allowing the differentiation of meningococci that can be used for epidemiological investigations [54]. The heterogeneity of meningococci considering class 1 and 2/3 OMPs appears to be limited [55]. In vitro bactericidal assays and animal protection studies with a collection of monoclonal antibodies demonstrate that antibodies against class 1 OMP are protective, as well as those against LPS. Antibodies against class 2/3 OMPs are hardly protective [56]. Growth conditions of meningococci before testing appears to be critical (Table I). These findings provide the framework for further vaccine development. Meningococci do

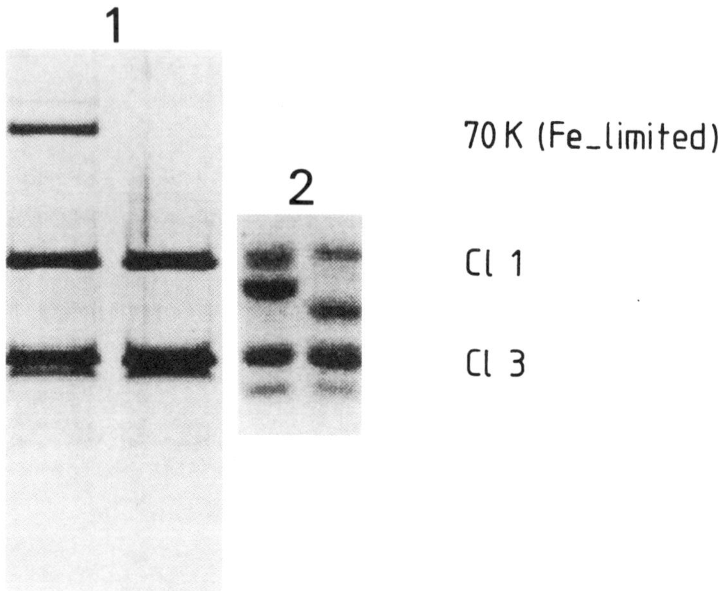


**Fig. 1.** Meningococcal OMPs. 1 = H44/76 OMP vaccine after cultivation without added $Fe(NO_3)_3$. 2 = H44/76 OMP vaccine after cultivation with added $Fe(NO_3)_3$.

express additional major OMPs that can be targets for bactericidal antibodies, for instance, the 70 kDa iron-limitation–induced OMP and major OMPs present in the area of class 1 and 2/3 OMPs [56]; see Figure 1 for an example.

The bactericical response is considered to be critical for protection against meningococcal diseases [57]. Meningococci appear to be relatively resistant to phagocytosis, because people with C6, C7, or C8 deficiencies and properdin deficiencies are extremely susceptible to meningococcal diseases. Properdin deficiency appears to result in aggressive meningococcal disease processes. These findings suggest that phagocytosis and bacteriolysis through the alternative complement pathway is the "normal" human defense mechanism. If bacteriolysis (C6, C7, or C8 deficiency) falls away, the host suffers as antibody-independent phagocytosis is not a very efficient process. If both phagocytosis and bacteriolysis through the alternative pathway fall away (properdin deficiency) the result is very bad for the host. Both categories of deficiencies probably benefit from vaccination with A, C, Y, and W-135 polysaccharides because antibody-dependent phagocytosis will then become active, most likely the protective mechanism behind capsular polysaccharide vaccines.

**TABLE II. Bactericidal Activity of Mice Immunized With Experimental Meningococcal OMP vaccines, Lots MP4 and MPC2**

| 2996(B:2b:P1.2) | H355(B:15:P1.15) | S3032(B:12:P1.16) | H44/76(B:15:P1.16) |
|---|---|---|---|
| MP-4 <1:20 | <1:20 | 1:1280 | 1:2560 |
| MPC-2 <1:20 | <1:20 | 1:1280 | 1:2560 |

Reprinted from Poolman et al. with permission of Kluwer Academic Publishers.

The mechanisms that will hopefully allow protection to be induced by OMP or LPS vaccines against group B meningococci (since capsular polysaccharide of group B is nonimmunogenic) are the antibody-mediated bacteriolysis and endotoxin (in case of LPS) neutralization processes. In this respect, our findings with class 1 OMP and LPS are stimulating.

### B. Meningococcal OMP Vaccine Development

Currently, first-generation meningococcal OMP vaccines are being tested for protection in Norway, Cuba, and Chile [58]. Such vaccines are prepared from isolated or extracted outer-membrane vesicles depleted of LPS to obtain acceptable pyrogenicity tests. Class 1, 2/3, 4, and 5 OMPS, and residual LPS, are present in such vaccines.

The vesicle structure is intact or the proteins are presented in micellar formulation with the group C capsular polysaccharide (thanks to the lipid tail) [59,60]. Such vaccines appear to induce strain-specific, serotype-dependent, bactericidal immunity [56] (Table II). A major question is how much cross-protection will be induced in the field. Mixing a number of such first-generation OMP vaccines is not possible because of the amount of protein, solubility, and pyrogenicity (LPS).

The aim of a meningococcal vaccine is the induction of broadly protective immunity by way of administration through the general child vaccination programs. Therefore we are currently developing second-generation vaccines that will allow the induction of broadly protective immunity.

### C. Second-Generation Meningococcal OMP/LPS Vaccines

Second-generation meningococcal OMP/LPS vaccines that are intended to induce cross-protective immunity can be constructed by way of the use of cross-protective surface epitopes or by way of combining serotype-specific epitopes by either peptide synthesis or rDNA technology. The combination of epitopes seems to be feasible since the presumably protective class 1 OMP and LPS molecules express limited serotypic heterogeneity [55,61]. As mentioned earlier, production of LPS vaccines coming from short-chain LPS will be very difficult to be carry out because the yields of the necessary oligosaccharides will be low after removal of lipid A. In addition, meningococcal LPS contains a tetraneolactose structure that is common to host structures [43,62]. A practical means to get around this problem is chemical synthesis of the minimum structure needed for protection. The first steps into this process have been made [44].

The construction of second-generation OMP vaccines has to take into account that the conformation, tertiary structure, of these proteins can sometimes be important and that the presentation or formulation to the immune system is important in order to direct the attention to the surface portions or loops of these proteins. These issues are clearly indicated by studies with gonococcal protein I, which demonstrate that the presentation of purified proteins is best achieved by incorporating the proteins into membrane, micelle, or liposome structures [63]. The outer-membrane vesicles themselves appear to be a logical natural presentation system that can be optimized by mutagenesis and upgrading of expression of the OMPs of interest. This principle has been shown to be valid by way of the first-generation meningococcal OMP vaccines. The LPS in these vesicles can be hidden, so such vaccines will not be pyrogenic and LPS will act as an adjuvant. The OMPs themselves can be manipulated by protein engineering in order to express several epitopes [64].

In the case of OMPs with porin function, the organization in the outer membrane can be predicted on the basis of membrane-spanning sheet structures and variable antibody or phage-binding surface loops [65]. These proteins form cylinders with a hole and the protein chain passes the outer membrane a number of times, so a number of surface loops are present. Regarding the meningococcal class 1 OMP, a number of epitopes defined by monoclonal antibodies recognize the SDS denatured protein and cyanogenic bromide fragments [66]. This opens up the way for the construction of synthetic peptide vaccines, thereby avoiding the immune presentation problems noted above. The published amino acid sequence of the meningococcal OMPs will be very helpful [67]. These B-cell epitopes, however, will then need to be conjugated to T-cell–stimulating carrier epitopes, for instance, a carrier protein, just as with oligo- or polysaccharides. T-cell epitopes of the homologous OMPs can also be used, if characterized and tested for HLA restrictiveness. Chemically synthesizing a number of B-cell and T-cell epitopes and conjugating these may become impractical; in such a case the whole OMP approach comes back into the picture. Whichever approach leads to a vaccine, the demonstration of good bactericidal activity and animal protection with class 1 OMP and LPS-specific monoclonal antibodies is promising, knowing the limited serotypic heterogeneity of these molecules among meningococci.

Multiepitopic vaccines seem feasible, but will take an enormous high-tech effort to be realized. The other options in the search for cross-reactive epitopes therefore still need considerable attention.

## VI. PILI, FLAGELLAE, AND M PROTEINS

Immunity on the basis of neutralization of proteinous surface appendages depends upon phagocytosis, inhibition of motility, and adhesion.

### A. Pili

Pili, or fimbriae, are involved in bacterial colonization processes, and prohibition of this will give protection to the infectious disease process taking place after colonization. Veterinary pilus vaccines have been used with success to induce colostral antibody that will give protection during the critical postnatal period [68]. In a human gonorrhea challenge model, a homologous single-pilus vaccine was highly protective [69]. However, such vaccine was not effective in a placebo-controlled field trial [70]. The straightforward explanation for this comes from the tremendous serological heterogeneity of gonococcal pili [71]. Although common epitopes among gonococcal pili with potential cross-protective potential have been suggested [72], this still needs serious proof. Problems similar to those described for gonococci can be expected to be critical in case of other bacteria such as *Haemophilus influenzae*]73] and *Pseudomonas aeruginosa* [74].

*Bordetella pertussis* pili [75] reveal very little heterogeneity and thus are interesting vaccine candidates, as are the bordetella OMPs. In case of *Bordetella pertussis,* the pertussis toxin is the only antigen proved to be effective in a field trial [76]. The chemically treated toxin, the toxoid, was proved to be protective, and the supposedly major adhesion factor of *B. pertussis,* filamentous hemagglutinin (FHA), added little protection to this. FHA is a proteinous surface appendage different from pili. Protection rates obtained with the toxoid were similar or a little less as compared to the whole-cell pertussis vaccine [77].

These results prove antitoxin immunity to be important, but give little support for anti-adhesion immunity to be induced by parenteral immunization. Clearly, more research is needed to determine if anti-adhesion (FHA, pili) or direct antibacterial (OMP, LPS) immunity can best be achieved by direct triggering of the mucosal immune mechanisms by oral or aerosol vaccines using live attenuated strains [39]. The protective antigens in such vaccines have to be elucidated in order to determine if cross-reactive immunity within species can be induced. LPS seems to be an important antigen in oral vaccines, and many bacterial species express several different LPS immunotypes.

### B. Flagellae

Flagella vaccines are able to protect against *Pseudomonas aeruginosa* infection [78]. A comparison with the protection induced by LPS is not provided here. Pseudomonas flagellae, however, occur in many antigenic forms [79], and the extent of cross-protection needs to be investigated.

### C. M Proteins

Streptococcal M proteins are protective antigens by way of phagocytosis processes. The type-specific and host–cross-reactive (heart-reactive) epitopes on M proteins can be separated, and synthetic peptide vaccines appear to be very promising [80]. A broadly protective vaccine needs to contain several type-specific M protein peptides, but maybe cross-reactive epitopes can be defined.

## VII. PREPARATION AND CONTROL OF MENINGOCOCCAL B/C VACCINE

### A. Preparation

**1. Introduction.** To study the immunogenicity of group B meningococcal OMPs, a vaccine has been prepared [56]. Zwittergent 3-14 extracted type 15:P1.16 OMPs containing only class 1, 3, and 4 proteins have been used as starting material. In the final vaccine, the OMPs are present in complexes formed with Zwittergent 3-14 or with meningococcal group C polysaccharide (CPS). $AlPO_4$ is added as adjuvant, merthiolate as preservative.

The group B *Neisseria meningitidis* strain H4476 has been cultivated in a semisynthetic medium, and supplemented with $Fe(NO_3)_3$ to suppress the formation of a 70 kDa Fe-binding protein. The fermentor-grown cells have been inactivated for 0.5 h at 56°C before extraction with Zwittergent 3-14. After the removal of cell debris, the supernatant fluid is purified by ethanol precipitation. Diafiltration is then used to eliminate ethanol and other low-molecular-weight products from the supernate and to concentrate it. Further purification of the OMPs is achieved by DEAE-Sepharose and Sephacryl S-300 chromatography.

In both steps only those fractions showing a favorable protein:endotoxin ratio are collected. The OMPs obtained after Sephacryl S-300 chromatography were combined with CPS, and this mixture was then dialyzed to remove the detergent and sterilized by membrane filtration. The solution was added to an $AlPO_4$ suspension. The final vaccine contains 100 μg of OMPs, 100 μg of CPS, 2 mg of $AlPO_4$, and 100 μg of thiomersal per milliliter.

The intended dose of the vaccine is 0.25 ml (25 μg OMP, 25 μg CPS) in a two-dose scheme. In all stages of the purification of the OMPs and in the preparation of the final vaccines, precautions have been taken to prevent bacterial contamination by maintaining a closed system. In the cases where this had to be broken, use was made of laminar air cabinets or, if their use was not possible, the solutions were sterilized by membrane filtration as soon as possible.

**2. Preparation of OMPs**

**a. Strain.** The group B *Neisseria meningitidis* strain H44/76 was isolated from a septicemia patient in Norway in 1976 by Holten [81]. It has since been used as a prototype 15:P1.16 strain by various investigators for the serotyping of other *N. meningitidis* [61] strains. The strain was stored, adsorbed onto glass beads. One primary glass bead is cultured on agar to produce a secondary culture that also is stored at −70°C. These secondary glass beads are used to inoculate agar plates, precultures, and fermentor cultures.

**b. Cultivation.** The strain is cultivated as has been described for the cultivation of group A and C meningococci [82]. The medium was slightly modified by the addition of 100 μM $Fe(NO_3)_3$ to suppress the production of a 70 kDa protein. During cultivation the pH and $pO_2$ were controlled and automatically adjusted to values of 7.0–7.2 and 10% air saturation, respectively. After the start of the stationary phase, the culture was heat inactivated (0.5 h, 56°C). The cells were harvested aseptically by centrifugation, washed with sterile 0.14 M saline, and, if not processed immediately, distributed in 0.5 liter bottles in 150 ml quantities and lyophilized.

**c. Extraction of OMPs.** The cells were extracted in a solution 0.5 M $CaCl_2$, 1% (w/v) Zwittergent 3-14 (Zw 3-14) and 0.14 M saline, pH 4.0 (100 ml per gram dry weight). After resuspension, the pH was adjusted to a value of 5–6. After stirring for 1 h at room temperature, the suspension was centrifuged (1 h, 3,000*g*) and the supernate collected aseptically.

**d. Ethanol fractionation.** Ethanol was added to the supernate to a final concentration of 20% (v/v). After stirring for 0.5 h, the suspension was transferred to sterile centrifuge bottles and centrifuged (0.5 h, 10,000*g*). After centrifugation the supernate was collected aseptically.

**e. Diafiltration.** $CaCl_2$ and ethanol were eliminated by diafiltration in an Amicon hollow fiber system (HIP 30-43; cut off, 30,000) sterilized beforehand with a 2% formalin solution. After rinsing the equipment with sterile 0.14 M saline, the supernate was concentrated and rediluted with buffer A (50 mM Tris, 10 mM EDTA, and 0.05% [w/v] Zw 3-14, pH 8.0). These steps were repeated five times. After the last concentration step the Amicon system was rinsed with a small volume of buffer A. The rinsing fluid was combined with the concentrate.

**f. Chromatography.** The OMPs were further purified by column chromatography on DEAE-Sepharose. The column was sterilized beforehand by treatment with 0.5% (v/v) formalin in 0.14 M saline and loaded with the

OMP concentrate. It was eluted with the same solution and with buffer A, after which a linear gradient was started from buffer A to buffer B (buffer A supplemented with 0.6 M NaCl). The eluate fractions were assayed for protein [83] and KDO [84]. The protein-containing fractions with a protein: KDO ratio ≥20 were combined.

**g. Concentration.** The combined fractions were concentrated in a sterile Amicon hollow fiber system. Ethanol was added to the concentrate up to a final concentration of 80% (v/v). The OMPs were isolated aseptically by centrifugation (1.5 h, 30,000*g*) and redissolved in buffer C (50 mM Tris, 0.5% [w/v] Zw 3-14 in 0.14 M saline, pH 9.0) to a protein concentration of 15 g/liter.

**h. Sephacryl S-300 chromatography.** A sterilized Sephacryl S-300 column was loaded with the protein concentrate and eluted with 0.05% (w/v) Zw 3-14 in 0.14 M saline solution. The fractions were analyzed for protein and KDO content as before. Protein-containing fractions with a protein:KDO ratio of at least 100 were combined and sterilized by Millipore GSWP filtration.

**3. Preparation of CPS.** For a detailed description of the procedures and media, see Tiesjema et al. [82].

**a. Strain** The *Neisseria meningitidis* strain C11 used for the preparation of the C polysaccharide was obtained from the Walter Reed Army Institute in the lyophilized state in June 1973.

**b. Cultivation.** Two successive precultures of the strain were made on sheep blood agar plates, the first for 18 h, the second for 12 h at 35°C in air with 10% $CO_2$. After checking the purity, a few colonies were then used for inoculation of 400 ml of liquid seed culture medium in 1 liter bottles. After incubation for 12 h at 35°C on a shaking machine and a check of the purity, 40 liters of liquid medium in a fermentor was inoculated with 400 ml of the liquid preculture. During the cultivation at 35°C, the pH and $pO_2$ were controlled. As soon as no substantial increase in optical density could be observed, the pH-drop leveled off and the $pO_2$ started rising from zero; the cultivation was stopped by cooling down the culture. The purity of the culture was checked.

**c. Cetavlon precipitation.** After centrifugation of the culture and collection of the supernatant liquid in two 20 liter bottles, Cetavlon was added to a final concentration of 0.1% (w/v). After centrifugation of the suspension (2,400*g*), the precipitate was dissolved in ~100 ml of cold 1 M $CaCl_2$.

**d. Ethanol fractionation.** Ethanol (96%) was added to this solution to a final concentration of 25% (v/v). After 1 h, the suspension was centrifuged (1 h, 50,000*g*), the clear supernatant was collected, and its ethanol concentration was increased to 80% (v/v). After 1 h, the suspension was centrifuged (20 min, 5,000*g*). The precipitate was washed with absolute ethanol, acetone, and diethylether and successively dried in a vacuum desiccator over $p_2O_5$ to constant weight. This crude polysaccharide was stored in a closed bottle at −18°C or below until further purification.

**e. Phenol extraction.** The crude polysaccharide was dissolved at a concentration of 20 mg/ml in 1:10 diluted saturated sodium acetate, pH 7.0. After stirring for 2–3 h to obtain a homogeneous suspension, an equal volume of phenol solution (100 g of phenol in 40 ml 0.5 M sodium acetate, pH 7.0) was added and the mixture stirred vigorously for 5 min. The milky suspension was centrifuged (20–60 min, 10,000*g*) and the upper water layer was collected. The extraction was repeated three times more, and the water layers were combined.

**f. Ultracentrifugation and ethanol precipitation.** After dialysis of the combined water layers for 16–24 h against three changes of 0.1 M $CaCl_2$, solution was centrifuged for 3–5 h at 100,000*g* and the clear supernatant was collected. Ethanol was added to a final concentration of 80% (v/v), and the precipitate was collected by centrifugation, washed with ethanol, acetone, and diethylether, and dried, as described above. The purified polysaccharide was stored in a closed bottle at −18°C or colder.

**4. Preparation of OMP/CPS vaccine**

**a. Combination of OMPs and CPS, dialysis.** To the Zwittergent 3-14–containing OMP solution was added meningococcal group C polysaccharide purified as described [82] to a final concentration equal to that of the OMP. The mixture was dialyzed aseptically against 0.14 M NaCl, containing 0.01% w/v thiomersal to eliminate the detergent, and sterilized by Millex-GV (thiomersal) filtration.

**b. Preparation of $AlPO_4$.** Ten liters of an $AlPO_4$ suspension were prepared aseptically by combining 0.625 liter of 1.05 M $AlCl_3$ and 5.05 liter of 0.14 M $Na_3PO_4$ solution and 1 liter of distilled water. The pH was increased from 4.5 to 7.0 by the addition of distilled water. The suspension prepared in this way has an $AlPO_4$ concentration of 8 g/liter (65.6 mM) and an NaCl concentration of 11.4 g/liter (196.8 mM).

**c. Preparation of the final bulk product.** The OMP vaccine complexed with group C polysaccharide was made by the combination of solutions of the polysaccharide-OMP complex, thiomersal, and NaCl and the $AlPO_4$ suspension to reach the intended composition of 100 μg of OMPs, 100 μg of group C polysaccharide, 100 μg of thiomersal, and 8.5 mg of NaCl and 2 mg of $AlPO_4$ per milliliter.

**d. Filling procedure.** The vaccine is filled in 3 ml quantities into 5 ml bottles.

**e. Dosage of the vaccine.** The dose of the vaccine is intended to be 0.25 ml for intramuscular use in a two-dose schedule.

**5. Composition of the vaccine.** One dose (0.25 ml ) of MPC-2 vaccine contains 25 μg OMPs (15:P1.16), 25 μg group C polysaccharide, 0.5 mg $AlPO_4$, and 25 μg thiomersal.

## B. Control

### 1. Purified OMP

1. The composition of the purified product is analyzed by SDS-polyacrylamide gel electrophoresis [85] for the presence of class 1 (P1.16 subtype), class 3 (serotype 15), and class 4 OMPs.
2. The activity of monoclonal antibodies with different specificities is tested against the purified 15:P1.16 OMPs [86]. The presence of the subtype P1.16 (class 1 OMP), serotype 15 (class 3 OMP), and the class 4 OMP is analyzed. Subtype P1.2 and serotype 2b specific monoclonal antibodies are used as negative controls.
3. The group B capsular polysaccharide is detected by ELISA [87].
4. Protein content of the purified OMP bulk product is determined according to Peterson [83].
5. Zwittergent content of the purified OMP bulk is measured after its binding to and elution from Extractie-gel D in an HPLC reversed-phase system or TLC. The detection limit of this assay is 10 μg/ml.
6. Endotoxin content is tested by KDO determination [84].
7. The bulk product is tested for sterility.

### 2. Purified C polysaccharide bulk

1. N-acetylneuramic acid content [88].
2. O-acetyl content [89].
3. Protein content: by 260 nm detection.
4. Molecular weight: by gel filtration.

### 3. Combined OMP/CPS

1. Protein content [83].
2. N-acetylneuraminic acid content [88].
3. Pyrogenicity according to WHO requirements [90,91], applying doses of 25 ng/kg and 250 ng/kg rabbit.
4. LAL test for endotoxin content [92].

### 4. Final vaccine

1. $Al^{3+}$ content.
2. Thiomersal content.

3. $Cl^-$ content.
4. Zwittergent-3,14 content.
5. pH.
6. The immunogenicity of the vaccine lots was tested in mice. The vaccine lots were injected s.c. with 2.5 μg protein and boosted after 4 weeks. The sera were assayed by ELISA using OMP or C polysaccharide as coat antigens. The bactericidal assay [93] was carried out with sera taken after the booster injection, tested with guinea pig complement.
7. Abnormal toxicity tests according to the WHO prescription [91] for the control of meningococcal polysaccharide vaccines was carried out.
8. Toxicity test in a mouse weight-gain test.
9. Sterility test.

## ACKNOWLEDGMENTS

Wilma Witkamp is gratefully thanked for her expert administrative help.

## REFERENCES


1. Heidelberger M, Avery OT: The soluble specific substance of pneumococcus I. J Exp Med 38:73–79, 1923.
2. MacLeod CM, Hodges RG, Heidelberger M, Bernhard WG: Prevention of pneumococcal pneumonia by immunization with specific capsular polysaccharides. J Exp Med 82:445, 1945.
3. Avery OT, Goebel WF: Chemo-immunological studies on conjugated carbohydrate-proteins II. Immunological specificity of synthetic sugarproteins. J Exp Med 50:521, 1929.
4. Peltola H, Kayhty H, Sivonen A, Makela PH: *Haemophilus influenzae* in children; a double-blind field study of 100,000 vaccines 3 months to 5 years of age in Finland. Pediatrics 60:730, 1977.
5. Beuvery EC, van Rossum F, Nagel J: Comparison of the induction of immunoglobulin M and G antibodies in mice with purified pneumococcal type 3 and meningococcal group C polysaccharides and their protein conjugates. Infect Immun 37:15, 1982.
6. Schneerson R, Barrera O, Sutton A, Robbins JB: Preparation, characterization and immunogenicity of *Haemophilus influenzae* type b polysaccharideprotein conjugates. J Exp Med 152:361, 1980.
7. Eskola J, Peltola H, Takola AK, Kayhty H, Hakulinen M, Karanko V, Kela E, Rekola P, Ronnberg P-R, Samuelson JS, Gordon LK, Makela PH: *Haemophilus influenzae* type b polysaccharide-diphtheria toxoid conjugate vaccine: Protective in infancy. N Engl J Med 317:717, 1987.
8. Sutton A, Schneerson R, Kendall-Morris S, Robbins JB: Differential complement resistance mediates virulence of Haemophilus influenzae type b. Infect Immun 35:95, 1982.
9. Humphrey JH, Parrott DMV, East J: Studies on globulin and antibody production in mice thymectomised at birth. Immunology 7:419, 1964.
10. Mosier DE, Feeney AJ, Scherle P: The physiology of B lymphocytes capable of generating anti-polysaccharide antibody response. In Bell R, Torrigiani G (eds): "Towards Better Carbohydrate Vaccines." Chichester: John Wiley and Sons, 1987, p 243.

11. Kabat EA: Carbohydrates and immunogens: Size, shape and nature of carbohydrate epitopes and idiotypes polysaccharide vaccines. In Bell R, Torrigiani G (eds): "Towards Better Carbohydrate Vaccines." Chichester: John Wiley and Sons, 1987, p 75.
12. Howard JG: T-cell-independent responses to polysaccharides: Their nature and delayed ontogeny. In Bell R, Torrigiani G (eds): "Towards Better Carbohydrate Vaccines." Chichester: John Wiley and Sons, 1987, p 221.
13. Mohgini PKA, Stein KE, Paul WE: T cell regulation of IgG subclass antibody production in response to T-independent antigens. J Exp Med 153:1, 1981.
14. Schrater AF, Goidl EA, Thorbecke J, Siskind GW: Production of auto-anti-idiotypic antibody during the normal immune response TNP-ficoll. III. Absence of nu/nu mice: Evidence for T-cell dependence of the anti-idiotypic antibody response. J Exp Med 150: 808, 1979.
15. Soderstrom T: Anti-idiotypes as surrogate polysaccharide vaccines. In Bell R, Torrigiani G (eds): "Towards Better Carbohydrate Vaccines." Chichester: John Wiley and Sons, 1987, p 119.
16. Geysen HM, MacFarlan R, Rodda SJ, Tribbick G, Mason TJ, Schoofs P: Peptides which mimic carbohydrate antigens. In Bell R, Torrigiani G (eds): "Towards Better Carbohydrate Vaccines." Chichester: John Wiley and Sons, 1987, p 103.
17. Lemieux RU: Applications of synthetic complex oligosaccharides to areas of molecular biology. In Bell R, Torrigiani G (eds): "Towards Better Carbohydrate Vaccines." Chichester: John Wiley and Sons, 1987, p 41.
18. Kabat EA, Liao J, Osserman EF, Gamian A, Michon F, Jennings HJ: The epitope associated with the binding of the capsular polysaccharide of the group B meningococcus and of Escherichia coli K1 to a human monoclonal macroglobulin, $IgM^{nov}$. J Exp Med 168: 699, 1988.
19. Crowle AJ: Immunization against tuberculosis: What kind of vaccine? Infect Immun 56:2769, 1988.
20. Handman E, McConville MJ, Goding JW: Carbohydrate antigens as possible parasite vaccines. Immunol Today 8:181, 1987.
21. Jennings JH, Roy R, Gamian A: Induction of meningococcal group B polysaccharide-specific IgG antibodies in mice by using an N-proionylated B polysaccharide-tetanus toxoid conjugate vaccine. J Immunol 137:1708, 1986.
22. Sinigaglia F, Guttinger M, Kilgus J, Doran DM, Matile H, Ettinger H, Trzeciak A, Gillenessen D, Pink JRL: A malaria T-cell epitope recognized in association with most mouse and human Mnc class II molecules. Nature 336:778, 1988.
23. Lai E, Kabat EA, Mobraaten L: Genetic and nongenetic control of the immune response of mice to a synthetic glycolipid stearylisomaltotetraose. Cell Immunol 92:172, 1985.
24. Osterholm MT, Rambeck JH, White KE, Jacobs JL, Pierson LM, Neaton JD, Hedberg CW, MacDonald KL, Granoff DM: Lack of efficacy of Haemophilus b polysaccharide vaccine in Minnesota. JAMA 260:1423, 1988.
25. Insel RA, Anderson PW: Response to oligosaccharide-protein conjugate vaccine against *Haemophilus influenzae* b in two patients with IgG2 deficiency unresponsive to capsular polysaccharide vaccine. N Engl J Med 315:499, 1986.
26. Makela O, Mattila P, Rautonen N, Seppala I, Eskola J, Kayhty H: Isotype concentrations of human antibodies to *Haemophilus influenzae* type b polysaccharide (Hib) in young adults immunized with the polysaccharide as such or conjugated to a protein (diphtheria toxoid). J Immunol 139:1999, 1987.
27. Weinberg GA, Granoff DM, Nahm MH, Shackelford PG: Functional activity of different IgG subclass antibodies against type b capsular polysaccharide of *Haemophilus influenzae*. J Immunol 136:4232, 1986.

28. Schneerson R, Barrera O, Sutton A, Robbins JB: Preparation, characterization and immunogenicity of *Haemophilus influenzae* type b polysaccharide-protein conjugates. J Exp Med 152:361, 1980.
29. Lepow ML, Samuelson JS, Gordon LK: Safety and immunogenicity of *Haemophilus influenzae* type b polysaccharide-diphtheria toxoid conjugate vaccine in adults. J Infect Dis 150:402, 1984.
30. Anderson P, Pichichero ME, Insel RA, Betts R, Eby R, Smith DH: Vaccines consisting of periodate-cleaved oligosaccharides from the capsule of *Haemophilus influenzae* type b coupled to protein carrier: Structural and temporal requirements for priming in the human infant. J Immunol 137:1181, 1986.
31. Marburg S, Jorn D, Tolman RL, Arison B, McCauly J, Kniskern PJ, Hagopian A, Vella PP: Bimolecular chemistry of macromolecules: Synthesis of bacterial polysaccharide conjugates with Neisseria meningitidis membrane protein. J Am Chem Soc 108:5282, 1986.
32. Hoogerhout P, Evenberg D, van Boeckel CAA, Poolman JT, Beuvery EC, van de Marel GA, van Boom JH: Synthesis of fragments of the capsular polysaccharide of *Haemophilus influenzae* type b, comprising two or three units. Tetrahedron Lett 28:1553, 1987.
33. Weinberg GA, Granoff DM: Polysaccharide-protein conjugate vaccine for the prevention of *Haemophilus influenzae* type b disease. J Pediatr 113:621, 1988.
34. Makela PH: Personal communication, 1987.
35. Ward JI, Brenneman G, Lepow M, Lum M, Burkhart K, Chiu CY: *Haemophilus influenzae* type b anticapsular immunity responses to PRP-pertussis and PRP-D vaccines in Alaska native infants. J Infect Dis 158:719, 1988.
36. Claesson BA, Trollfors B, Lagergard T, Taranger J, Bryla D, Otterman G, Cramton T, Younghong R, Reimer CB, Robbins JH, Schneerson R: Clinical and immunological responses to the capsular polysaccharide of *Haemophilus influenzae* type b alone or conjugated to tetanus toxoid in 18- to 23-month-old children. J Pediatr 112:695, 1988.
37. Cryz SJ Jr, Furer E, Sadoff JC, Germanier R: *Pseudomonas aeruginosa* immunotype 5 polysaccharide-toxin A conjugate vaccine. Infect Immun 52:161, 1986.
38. Svenson SB, Lindberg AA: Coupling of acid labile *Salmonella* specific oligosaccharides to macromolecular carriers. J Immunol Methods 120:1750, 1979.
39. Hooke AM: Attenuated live bacteria: Vaccine efficacy and shortcomings. In Bell R, Torrigiani G (eds): "Towards Better Carbohydrate Vaccines." Chichester: John Wiley and Sons, 1987, p 167.
40. Saukkonen K, Leinonen MP, Kayhty H, Abdillahi H, Poolman JT: Monoclonal antibodies to the rough lipopolysaccharide of *Neisseria meningitidis* protect infant rats from meningococcal infection. J Infect Dis 158:209, 1988.
41. Griffis J, Brandt BL, Broud DD, Goroff DK, Baker CJ: Immune response of infants and children to disseminated *Neisseria meningitidis* infection. J Infect Dis 150:71, 1984.
42. Parsons NJ, Patel PV, Martin PMV, Tan EL, Nairn CA, Goldner M, Brogden KA, Smith H: Further studies on the host and bacterial determinants of gonococcal resistance to killing by human serum and by phagocytes. In Poolman JT, Zanen HC, Meyer TF, Heckels JE, Mähelä PRH, Smith H, Beuvery EC (eds): "Gonococci and Meningococci." Dordrecht: Kluwer Academic, 1988, p 775.
43. Jennings HJ, Beurret M, Gamian A, Michon F: Structure and immunochemistry of meningococcal lipopolysaccharides. In Poolman JT, et al. (eds): "Gonococci and Meningococci." Dordrecht: Kluwer Academic, 1988, p 553.
44. Boons GJPH, van der Klein PA, van der Marel GA, van Boom JH: Synthesis of L-glycero-D-manno-heptose. Red Trav Chim Pays-Bas 107:507, 1988.
45. Sawada S, Suzuki M, Kawamura T, Fujinaga S, Masuho Y, Tomibe K: Protection against

infection with *Pseudomonas aeruginosa* by passive transfer of monoclonal antibodies to lipopolysaccharides and outer membrane proteins. J Infect Dis 150:570, 1984.
46. Saukkonen K, Abdillahi H, Poolman JT, Leinonen M: Protective efficacy of monoclonal antibodies to class 1 and class 3 outer membrane proteins of *Neisseria meningitidis* B;15: P1.16 in infant rat infection model: New prospects for vaccine development. Microbiol Pathogen 3:261, 1987.
47. Granoff DM, Munson RS: Prospects for prevention of *Haemophilus influenzae* type b disease by immunization. J Infect Dis 153:448, 1986.
48. Murphy TF, Campagneri AA, Nelson B, Apicella MA: Somatic antigens of *Haemophilus influenzae* as vaccine components. Pediatr Infect Dis J 8, S66–S69, 1989.
49. Lynch EC, Blake MS, Gotschlich EC, Mauro A: Studies of porins: Spontaneously transferred from whole cells and reconstituted from purified proteins of Neisseria gonorrhoeae and Neisseria meningitidis. Biophys J 45:104, 1984.
50. Lambden PR, Heckels JE, James LT, Watt PJ: Variations in surface protein composition associated with virulence properties in opacity types of *Neisseria gonorrhoeae*. J Gen Microbiol 114:305, 1979.
51. Poolman JT, de Marie S, Zanen HC: Variability of low-molecular-weight, heat-modifiable outer membrane proteins of *Neisseria meningitidis*. Infec Immun 30:642, 1980.
52. Poolman JT, Hopman CTP, Zanen HC: Immunogenicity of meningococcal surface antigens as detected in patient sera. Infec Immun 40:398, 1983.
53. Rice PA, Blake MS, Joiner KA: Mechanisms of stable serum resistance of *Neisseria gonorrhoeae*. In Poolman JT, Zanen HC, Meyer TF, Heckels JE, Mähelä PRH, Smith H, Beuvery EC (eds): ''Gonococci and Meningococci.'' Dordrecht: Kluwer Academic, 1988, p 789.
54. Frasch CE, Zollinger WD, Poolman JT: Serotype antigens of *Neisseria meningitidis* and a proposed scheme for designation of serotypes. Rev Infect Dis 7:504, 1985.
55. Abdillahi H, Poolman JT: *Neisseria meningitidis* group B serosubtyping using monoclonal antibodies in whole-cell ELISA. Microbiol Pathogen 4:27, 1988.
56. Poolman JT, Timmermans JAM, Hopman CTP, Teerlink T, van Vught PAM, Witvliet MH, Beuvery EC: Comparison of meningococcal outer membrane protein vaccines, solubilized with detergent of C polysaccharide. Antonie Van Leeuwenhoek 53:413, 1987.
57. Goldschneider I, Gotschlich EC, Artenstein MS: Human immunity to the meningococcus. I. The role of humoral antibodies. J Exp Med 129:1307, 1969.
58. Froholm LO, Berdal BP, Bovre K, Gaustad P, Halstensen AI, Harboe A, Harthug S, Holten E, Hoiby EA, Lystad A, Omland T, Rosenqvist E, Viko G, Frasch CE, Zollinger WD: Preliminary results from a clinical trail with a meningococcal vaccine containing serotype 2b and 15 antigens in complex with mixed A, C, Y and W135 polysaccharides. Antonie Van Leeuwenhoek 52:239, 1986.
59. Frasch CE, Zahradnik JM: Human immune response to an aluminum hydroxide absorbed *Neisseria meningitidis* serotype 2b protein vaccine. Antonie Van Leeuwenhoek 52:229, 1986.
60. Zollinger WD, Boslego W, Brandt B, Moran EE, Ray J: Safety and antigenicity studies of a polyvalent meningococcal protein-polysaccharide vaccine. Antonie Van Leeuwenhoek 52:225, 1986.
61. Poolman JT, Hopman CTP, Zanen HC: Problems in the definition of meningococcal serotypes. FEMS Lett 13:339, 1982.
62. Mandrell RE, Griffiss JM, Macher BA: Lipooligosaccharides (LOS) of *Neisseria gonorrhoeae* and *Neisseria meningitidis* have components that are immunochemically similar to precursors of human blood group antigens. J Exp Med 168:107, 1988.
63. Kersten GF, Teerlink T, Derks HJGM, Verkleij AJ, Van Wezel TL, Crommelin DJA,

Beauvery EC: Incorporation of the major outer membrane protein of *Neisseria gonorrhoeae* in saponin-lipid complexes (iscoms): Chemical analysis, some structural features, and comparison of their immunogenicity with three other antigen delivery systems. Infect Immun 56:432, 1986.

64. Agterberg M, Adriaanse H, Tommassen J: Use of outer membrane protein pho E as a carrier for the transport of a foreign antigenic determinant to the cell surface of Escherichia coli U12. J Gene 59:145, 1987.
65. van der Leij P, Struyve M, Tommassen J: Topology of outer membrane pore proteins of Escherichia coli. Identification of cell surface-exposed amino acids with the aid of monoclonal antibodies. J Biol Chem 261:12222, 1986.
66. Poolman JT, Timmermans JAM, Teerlink T, Seid RC Jr: Purification cyanogen bromide cleavage and amino terminus sequencing of class 1 and 3 outer membrane proteins of meningococci. Infect Immun, 1989 (in press).
67. Barlow AK, Heckels JE, Clarke IN: The class 1 outer membrane protein of *Neisseria meningitidis:* Gene sequence and structural and immunological similarities to gonococcal porins. Mol Microbiol, 1989 (in press).
68. Rutter JM, Jones GW: Protection against enteric disease caused by *Escherichia coli*—Model for vaccination with a virulence determinant. Nature 242:531, 1973.
69. Brinton CC Jr, Wood SW, Brown A, Labik AM, Lee JR, Polen SW, Tramont EC, Sadoff JC, Zollinger WD: The development of a neisserial pilus vaccine for gonorrhea and meningococcal meningitis. In Robbins JB, Weinstein L, Fields BN, Hill JC, Sadoff JC (eds): "Bacterial Vaccines." New York: Thieme-Stratton, 1982, p 140–150.
70. Tramont EC, Boslego JW, Chung R, McChesney D, Ciak J, Sadoff J, Piziak M, Brinton CC, Wood S, Bryan J: Parenteral gonococcal pilus vaccine. In Schoolnik GK, Brooks GF, Falkow S, Frasch CE, Knapp JS, McCutchan JA, Morse SA (eds): "The Pathogenic Neisseriae." Washington, DC: American Society for Microbiology, 1985, p 316.
71. Buchanan TM: Antigenic heterogeneity of gonococcal pili. J Exp Med 141:1470, 1975.
72. Rothbard JB, Fernandez R, Schoolnik GK: Strain-specific and common epitopes of gonococcal pili. J Exp Med 160:208, 1984.
73. Brinton CC, Carter MJ, Derber DB, Kar S, Kramarik JA, To AC-C, To SC-M, Wood SW: Design and development of pilus vaccines for *Haemophilus influenzae* diseases. Pediatr Infect Dis J 8:S54, 1989.
74. Sastry PA, Finlay BB, Pasloske BL, Paranchych W, Pearlstone JR, Smillie LB: Comparative studies of the amino acid and nucleotide sequences of pilin derived from *Pseudomonas aeruginosa* PAK and PAO. J Bacteriol 164:571, 1985.
75. Zhong ML, Brennan MJ, David JL, Carter PH, Cowell JL, Manclarck CR: Comparison of type 2 and type 6 fimbriae of *Bordetella pertussis* by using agglutinating monoclonal antibodies. Infect Immun 56:3184, 1988.
76. Ah hoc group for studies of pertussis vaccines. Placebo controlled field trial of 2 acellular pertussis vaccines in Sweden—protective efficacy and adverse events. The Lancet 1, 955–960, 1989.
77. Blennow M, Olin P, Granstrom M, Bernier RH: Protective efficacy of a whole cell pertussis vaccine. Br Med J 296:1570, 1988.
78. Drake D, Montie TC: Protection against *Pseudomonas aeruginosa* infection by passive transfer of anti-flagellar serum. Can J Microbiol 33:755, 1987.
79. Allison JS, Dawson M, Drake D, Montie TC: Electrophoretic separation and molecular weight characterization of *Pseudomonas aeruginosa* H-antigen flagellins. Infect Immun 49:770, 1985.
80. Beachey EH: Protective immunogenicity of chemically synthesized peptide fragments of

group A streptococcal M proteins. In Jackson GG, Thomas H (eds): "The pathogenesis of bacterial infections." Berlin: Springer-Verlag, 1985, p 208.
81. Holten E: Serotypes of *Neisseria meningitidis* isolated from patients in Norway during the first six months of 1978. J Clin Microbiol 9:186, 1979.
82. Tiesjema RH, van Hemert PA, van den Heuvel N, te Pas BJ, Smid P: Productie en controle van een gevriesdroogd groep A en C meningococcen polysaccharide vaccin. RIVM rapport 44/76, 1976.
83. Peterson GL: A simplification of the protein assay method of Lowry et al which is generally applicable. Anal Biochem 83:346, 1977.
84. Karkahnis YD, Zeltner JY, Jackson JJ, Carlo J: A new and improved microassay to determine 2-keto-3-deoxyoctonate in lipopolysaccharide of gram-negative bacteria. Anal Biochem 85:680–685, 1978.
85. Laemmli UK: Cleavage of structural proteins during the assembly of the head of the bacteriophage T4. Nature 227:680, 1970.
86. Beuvery EC, Miedema F, Van Delft RW, Haverkamp J, Leussink AB, Te Pas BJ, Teppema JS, Tiesjema RH: Preparation and physiochemical and immunological characterization of polysaccharide outer membrane protein complexes of *Neisseria meningitidis*. Infect Immun 40:369, 1983.
87. Beuvery EC, Van Delft RW, Tiesjama RH, Nagel J: The enzyme-linked immunosorbent assay of meningococcal and some related *Escherichia coli* polysaccharides. J Biol Stand 11:195, 1983.
88. Svennerholm L: Quantitative estimation of sialic acids. II. A colorimetric resorcinol-hydrochloric acid method. Biochim Biophys Acta 24:604, 1957.
89. Hestrin S: The reaction of acetylcholine and other carboxylic acid dervates with hydroxylamine and its analytical application. J Biol Chem 180:249, 1949.
90. European Pharmacopoeia, 2nd edition: Test for non-pyrogenicity.
91. Requirements for meningococcal polysaccharide vaccine. WHO Tech Rep Ser 594:50, 1976.
92. Kreeftenberg JG, Loggen HG, Ramshost JD, en Beuvery EC (1977). The limulus amebocyte lysate test; micromethod and application in the control of sera and vaccines. Develop. Biol. Standard. 34, 15–20 (S. Karger, Basel).
93. Poolman, J.T., Wientjes, F.B., Hopman, C.T.P., Zanen, H.C. (1985). Influence of the length of lipopoly saccharide molecules on the surface exposure of class 1 and 2 outer membrane proteins of *Neisseria meningitidis* (B:2b:P1.2). The pathogenic Neisseria, proc. IVth Int. Symp., (American society for Microbiology, Washington, D.C.). pp. 562–570.

Bacterial Vaccines, pages 87–103


# *Klebsiella* Polysaccharide Vaccines

**S.J. Cryz, Jr.**

Swiss Serum and Vaccine Institute, 3001 Berne, Switzerland



## I. INTRODUCTION

### A. Causative Agent

The genus *Klebsiella* is a member of the family Enterobacteriaceae. Two species of *Klebsiella* are pathogenic for man: *K. pneumoniae,* including subspecies *pneumoniae, ozaenae,* and *rhinoscleromatis,* and *K. oxytoca. Klebsiella* spp. are nonmotile gram-negative rods. They are facultative anaerobes and not nutritionally fastidious. Almost all clinical isolates of *Klebsiella* are encapsulated. The amount of capsule produced is dependent upon both innate strain characteristics and growth media. It is not uncommon, therefore, to isolate "mucoid" strains of *Klebsiella* from clinical specimens. Upon passage on solid media, colonial variants that produce markedly less capsule can be found.

### B. Epidemiology

The majority of life-threatening infections caused by *Klebsiella* are hospital acquired [1–3]. Numerous epidemiological investigations have focused upon *Klebsiella* within the hospital environment because of its propensity to cause epidemics. Several methods, including capsular serotyping [4], bacteriocins [5], and biochemical profiles [6], have been used to study the epidemiology of *Klebsiella*. Of these, capsular serotyping has gained the most favor because of the large number of distinct serotypes and its proven ability to track strains within a hospital.

Both the hospital environment and its personnel can serve as a reservoir for *Klebsiella* [2,3]. Although *Klebsiella* have been isolated from inanimate objects, especially associated with moist areas such as sinks, there is no direct evidence to implicate such isolates with disease. In contrast, there are considerable data to indicate that the hospital staff plays a crucial role in transmission of *Klebsiella* to patients. For example, *Klebsiella* can often be cultured from the hands of nurses or other hospital staff [7,8]. Casewell and Phillips [9] have shown conclusively that patients can readily acquire *Klebsiella* after handling by carriers.

Normal healthy adults can harbor *Klebsiella* among their intestinal flora. However, it is difficult to say what percentage of the population are carriers and how long carriage can persist [2]. What is known for certain is that hospitalized patients are more likely to harbor *Klebsiella* in their bowels than healthy individuals. Several investigators have linked this trend to antibiotic usage [10–13]. For example, carriage rates increased from 11 to 34% in patients receiving antibiotics, while the rate remained unchanged for hospitalized patients not receiving antibiotics [11]. Intestinal colonization has also been shown to predispose patients to subsequent infection. Therefore, while 14 of 31 patients who became colonized during hospitalization developed an infection, only 11 of 101 noncolonized patients did so. Colonization of neonates by *Klebsiella* occurs more readily than in children or adults and is not markedly influenced by antibiotic usage. During *Klebsiella* outbreaks among closed nursery units, carriage rates can approach 100%, illustrating the propensity of *Klebsiella* to rapidly spread among a susceptible population [14]. *Klebsiella* can also colonize the nasopharynx and skin, although the relevance to subsequent transmission and disease is unknown [2].

Considerable effort has been devoted to the seroepidemiology of *Klebsiella*. These studies have attempted to define the overall prevalence of given capsular serotypes among clinical specimens in order to show the predilection of a given serotype for an anatomical site or an increase in the virulence of a given serotype. At present, 77 distinct *Klebsiella* capsular serotypes have been identified [15]. Most investigators find that approximately 90% of *Klebsiella* clinical isolates are typeable by a given technique or combination of techniques [16–18].

Most seroepidemiological studies involving large numbers of *Klebsiella* clinical isolates have shown that ≥50% of the known capsular types will be found [6,16–18]. However, certain serotypes were observed at a higher overall frequency than others, while other serotypes showed a predilection for certain anatomical sites. Martin et al. [17] found that, of 213 isolates studied, serotypes 35 and 2 were the most common and that serotype 2 clearly predominated in the urine. In a study involving 357 isolates from 20 hospitals, Casewell and Talsania [18] found that serotypes 21, 2, and 9 accounted for 42, 16.5, and 7.3%, respectively, of all strains. Simoons-Smit et al. [6] noted that although serotypes 55, 18, 69, and 8 accounted for nearly 50% of neonatal isolates, serotypes 33, 21, and 24 predominated among isolates from other hospital units. In contrast, some studies have seen no clear predominance of a given serotype or serotypes [19,20].

The lower capsular types (serotypes 1–6) appear to be most commonly associated with the respiratory tract [21,22], although serotype 2 is also frequently isolated from urine. Serotypes 9, 10, and 24 are also found more frequently in urine [21,23,24]. Similarly, serotype 8 showed a predilection for the skin, whereas serotype 68 was most commonly found in the stool [21]. Several authors have postulated that these serotypes possess some innate advantage, probably due to capsular type and/or other as-yet-unidentified factors, that allows them to more readily colonize a patient and, therefore, cause disease [18,21,25].

In most of the above-cited studies, *Klebsiella* isolates were sequentially isolated from routine specimens submitted to the hospital's microbiology laboratory. Hence, their causative relationship to a disease state can not be assumed because of the ability of *Klebsiella* to colonize various anatomical sites without causing overt infection. In an attempt to address the question of which capsular types are associated with infections of clinical significance, several investigators have restricted their analyses to bacteremic isolates [16,26–28]. In the largest of these studies, Cryz et al. [16] serotyped 703 *Klebsiella* blood isolates from 13 hospitals in North America and Europe. A majority (90%) of the 77 serotypes were identified. Three serotypes—2, 21, and 55—were found at a significantly ($P < 0.05$) higher frequency than all other serotypes together and represented 20.5% of all isolates. The remainder of the serotypes occurred at a frequency ranging from 0.16 to 2.8%. Approximately half of the capsular serotypes found appeared at an individual frequency of <0.5%. Similar results were obtained in three prior studies where serotypes 2, 16, 18, 21, and 24 were most frequently seen [26–28].

The above seroepidemiological studies have important implications for vaccine development based upon serospecific capsular polysaccharides [16] (see below). Any such vaccine must be multivalent in order to "cover" a substantial percentage of clinical isolates. However, selection of vaccine serotypes should not only be based upon overall frequency of occurrence. It is also critical to know the distribution of serotypes among various hospitals in order to include broadly distributed serotypes [16]. Attention must also be given to the fact that there is considerable immunological cross-reactivity between several serotypes. For certain pairs of serotypes, such as 2 and 13,

**TABLE I. Factors Predisposing to *Klebsiella* Infection**

| | |
|---|---|
| Leukemia | Surgical incisions |
| Malignancy | Pulmonary intubation |
| Reduced pulmonary function | Urinary catherization |
| Low birth weight | Ankylosing spodylitis |

3 and 68, 14 and 64, and 7 and 10, immunization with a purified capsular polysaccharide of one serotype engenders an immune response of nearly equal magnitude to the other capsular antigen [25] (author's unpublished observations).

## II. CLINICAL SIGNIFICANCE

The potential for *Klebsiella* to cause life-threatening disease was initially noted in 1882, when Frieländer first isolated the bacillus from pneumonia patients. Until the antibiotic era, *Klebsiella* was rarely associated with infections of any consequence. Several investigators first noticed an increase in hospital-acquired *Klebsiella* infections during the 1950s [29,30]. *Klebsiella* is now known to be a leading cause of nosocomial infections, resulting in approximately 20 infections per 10,000 patient discharges [3]. *Klebsiella* is second only to *Escherichia coli* as the overall cause of gram-negative bacteremia [31–33]. Numerous patient populations are at an increased risk of acquiring a serious *Klebsiella* infection, usually because of an underlying host defect or manipulative procedure (Table I).

*Klebsiella* bacteremia carries a high attendant mortality rate (Table II). This can partially be attributed to the seriousness of the underlying host defect. However, the recent occurrence of endemic hospital strains of *Klebsiella* that are multiply resistant to antibiotics has greatly complicated therapeutic regimens. Several investigators have documented an increased mortality rate for infection in which the strain of *Klebsiella* was resistant to the therapeutic agent employed [3,39,40]. Numerous outbreaks of *Klebsiella* in the hospital setting have been described in the 1970s, in most cases caused by strains resistant to aminoglycosides and/or early-generation cephalosporins [41–43]. The rapid spread of aminoglycoside-resistant strains within a hospital and the transfer of resistance from *Klebsiella* to other gram-negative nosocomial pathogens by transmissible R factors has been well documented [43–45]. Therefore, *Klebsiella* can serve as the "hub" in the spread of multiple resistance.

Until recently, most clinical isolates of *Klebsiella* were uniformly sensitive to third-generation cephalosporins [46,47]. However, several reports have now described emerging resistance among *Klebsiella* to these agents [40,48–50]. This is of particular concern because cephalosporins together

**TABLE II. Mortality Rates for *Klebsiella* Bacteremia**

| Study years | No. of cases | Fatality rate (%) | Reference |
|---|---|---|---|
| 1956–1966 | 30 | 43 | 34 |
| 1965–1974 | 74 | 24 | 33 |
| 1972–1977 | 117 | 38 | 35 |
| 1975–1977 | 25 | 48 | 36 |
| 1977–1982 | 100 | 25 | 1 |
| 1982–1983 | 20 | 30 | 37 |
| 1983–1986 | 17 | 35 | 38 |

with an aminoglycoside have become the treatment regimen of choice for *Klebsiella* sepsis. Concurrent aminoglycoside and cephalosporin resistance has also been noted and appears to be mediated by a transmissible R factor [40]. This R factor can be acquired by other Enterobacteriaceae and has been shown to spread rapidly within the hospital environment [40]. A particularly vexing problem is the emergence of resistant strains in vivo during cephalosporin treatment [40,49].

A combination of increased clinical significance, multiply resistant strains, and a persistently high mortality rate has stimulated research toward the development of immunological agents for the control of *Klebsiella* infections.

## III. VIRULENCE FACTORS OF *KLEBSIELLA*

The polysaccharide capsule (CPS) of *Klebsiella* is a critical virulence factor. As noted above, most, if not all, *Klebsiella* clinical isolates are encapsulated. *Klebsiella* CPSs are acidic polysaccharides composed of repeating units consisting of galacturonic or glucuronic acid linked to hexoses and 6-deoxyhexoses [51,52]. Acetyl, formyl, or keto-linked pyruvate groups can also be found on some serotypes.

It has been well documented that nonencapsulated strains or variants possessing a smaller capsular phenotype are far less virulent in experimental infection models [53,54] (author's unpublished observations). Such strains are unable to establish an infection and/or persist in vivo [53,54]. Several mechanisms have been proposed to explain this occurrence. The most likely is that *Klebsiella* CPS interferes with opsonization mediated by either anti-CPS or anti-lipopolysaccharide (LPS) antibody [55–57]. This could occur at several different levels. First, CPS could physically detach from the bacterial cell upon antibody binding. Second, CPS released from the invading bacteria could serve to bind anti-CPS antibody away from the cell, thereby negating its protective effect. Third, CPS could "mask" other surface components, such as LPS or outer-membrane proteins, from circulating antibody. Studies by Kadurugamuwa and coworkers [57] have shown that subinhibitory amounts of cephalosporins, which reduce capsule formation by *Klebsiella*, allow bacteria to be agglutinated by anti-LPS antisera. However, full expression of the capsule prevents this effect. *Klebsiella* CPS also appears to play a role in conferring serum resistance [56]. Trautmann et al. [58], using a K2-specific monoclonal antibody, have shown that the capsule interferes with the deposition of complement onto the bacterial cell surface. Hence, the critical step of complement activation is aborted.

*Klebsiella* CPS also may be able to induce a state of immune paralysis [59–61]. Ørskov [59] noted that when rabbits were immunized with a heavily encapsulated strain of *Klebsiella* serotype 5, little anti-CPS antibody was produced. In contrast, a K5 strain that possessed a much smaller capsule induced good levels of anti-CPS antibody. Paralysis by injecting crude CPS was dose dependent, i.e., small quantities of CPS induced protective levels of antibody whereas large quantities did not [60,61]. Immunological nonresponsiveness to serospecific CPS lasted approximately 3 months and at that time could be overcome by administering small doses of antigen. Yokochi et al. [62,63], using an in vitro cultivation system, demonstrated that *Klebsiella* CPS could act to block the generation of macrophages from precursor cells. This effect was not due to toxicity associated with the CPS preparation.

There appears to be a correlation between animal virulence and *Klebsiella* capsular serotype. Mizuta et al. [64] noted that of 82 *Klebsiella* isolates of the 01 LPS serotype, only those possessing either K1 or K2 capsule were highly virulent for mice when administered intraperitoneally. However, not all 01:K1 and 01:K2 strains were virulent, indicating the need for additional virulence factors. The authors found that strain virulence did not depend upon the site from which the clinical isolates were obtained. Using a murine model whereby virulence was defined as the ability to persist and multiply in skin, strains expressing the K1, K2, K4, and K5 serotypes were far more virulent than all other serotypes tested [65]. Again, there was no correlation between virulence and site of isolation. Similarly, we have found that only *Klebsiella* K1 or K2 strains are virulent in a variety of animal models, including a murine burn-wound sepsis model [53], a lobar pneumonia model [66], and an immunosuppression model [67] (D. Woods, personal communication and author's unpublished observations).

Evidence to support an important role for *Klebsiella* CPS in the course of human disease was presented by Pollack [68]. There was a significantly higher mortality rate in bacteremic patients with CPS antigenemia than in patients without antigenemia. Antigenemia and antigenuria also correlated with a more severe disease state and poorer prognosis. The precise role of the free antigen was not defined, but evidence suggests that it may bind preexisting anti-CPS antibody away from the invading organisms or the infectious foci.

LPS has been shown to contribute to the serum-resistant phenotype of *Klebsiella* clinical isolates [56,69,70]. Of critical importance is the length of the O-polysaccharide side chains. Only ''smooth'' LPS was able to completely inhibit complement-mediated bacteriolysis. LPS may also allow longer intracellular survival of *Klebsiella* subsequent to phagocytosis [55]. Straus and coworkers [71,72] have identified an ''extracellular toxic complex'' (ETC) produced by *Klebsiella* that is composed of CPS, LPS, and protein. ETC was able to kill mice when injected intraperitoneally and to induce pulmonary pathology similar to that seen following experimental *Klebsiella* pneumonia when instilled into the lungs of rats. Toxicity of ETC is attributed to the LPS component and can be neutralized by anti-LPS antibody.

The ability of *Klebsiella* to establish an infection depends upon the ability of the strain to colonize a mucosal surface. Two mechanisms for attachment have been described for *Klebsiella*. *Klebsiella* produce a type 1 pili that has been shown to mediate attachment to the surface of rat bladders [73]. Nonpiliated variants caused far less renal scarring than a wild-type strain in a rat pyelitis model [74]. Nonencapsulated strains of *Klebsiella* can bind to epithelial cells through a mannose-inhibitable adherence (MIAT) mechanism involving outer-membrane proteins [75–77]. MIAT-positive strains were more virulent for mice than MIAT-negative variants. This was attributed to the fact that MIAT strains are more resistant to both phagocytosis and intracellular killing by polymorphonuclear leukocytes [77]. The role of MIAT-mediated binding in human disease, where most isolates are encapsulated, is unknown.

As noted above, not all K1- and K2-bearing strains of *Klebsiella* are equally virulent. It has been recently shown that all K1 and K2 strains studied produced the siderophore enterochelin. Only those strains that are highly virulent also produce aerobactin [78]. The gene encoding aerobactin and its receptor protein was found on a 180 kilobase plasmid. Avirulent *Klebsiella* K1 or K2 strains that received the plasmid showed a marked increase in virulence.

## IV. IMMUNITY TO *KLEBSIELLA:* EXPERIMENTAL ANIMAL MODELS

For reasons described above, most attempts to protect against experimental *Klebsiella* infections have centered around the use of CPS-containing vaccines. Ribosomal vaccines prepared from *Klebsiella* K1 or K2 strains were able to protect mice in a serospecific manner [79]. Protection could also be obtained by passive transfer of immune sera. Further studies demonstrated that CPS contained in the ribosomal vaccine was the protective moiety [80]. Therefore, ribosomal preparations extracted from a nonencapsulated variant were nonprotective. In addition, treatment of the K2 vaccine with K2-specific bacteriophage glycanase destroyed its protective capacity, but had no effect on a K1 ribosomal vaccine. The O serotype of the strain from which vaccines were prepared had no bearing on their protective capacities. Crude cell surface preparations (CSPs) were made by extracting cells with lithium acetate [81]. CSP, which contained both CPS and LPS, afforded protection against a live challenge in a serospecific manner as related only to the capsular component. Surprisingly, purified CPS, when employed as an active immunogen, provided only minimal protection against a live challenge [81]. The authors speculate that those results are due to the poor immunogenicity of purified CPS. It was postulated that the ability of CPS to evoke a protective immune response is dependent upon its presentation either on the bacterial cell surface or as part of ribosomal extracts.

Jones and Roe [82,83] have produced crude CPS preparations from various serotypes of *Klebsiella*. Surprisingly, a given monovalent vaccine was able to afford protection in mice against as many as 30 different serotypes. A 12-valent vaccine was effective against most (71) of the 77 known serotypes. The authors concluded that *Klebsiella* CPSs can induce protective antibodies of a broadly cross-reactive nature. However, a number of points concerning these studies remain unclear. Most importantly, the vaccines were not characterized as to relative content (CPS, LPS, protein, etc.). Given the purification procedure used, these vaccines surely contained a significant amount of LPS. Given the limited number of *Klebsiella* LPS serotypes [25], the protection seen may be due to anti-LPS antibody, since antibody levels to neither CPS nor LPS were not correlated with protection. An additional concern is the fact that a single dose of vaccine protected against a challenge 4 days later. This interval would not be expected to allow generation of specific antibody in unprimed animals. A distinct possibility is that such LPS-containing vaccines stimulated a nonspecific immunity mediated by the release of lymphokines or other families of immune modulators.

Cryz et al. [84,85] have been able to purify numerous *Klebsiella* CPSs essentially to homogeneity by use of detergent precipitation, ethanol precipitation, ultracentrifugation, and extraction in organic solvents. *Klebsiella* CPSs purified in such a manner contained only trace elements of nucleic acids, proteins, and LPS, were of a high molecular weight, were nonpyrogenic, and were nontoxic when 1 mg was administered intraperitoneally to mice. Immunization of mice with K1 CPS provided a high degree of protection against fatal *Klebsiella* K1 burn-wound sepsis [86]. Hyperimmune rabbit anti-K1 CPS immunoglobulin G (IgG) was also capable of protecting mice when passively transferred [85]. Similarly, rats immunized with K2 CPS were protected against fatal *Klebsiella* K2 pneumonia [87]. Protection afforded by purified *Klebsiella* CPS in the above studies was serospecific. These results indicated that these highly purified, high-molecular-weight, nontoxic *Klebsiella* CPSs warranted evaluation as human vaccine candidates.

## V. HUMAN VACCINE STUDIES WITH *KLEBSIELLA* POLYSACCHARIDES

As noted above, any potentially useful *Klebsiella* vaccine based upon serospecific CPSs would have to be multivalent (>20 serotypes). Preliminary studies in our laboratory showed that most purified *Klebsiella* CPS preparations possessed a minimal pyrogenic dose (defined as a ≥0.5%°C increase in body temperature) of ≥0.5 μg/kg when intravenously administered to rabbits. While monovalent preparations would be expected to be safe when parenterally administered to humans, a polyvalent (>20 antigens) vaccine would probably contain a sufficient amount of LPS to evoke unwanted side reactions. Attempts to remove the small quantity of LPS in preparations of purified CPS by a variety of physical means was unsuccessful. We found, however, that residual LPS could be detoxified by treatment of the CPSs in a solution of 95% ethanol containing 0.1 N NaOH [84,88]. This procedure, which removed fatty acids from the lipid A portion of LPS, had a negligible effect on the size of the polysaccharides and did not greatly diminish their immunogenicity in mice.

As a first step in the evaluation of *Klebsiella* CPSs as human vaccines, healthy adult volunteers were immunized subcutaneously with either 25 or 50 μg of native K1 CPS or NaOH-treated K1 CPS [88]. Reactions following vaccination with the NaOh-treated vaccine were far less frequent and milder in nature than with native K1 CPS. Both vaccines were equally immunogenic, inducing a significant rise (fourfold or greater) in IgG antibody titer in all vaccinees. A second dose of vaccine given 5 weeks later did not result in a further rise in anti-CPS antibody levels. Vaccination engendered an antibody response that recognized only K1 CPS. IgG purified from the serum of vaccinated individuals was highly effective at preventing fatal *Klebsiella* K1 burn-wound sepsis in mice. This study showed that NaOH-treated *Klebsiella* CPS is well tolerated, immunogenic, behaves as a T-independent antigen and induces protective antibodies.

A hexavalent *Klebsiella* CPS vaccine containing serotypes often associated with bacteremia (K2, K3, K10, K21, K30, K55) was then formulated [89]. The vaccine was tested for safety and immunogenicity in healthy adult human volunteers who received either 25 or 50 μg of NaOH-treated capsular antigen (150 and 300 μg of total CPS, respectively). The vaccine was well tolerated at both doses. The immune response was slightly better at the higher vaccine dose, but most vaccinees responded with a fourfold or higher rise in IgG titer to all six vaccine antigens. Vaccination engendered cross-reactive antibody that recognized K7, K61, and K68 CPSs. This is not surprising since the K10 vaccine antigen is known to cross-react with K7 and K61, and K3 vaccine antigen with K68. IgG isolated from postimmunization sera was able to provide significant protection against five K2 strains of *Klebsiella* when tested in a murine burn-wound sepsis model.

A substantial portion of patients at increased risk to acquiring a *Klebsiella* infection would not be prime candidates for active immunization, because they are either immunocompromised or would be at an increased risk for a considerable time before obtaining significant levels of vaccine-induced antibodies. Under these conditions, a hyperimmune gamma globulin for intravenous use (IVIG) would be ideal. To explore this possibility, an experimental IVIG was produced from the sera of volunteers immunized with the hexavalent *Klebsiella* vaccine described above and compared with several commercial IVIGs for anti-*Klebsiella* activity [90]. The hyperimmune *Klebsiella* IVIG possessed antibody titers that were significantly higher (32- to 64-fold) against the six vaccine serotypes than the commercial IVIGs. In addition, only the hyperimmune preparation was able to promote the uptake and killing of all six serotypes of test organisms by human polymorphonuclear leukocytes. When administered prophylactically, as little as 5 mg/kg of hyperimmune IVIG was able to provide significant protection against experimental *Klebsiella* sepsis. To obtain a similar level of protection, 100-fold more normal IVIG had to be used. When used therapeutically (IVIG given at 24 h and 48 h postinfection), only the hyperimmune preparation was effective at reducing mortality. Similarly, only the hyperimmune product acted synergistically with antibiotics in a therapeutic mode.

## VI. CURRENT STUDIES AND FUTURE PROSPECTS

At present, a 24-valent CPS vaccine has been produced (Table III). Selection of vaccine serotypes was based upon the serotype distribution of *Klebsiella* bacteremic isolates [16,27,91] and cross-reactions noted between serotypes. In total, this vaccine will "cover" roughly 70% of *Klebsiella* bacteremic strains. The vaccine is composed of 89.7% carbohydrate, 1% protein, 0.84% nucleic acids, and 5.8% residual moisture. The average molecular weight of the vaccine polysaccharide is $\geq 2 \times 10^6$ as determined by gel filtration. The vaccine was nonpyrogenic when 24 μg/kg was administered intravenously to rabbits. The vaccine is nontoxic for mice and guinea pigs.

**TABLE III. Serotypes Contained in Polyvalent *Klebsiella* Polysaccharide Vaccine and Cross-Reactive Nonvaccine Serotypes**

| Vaccine serotype | Cross-reacting serotype |
|---|---|
| 2 | 13, 69 |
| 3 | 12, 68 |
| 5 | 7, 46 |
| 9 | |
| 10 | 7 |
| 15 | |
| 16 | |
| 17 | |
| 18 | |
| 21 | 11, 26 |
| 22 | 37 |
| 25 | |
| 28 | |
| 30 | |
| 35 | |
| 43 | |
| 52 | |
| 53 | |
| 55 | |
| 60 | 31 |
| 61 | |
| 62 | |
| 63 | |
| 64 | 14 |

Initial phase I studies in human volunteers entailed administering 25, 50, or 75 μg per vaccine serotype to a small number of healthy adults. All three vaccine doses were well tolerated. The only reactions of note consisted of an area of nonpainful erythema at the injection site in subjects who received the 75 μg per serotype dose. Although the number of vaccinees were small, the immune response to most antigens was excellent and was comparable between the three doses. Vaccination also engendered an IgG antibody response to the 11 cross-reactive serotypes shown in Table III.

Based upon the above results, a phase II trial was conducted in which 33 volunteers received either 25 or 50 μg per vaccine serotype. No systemic reactions were noted. Local reactions were infrequent, mild, and transient. To date, the immune response to the 50 μg dose has been analyzed. Significant rises (fourfold or greater) in IgG antibody levels were seen in 90% of vaccinees for 9 of 24 serotypes, in 70% of vaccinees for 18 of 24 serotypes, and in 60% of vaccinees for 21 of 24 serotypes. Only 20 to 40% of subjects responded to serotypes 35, 53, and 63. It is important to note that in at least 50% of the cases where the vaccinees did not mount a significant rise in titer, there was ≥5 μg/ml of specific IgG in their preimmunization serum. This would indicate that high levels of preexisting anti-CPS antibody can act to suppress the immune response following vaccination. Also, 73% of the vaccinees mounted a fourfold response or higher to ten cross-reactive serotypes not included in the vaccine (Table III). The IgG antibody engendered by vaccination was long lived. Most volunteers still had significantly elevated titers 6 months postimmunization. Vaccination also engendered significant increases in opsonic antibody to 21 of the 24 vaccine serotypes.

In addition to the IgG antibody, vaccination also engendered a vigorous IgA antibody response. The IgG antibody response was distributed among both the IgG1 and IgG2 subclasses. Similarly significant rises in both IgA1 and IgA2 antibodies were noted.

All studies to date indicate that this polyvalent *Klebsiella* polysaccharide vaccine is very well tolerated and immunogenic. The next major question is what is the best way to evaluate its clinical efficacy. We know from ongoing studies that good levels of anti-CPS antibody are not obtained until 2 to 3 weeks postvaccination. Given this, plus the other constraints mentioned above, the vaccine by itself would not have the ability to prevent nosocomial infections due to *Klebsiella* simply by immunizing at-risk patients. A much more promising approach would be to use this vaccine to immunize donors whose plasma could then be processed to obtain a hyperimmune IVIG. Given the number of gram-negative nosocomial pathogens, a monovalent *Klebsiella* IVIG would be of little utility in the hospital setting. Our plans envision the use of the currently described vaccine in combination with vaccines against two other leading nosocomial pathogens, *Pseudomonas aeruginosa* [92] and *Escherichia coli*. Such a trivalent IVIG would cover those agents responsible for approximately 70% of the deaths due to gram-negative infections.

## REFERENCES

1. de la Torre MG, Romero-Vivas J, Martinez-Beltrán J, Guerro A, Meseguer M, Bouza E: Klebsiella bacteremia: An analysis of 100 episodes. Rev Infect Dis 7:143, 1985.
2. Montgomerie JZ: Epidemiology of Klebsiella and hospital-associated infections. Rev Infect Dis 1:736, 1979.
3. Jarvis WR, Munn VP, Highsmith AK, Culver DH, Hughes JM: The epidemiology of nosocomial infections caused by Klebsiella pnuemoniae. Infect Control 6:68, 1985.
4. Riser E, Noone P, Howard FM: Epidemiological study of Klebsiella infection in the special care baby unit of a London hospital. J Clin Pathol 33:400, 1980.
5. Bauernfeind A: Epidemiological typing of Klebsiella by bacteriocins. Methods Microbiol 16:213, 1984.

6. Simoons-Smit AM, Verwey-Van Vught AMJJ, De Vries PMJM, MacLaren DM: Comparison of biochemical and serological typing results and antimicrobial susceptibility patterns in the epidemiological investigation of *Klebsiella* spp. Epidemiol Inf 99:625, 1987.
7. Adler JL, Shulman JA, Terry PM, Feldman D, Skaley P: Nosocomial colonization with kanamycin-resistant Klebsiella pneumoniae, types 2 and 11, in a premature nursery. J Pediatr 77:376, 1970.
8. Salzman TC, Clark JJ, Klemm J: Hand contamination of personnel as a mechanism of cross-infection in nosocomial infections with antibiotic-resistant Escherichia coli and Klebsiella-Aerobacter. Antimicrob Agents Chemother 7:97, 1967.
9. Casewell MW, Phillips I: Aspects of the plasmid-mediated antibiotic resistance and epidemiology of Klebsiella species. Am J Med 70:459, 1981.
10. Montgomerie JZ, Doak PB, Taylor DEM, North JDK, Marlim WJ: Klebsiella in faecal flora of renal-transplant patients. Lancet 2:787, 1970.
11. Rose HD, Schreier J: The effect of hospitalization and antibiotic therapy on the gram-negative fecal flora. Am J Med Lab Sci 255:228, 1968.
12. Smith RF, Dayton SL, Chipps DD, Blasi D: Intestinal carriage of Klebsiella and Pseudomonas in burned children and their comparative role in nosocomial infection. Health Lab Sci 10:173, 1973.
13. Selden R, Lee S, Wang WLL, Bennett JV, Eickhoff TC: Nosocomial klebsiella infections: Intestinal colonization as a reservoir. Ann Intern Med 74:657, 1971.
14. Eisenach KD, Reber RM, Eitzman DV, Baer H: Nosocomial infections due to kanamycin-resistant, (R)-factor carrying enteric organisms in an intensive care nursery. Pediatrics 50:395, 1972.
15. Ørskov I, Fife-Asbury MA: New Klebsiella capsular antigen, K82, and the deletion of five of those previously assigned. Int J System Bacteriol 27:386, 1977.
16. Cryz SJ Jr, Mortimer PM, Mansfield V, Germanier R: Seroepidemiology of Klebsiella bacteremic isolates and implications for vaccine development. J Clin Microbiol 23:687, 1986.
17. Martin WJ, Yu PKW, Washington JA II: Epidemiologic significance of Klebsiella pneumoniae. A 3-month study. Mayo Clin Proc 46:785, 1971.
18. Casewell M, Talsania HG: Predominance of certain klebsiella capsular types in hospitals in the United Kingdom. J Infect 1:77, 1979.
19. Ullman U: The distribution of *Klebsiella pneumoniae* serotypes from different sources and their sensitivity to cephalosporins. Infection 11(Suppl):S28, 1983.
20. Rennie RP, Duncan IBR: Combined biochemical and serological typing of clinical isolates of Klebsiella. Appl Microbiol 28:534, 1974.
21. Riser E, Noone P: Klebsiella capsular type versus site of isolation. J. Clin Pathol 34:552, 1981.
22. Weiss, W, Eisenberg GM, Spirack A, Nadel J, Kayser HL, Sathavara S, Flippin HF: Klebsiella in respiratory disease. Ann Intern Med 45:1010, 1956.
23. Brooke MS: Further capsular antigens of Klebsiella strains. Acta Pathol Microbiol Scand 28:313, 1951.
24. Steinhauer BW, Eickhoff TC, Kislack JW, Finland M: The Klebsiella-Enterobacter-Serratia division. Clinical and epidemiological characteristics. Ann Intern Med 65:1180, 1966.
25. Ørskov I, Ørskov F: "Serotyping of Klebsiella." London: Academic Press, 1984, p 143.
26. Maiztegui JI, Biegeleisen JZ Jr, Cherry WB, Kass EH: Bacteremia due to gram-negative rods. A clinical, bacterologic, serologic and immunofluorescent study. N Engl J Med 272:222, 1965.
27. Montgomerie JZ, Ota JK: Klebsiella bacteremia. Arch Intern Med 140:525, 1980.

28. Kiseleva BS, Krasnogolovets VN: Role of Klebsiella pneumoniae in the etiology of bacterial sepsis. Zh Mikrobiol Epidemiol Immunbiol 2:20, 1983.
29. McGowan JE, Barnes MW, Finland M: Bacteremia at Boston City Hospital: Occurrence and mortality during 12 selected years (1935–1972), with special reference to hospital-acquired cases. J Infect Dis 132:316, 1975.
30. Ørskov I: Nosocomial infection with *Klebsiella* in lesions of the urinary tract. Acta Pathol Microbiol Scand 93:257, 1952.
31. Bryan CS, Reynolds KL, Brenner ER: Analysis of 1,186 episodes of gram-negative bacteremia in non-university hospitals: The effects of antimicrobial therapy. Rev Infect Dis 5:629, 1983.
32. McGowan JE Jr: Changing etiology of nosocomial bacteremia and fungemia and other hospital-acquired infections. Rev Infect Dis 7(Suppl):S357, 1985.
33. Kreger BE, Craven DE, Carling PC, McCabe WR: Gram-negative bacteremia. III. Reassessment of etiology, epidemiology and ecology in 612 patients. Am J Med 68:332, 1980.
34. Freid MA, Vosti KL: The importance of underlying disease in patients with gram-negative bacteremia. Arch Intern Med 121:428, 1968.
35. Michel MF, Priem CC: Positive blood cultures in a university hospital in The Netherlands. Infection 9:283, 1981.
36. Weinstein MP, Reller LB, Murphy JR, Lichtenstein KA: The clinical significance of positive blood cultures: A comprehensive analysis of 500 episodes of bacteremia and fungemia in adults. I. Laboratory and epidemiologic observations. Rev Infect Dis 5:35, 1983.
37. Elhag KM, Mustafa AK, Sethi SK: Septicemia in a teaching hospital in Kuwait. I. Incidence and etiology. J Infect 10:17, 1985.
38. Haddy RI, Klimberg S, Epting RJ: A two center review of bacteremia in the community hospital. J Fam Pract 24:253, 1987.
39. Ma MY, Mulligan ME, Friedman MH, Anderson MS: Prognosis of klebsiella bacteremia: Possible association with antimicrobial usage and bacterial resistance. In Easmon CSF, Gaya H (eds): "Second International Symposium on Infections in the Immunocompromised Host." New York: Academic Press, 1983, p. 211.
40. Brun-Buisson C, Legrand P. Philippon A, Montravers F, Ansquer M, Duval J: Transferable enzymatic resistance to third-generation cephalosporins during nosocomial outbreak of multiresistant Klebsiella pneumoniae. Lancet 2:302, 1987.
41. Thomas FE, Jackson RT, Melley MA, Alford RH: Sequential hospitalwide outbreaks of resistant serratia and klebsiella infections. Arch Intern Med 137:581, 1977.
42. Courtney M-A, Miller JR, Summersgill J, Melo J, Raff MJ, Streips UN: R-factor responsible for an outbreak of multiply antibiotic-resistant Klebsiella pneumoniae. Antimicrob Agents Chemother 18:926, 1980.
43. Markowitz SM, Veazey JM Jr, Macrina FL, Mayhall CG, Lamb VA: Sequential outbreaks of infection due to Klebsiella pneumoniae in a neonatal intensive care unit: Implication of a conjugative R plasmid. J Infect Dis 142:106, 1980.
44. Cross AS, Opal S, Kopecko DJ: Progressive increase in antibiotic resistance of gram-negative bacterial isolates. Walter Reed Hospital, 1976–1980: Specific analysis of gentamicin, tobramycin, and amikacin resistance. Arch Intern Med 143:2075, 1983.
45. Woloj M, Tolmasky ME, Roberts MC, Crosa JH: Plasmid-encoded amikacin resistance in multiresistant strains of Klebsiella pneumoniae isolated from neonates with meningitis. Antimicrob Agents Chemother 29:315, 1986.
46. Neu HC: Structure-activity relations of new $\beta$-lactam compounds and in vitro activity against common bacteria. Rev Infect Dis 5(Suppl):S319, 1983.

47. Donwitz GR, Mandell GL: Beta-lactam antibiotics. N Engl J Med 318:490, 1988.
48. Sanders CC, Sanders WE Jr: Microbial resistance to newer generation β-lactam antibiotics: Clinical and laboratory implications. J Infect Dis 151:399, 1985.
49. Then RL, Glauser MP, Angehrm P, Arisawa M: Cephalosporin resistance in strains of Klebsiella oxytoca isolated during antibiotic therapy. Zentralbl Bakteriol Hyg 254:469, 1983.
50. Kliebe C, Nies BA, Neyer JF, Tolxdorff-Neutzling RM, Wiedeman B: Evolution of plasmid-coded resistance to broad-spectrum cephalosporins. Antimicrob Agents Chemother 28:302, 1985.
51. Heidelberger M, Nimmich W: Immunochemical relationships between bacteria belonging to two separate families: Pneumococci and *Klebsiella*. Immunochemistry 13:67, 1976.
52. Kenne L, Lindberg B: Bacterial polysaccharides. In Aspinall GO (ed): ''The Polysaccharides.'' New York: Academic Press, 1983, Vol. 2, p 287.
53. Cryz SJ Jr, Fürer E, Germanier R: Experimental Klebsiella pneumoniae burn wound sepsis: Role of capsular polysaccharide. Infect Immun 43:440, 1984.
54. Domenico P, Johanson WG Jr, Straus DC: Lobar pneumonia in rats produced by clinical isolates of Klebsiella pneumoniae. Infect Immun 37:327, 1982.
55. Allen PM, Fisher D, Saunders JR, Hart CA: The role of capsular polysaccharide K21b of Klebsiella and of the structurally related colanic-acid polysaccharide of Escherichia coli in resistance to phogocytosis and serum killing. J Med Microbiol 24:363, 1987.
56. Williams P, Lambert PA, Brown MRW, Jones RJ: The role of the O and K antigens in determining the resistance of Klebsiella aerogenes to serum killing and phagocytosis. J Gen Microbiol 129:2181, 1983.
57. Kadurugamuwa JL, Anwar H, Brown MRW, Zak O: Effect of subinhibitory concentrations of cephalosporins on surface properties and siderophore production in iron-depleted Klebsiella pneumoniae. Antimicrob Agents Chemother 27:220, 1985.
58. Trautmann M, Cryz SJ Jr, Sadoff JC, Cross AS: A murine monoclonal antibody against *Klebsiella* capsular polysaccharide is opsonic in vitro and protects against experimental Klebsiella pneumoniae infection. Microb Pathogen 1988 (in press).
59. Ørskov I: ''Immunological paralysis'' induced in rabbits by a heavily capsulated Klebsiella strain. Acta Pathol Microbiol Scand 38:375, 1956.
60. Batshon BA, Baer H, Shaffer MF: Immunological paralysis in mice by Klebsiella pneumoniae type 2 polysaccharide. J Immunol 90:121, 1963.
61. Baer H, Bringaze JK, McNamee M: The effect of the doses of bacterial polysaccharide antigen in antibody production in mice. J Bacteriol 67:123, 1954.
62. Yokochi T, Nakashima I, Kato N: Effect of capsular polysaccharide of Klebsiella pneumoniae on the differentiation and functional capacity of macrophages cultured in vitro. Microbiol Immunol 21:601, 1977.
63. Yokochi T, Nakashima I, Kato N: Further studies on generation of macrophages in in vitro cultures of mouse spleen cells and its inhibition by the capsular polysaccharide of Klebsiella pneumoniae. Microbiol Immunol 23:487, 1979.
64. Mizuta K, Ohta M, Mori M, Hasegawa T, Nakashima I, Kato N: Virulence for mice of Klebsiella strains belonging to the 01 group: relationship to their capsular (K) types. Infect Immun 40:56, 1983.
65. Simoons-Smit AM, Verwey-Van Vught AMJJ, Kanis IYR, MacLaren DM: Virulence of Klebsiella strains in experimentally induced skin lesions in the mouse. J Med Microbiol 17:67, 1984.
66. Cryz SJ Jr, Fürer E, Germanier R: Immunization against fatal experimental Klebsiella pneumoniae pneumonia. Infect Immun 54:403, 1986.

67. Cryz SJ Jr, Fürer E, Germanier R: Simple model for the study of Pseudomonas aeruginosa infections in leukopenic mice. Infect Immun 39:1067, 1983.
68. Pollack M: Significance of circulating capsular antigen in klebsiella infections. Infect Immun 13:1543, 1976.
69. Tomás JM, Benedi VJ, Ciurana B, Jofre J: Role of capsule and O antigen in resistance of Klebsiella pneumoniae to serum bactericidal activity. Infect Immun 54:85, 1986.
70. Ciurana B, Tomás JM: Role of lipopolysaccharide and complement in susceptibility of Klebsiella pneumoniae to nonimmune serum. Infect Immun 55:2741, 1987.
71. Straus DC, Atkisson DL, Garner CW: Importance of a lipopolysaccharide-containing extracellular toxic complex in infections produced by Klebsiella pneumoniae. Infect Immun 50:787, 1985.
72. Straus DC: Production of an extracellular toxic complex by various strains of Klebsiella pneumoniae. Infect Immun 55:44, 1987.
73. Fader RC, Davis CP: Effect of piliation on Klebsiella pneumoniae infection in rat bladders. Infect Immun 30:554, 1980.
74. Fader RC, Davis CP: Klebsiella pneumoniae induced experimental pyelitis: The effect of piliation on infectivity. J Urol 128:197, 1982.
75. Pruzzo C, Valisena S, Satta G: Laboratory and wild-type Klebsiella pneumoniae strains carrying mannose-inhibitable adhesions and receptors for adiphages T3 and T7. Infect Immun 39:520, 1983.
76. Pruzzo C, Deblia EA, Satta G: Identification of the major adherence ligand of Klebsiella pneumoniae in the receptor for coliphage T7 and alteration of Klebsiella adherence properties by lysogenic conversion. Infect Immun 30:562, 1980.
77. Pruzzo C, Deblia E, Satta G: Mannose-inhibitable adhesions and T3-T7 receptors of Klebsiella pneumoniae inhibit phagocytosis and intracellular killing by human polymorphonuclear leukocytes. Infect Immun 36:949, 1982.
78. Nassif X, Sansonetti P: Correlation of the virulence of Klebsiella pneumoniae K1 and K2 with the presence of a plasmid encoding aerobactin. Infect Immun 54:603, 1986.
79. Riottot M-M, Fournier J-M, Pillot J: Capsular serotypic specificity of the protection conferred on mice by Klebsiella pneumoniae ribosomal preparations. Infect Immun 24: 476, 1979.
80. Riottot M-M, Fournier J-M, Jouin H: Direct evidence for the involvement of capsular polysaccharide in the immunoprotective activity of Klebsiella pneumoniae ribosomal preparations. Infect Immun 31:71, 1981.
81. Fournier J-M, Jolivet-Reynaud C, Riottot M-M, Jouin H: Murine immunoprotective activity of Klebsiella pneumoniae cell surface preparations: Comparative study with ribosomal preparations. Infect Immun 32:420, 1981.
82. Jones RJ, Roe EA: Vaccination against 77 capsular types of Klebsiella aerogenes with polyvalent Klebsiella vaccines. J Med Microbiol 18:413, 1984.
83. Roe EA, Jones RJ: Vaccination against Klebsiella aerogenes. J Hyg (Camb) 93:355, 1984.
84. Cryz SJ Jr, Fürer E, Germanier R: Purification and vaccine potential of Klebsiella capsular polysaccharides. Infect Immun 50:225, 1985.
85. Cryz SJ Jr, Fürer E, Germanier R: Protection against fatal *Klebsiella pneumoniae* burn wound sepsis by passive transfer of anticapsular polysaccharide. Infect Immun 45:139, 1984.
86. Cryz SJ Jr, Fürer E, Germanier R: Prevention of fatal experimental burn-wound sepsis due to Klebsiella pneumoniae KP1-O by immunization with homologous capsular polysaccharide. J Infect Dis 150:817, 1984.
87. Cryz SJ Jr, Fürer E, Germanier R: Immunization against fatal experimental Klebsiella pneumoniae pneumonia. Infect Immun 54:403, 1986.

88. Cryz SJ Jr, Fürer E, Germanier R: Safety and immunogenicity of Klebsiella pneumoniae K1 capsular polysaccharide vaccine in humans. J Infect Dis 151:665, 1985.
89. Cryz SJ Jr, Mortimer P, Cross AS, Fürer E, Germanier R: Safety and immunogenicity of a polyvalent Klebsiella capsular polysaccharide vaccine in humans. Vaccine 4:15, 1986.
90. Cryz SJ Jr, Cross AS, Fürer E, Chariatte N, Sadoff JC, Germanier R: Activity of intravenous immune globulins against Klebsiella. J Lab Clin Med 108:182, 1986.
91. Terman JW, Alford RH, Bryant RE: Hospital-acquired Klebsiella bacteremia. Am J Med Sci 264:191, 1972.
92. Cryz SJ Jr, Fürer E, Cross AS, Wegmann A, Germanier R, Sadoff JC: Safety and immunogenicity of a Pseudomonas aeruginosa O-polysaccharide toxin A conjugate vaccine in humans. J Clin Invest 80:51, 1987.

Bacterial Vaccines, pages 105–122


# Anthrax Vaccine Development: A Continuing Story

**Peter Hambleton and Peter C.B. Turnbull**

Division of Biologics, Public Health Laboratory Service, Centre for Applied Microbiology and Research, Porton Down, Salisbury, Wiltshire, SP4 OJG, England



## I. INTRODUCTION

Anthrax is a disease of some historical significance in microbiology: the antiquity of the disease is apparent from biblical [1] and classical [2] references to diseases of man and animals that are clearly recognizable as being descriptions of various forms of anthrax. The disease remained a scourge [3] until during the last century, when the development of hygienic, prophylactic, and therapeutic measures brought the disease into decline. Anthrax was in the forefront of the great leap forward in microbiology that occurred during the latter half of the nineteenth century. At this time the bacterial etiology of the disease was clearly established by various workers, including Koch, whose successful cultivation in vitro of the anthrax bacillus [4] led to the establishment of Koch's postulates, which, even today, form the basis for establishing the etiology of diseases of microbial origin. Shortly afterward, the involvement of *Bacillus anthracis* as the causative organism of an industrial illness, wool-sorters' disease, was identified [5,6]. These findings exemplify the significant developments in the understanding of microbial causes of disease that were taking place at this time. Of equal significance, in terms of health care, were early attempts to develop vaccines against anthrax [7,8], since these represent the first recognized attempts at prophylaxis for bacterial infections.

As with other diseases of bacterial origin [9], antibiotic therapy [10] and improved environmental hygiene [11] have contributed to the global decline of anthrax. Immunization has played the major role in controlling anthrax in animals and, in consequence, has obviously contributed to a decline in the disease in humans. The role of the human vaccine in protecting persons at risk is less easily quantified, but, as discussed below, it clearly has some effect.

Despite its increasing rarity, anthrax remains an important disease, because of its continuing incidence worldwide in animal populations [12] and the risk of breakdown in public health cover allowing significant human outbreaks [13]. For this reason, studies on the pathogenesis of anthrax have remained of scientific interest, and there is a continuing effort to apply new understanding of pathogenic mechanisms to improving the efficacy of vaccine prophylaxis. Indeed, anthrax represents an excellent example of the rational approach to vaccine development, a concept reviewed elsewhere [9,14]. The pathogenic mechanisms of *B. anthracis* and the development of anthrax vaccines have been the subject of previous comprehensive reviews [15–20]; this chapter examines progress in the understanding of the pathogenesis of anthrax as related to the development of vaccines of improved safety and efficacy.

## II. THE DISEASE

### A. Etiology

The aerobic, spore-forming, gram-positive rod *B. anthracis* is the causative agent of anthrax. The organism can infect all mammals and even birds and possibly reptiles, but anthrax is primarily a disease of herbivores. A comprehensive survey of the disease in animals can be found elsewhere [10], but it would appear that susceptible domestic livestock, such as sheep, cattle, or goats, and wild herbivores contract the disease primarily by ingesting *B. anthracis* spores present in contaminated soil. Human disease usually results from contact with infected animals or *B. anthracis*-contaminated animal products. The cycle of anthrax is dependent on the stage of sporulation outside the animal host [10], and the ability of *B. anthracis* spores to survive in the environment for extended periods of time is well documented. It is not known how long spores can remain viable in soil, but contamination may persist for decades at least [21].

### B. Anthrax in Animals

The disease usually occurs as a septicemia, and in the peracute form animals may not show signs of illness until a few hours before death. The animal then may develop signs of toxemia that develop rapidly into coma and death. During these later stages there is severe mucosal congestion that gives rise to the characteristic terminal hemorrhaging from body orifices. Vegetative *B. anthracis* present in the blood in high numbers subsequently sporulate; this is the usual natural source of infective spores. A slower, acute form of the disease usually runs a course of 2 to 3 days before death. Subacute infection is sometimes observed in carnivores and pigs; this may involve a milder intestinal form of disease or swelling of glands of the mouth and throat.

### C. Anthrax in Humans

The incidence of human anthrax is generally low [18,22]. This may well be because, in comparison with herbivores, humans would appear to be relatively resistant to *B. anthracis* infection [10,23]. The disease takes three basic forms, two of which, cutaneous and pulmonary anthrax, are of particular relevance to developed countries.

**1. Cutaneous anthrax.** The vast majority of human anthrax cases (>95%) are of the cutaneous form [10,23,24]; if untreated, some 10–20% of cases may progress to a generalized, fatal septicemia [23,24]. This figure decreases significantly (0–3%) where hospital treatment is available. Infection occurs through a cut, abrasion, or insect bite; following an incubation period of 2–3 days, a small red pimple develops. A ring of vesicles forms around the central papule, which ulcerates and then dries to form an eschar that subsequently blackens and covers the vesicles as they dry up. This characteristic lesion generally remains small (~2 cm diameter) but may become extensive (>10 cm diameter). It is always accompanied by edema extending some distance from the lesion itself. In uncomplicated cases the bacteria remain localized in the lesion.

**2. Pulmonary anthrax.** Pulmonary anthrax is rarely encountered outside the industrial context, where the disease may follow inhalation of anthrax spores during the handling of contaminated animal products. A proportion of the inhaled spores are phagocytosed and transported to lymph nodes, where spore germination and growth of vegetative organisms occur [11,25] with subsequent fatal toxemia.

**3. Intestinal anthrax.** The ingestion of meat contaminated with *B. anthracis* spores can give rise to what is essentially a form of cutaneous anthrax that develops on the intestinal mucosa. It only occurs in countries where infected animals are slaughtered for meat [26] and is not seen in the social and climatic conditions of developed countries. The course of the disease resembles that of cutaneous anthrax, causing acute gastroenteritis, abdominal pain, and prostration, sometimes with high mortality rates [23].

**4. Anthrax meningitis.** A rare clinical complication of anthrax septicemia in man, anthrax meningitis has a high mortality and features a leukocytosis with hemorrhagic and purulent cerebrospinal fluid [23,27].

## III. PATHOGENIC MECHANISMS

Although the bacterial etiology of anthrax was established over a century ago, the nature of the virulence determinants of *B. anthracis* and the fundamental pathogenesis of anthrax remained obscure until the 1950s. It is now known that the virulence of *B. anthracis* vegetative organisms is determined by two distinct factors, the polypeptide (poly D-glutamic acid) capsule and a three-component exotoxin [28], which are encoded on separate plasmids [29–33]. These virulence factors each possess unique toxic and immunological activities (Table I); for virulence, all components are required to be expressed by *B. anthracis,* from which they act synergistically in the infected host to produce clinical disease. It was first demonstrated by Bail and Weil [35] that anthrax bacilli produced substances in anthrax lesions that were capable of interfering with host defense mechanisms. It was later suggested that the bacilli were protected from phagocytosis by substances associated with capsule formation [36]. Virulent strains only of *B. anthracis* were found to possess a capsule containing a polypeptide (poly D-glutamic acid) [37] that functions by inhibiting phagocytosis and opsonization of the bacilli [38]. The gene coding for synthesis of this capsular polypeptide resides on an extrachromosomal element (plasmid) designated pX02 [29,30].

While the capsular properties highlighted the mechanism whereby *B. anthracis* could be seen to overcome host defense mechanisms, the factors producing the lethal systemic effects only became apparent following a pe-

**TABLE I. Toxic and Immunological Activities of *Bacillus anthracis* Virulence Determinants**

| Factor | Toxin action | Immunological activity |
|---|---|---|
| Capsule (poly D-glutamate) | Antiphagocytic | Weakly antigenic, nonimmunizing |
| Edema factor (EF) | Inactive | Antigenic, nonimmunizing |
| Protective antigen (PA) | Inactive | Strongly immunogenic |
| Lethal factor (LF) | Inactive | Antigenic, weakly immunogenic |
| EF + PA | Local edema | Strongly immunogenic |
| EF + LF | Inactive | Weakly immunogenic |
| PA + LF | Lethal | Immunogenic |
| EF + PA + LF | Edema, lethal | Immunogenic |

Abstracted in part from references 16, 18, and 34.

riod of intensive research carried out in the 1950s. It had been considered that the large number of organisms involved in the septicemia might cause deoxygenation of the blood [28] or occlusion of capillaries [39]. However, the elegant studies of Smith and his colleagues at Porton Down, England [28,40–43], clearly demonstrated the toxigenic component of anthrax. It was shown that once bacteremia in guinea pigs had reached a critical level, termination of the bacteremia and the infection by streptomycin therapy served only to delay the time to death of the animal. Furthermore, sterile plasma from moribund guinea pigs with anthrax was lethal to other animals when administered by intravenous injection. Although these workers attempted to analyze the chemical nature of the toxic components in the lethal serum, the nature of the toxic factors was only understood after the successful production of toxin in and isolation from in vitro cultures [44–46].

The exotoxin of *B. anthracis* is known to comprise three proteins, edema factor (EF), protective antigen (PA), and lethal factor (LF), all between 80 and 90 kDa [20] and each having distinct biological and serological activities (see Table I). All three components are encoded on a second plasmid (pX01) [31–33] and act synergistically to produce the systemic effects of anthrax. Individually, the pure proteins have no demonstrable toxic activity (Table I) but, when injected intravenously in various combinations, induce significant pathological responses. Thus, combinations of PA and LF injected intravenously are lethal to mice, rats, and guinea pigs [47], while intradermal injection of PA plus EF elicits edematous lesions in guinea pigs and rabbits [16]. The biochemical basis of the action of the toxin components is becoming increasingly clear. For example, it appears that PA binds avidly to the surface of eukaryotic cells to as-yet-uncharacterized receptor(s). The bound PA is cleaved proteolytically to reveal a high-affinity site for which LF and EF compete [48]. The resulting complex(es) then probably enters the host cell by endocytosis. EF has been identified by Leppla and coworkers [47,48] as a calmodulin-dependent adenyl cyclase that, when activated, generates high levels of cyclic AMP in the cytoplasm of eukaryotic cells. The mode of action of LF remains unknown.

Although the molecular basis for toxicity is now better understood, several aspects of the pathogenesis of anthrax remain unresolved, in particular the action of LF, as mentioned above, and the relationship between toxin production and the onset of the bacteremic phase. Finally, the primary cause of death from anthrax remains unclear. It seems most likely that death results from oligemic shock [16,34,49]. In primates, the toxin depresses electrical activity in the cerebral cortex and, by affecting the respiratory center, may contribute to anoxic hypertension, cardiac collapse, shock, and sudden death [23].

## IV. DEVELOPMENT OF PRESENT VACCINES

The elucidation of the bacterial etiology of anthrax opened the way for effective control of the disease. Although a significant reduction in the incidence of anthrax as an industrial disease in the leather and wool industries resulted from the introduction of disinfection and improved working conditions [5,6], undoubtedly live spore vaccines have been fundamental in controlling anthrax in livestock and hence reducing the incidence of human disease. Cell-free vaccines have also contributed to disease control by affording protection to humans at risk of exposure to *B. anthracis* spores.

Although early vaccines were derived essentially on empirical bases, knowledge of the virulence factors of *B. anthracis* and their role in the pathogenesis of anthrax has been applied in the development of existing human vaccines. It is nevertheless the case that, although vaccines against anthrax have been in use for a century now, the mechanisms of immunity to the disease are still not fully understood. Recent progress in elucidating the molecular biology of the virulence determinants of *B. anthracis* may soon find application to the development of improved prophylactics.

### A. Early Vaccines

Pasteur [8] is generally accredited with having developed the first effective bacterial vaccine, although others [7] were actively pursuing this goal. Pasteur's vaccines consisted of vegetative cultures of *B. anthracis* that had been attenuated by growth at elevated temperatures (42–43°C) and that served as the primary inoculation. These cultures (Pasteur I strain), used for the primary immunizing dose, had diminished virulence, being pathogenic only to small mammalian species. A second type of culture, incubated at high temperatures for a shorter period of time and consequently rather less attenuated (Pasteur II strain), was used for the second immunization of the schedule. The probable mechanism of the attenuation process used by Pasteur was described by Mikesell et al. [50], who showed that growing pathogenic *B. anthracis* at 42°C cures the organisms of the pX01 plasmid, which codes for the toxin components.

Although the Pasteur duplex vaccine was generally effective and remained in use until about 1935, the unpredictability of the attenuation process gave it limited and variable effectiveness. Modified heat-attenuated vaccines were introduced [50,51], but they too suffered from variable attenuation and were superseded by attenuated spore and accellular vaccines.

## B. Attenuated Spore Vaccines

Perhaps the most significant event in the control of anthrax was the development by Sterne [52–55] of a live attenuated spore vaccine derived from an avirulent noncapsulated variant of *B. anthracis,* later shown to be lacking the plasmid pX02 encoding the poly D-glutamic acid capsule [29,30]. This vaccine and its derivatives [56] have proved very effective as veterinary vaccines and have had a pronounced effect on limiting economic losses due to anthrax [57]. Furthermore, control of the disease in livestock has unquestionably been the major factor responsible for the reduced human incidence [10].

The main limitation of the Sterne-type vaccine is one of safety. The attenuated strain(s) retains a measurable, albeit low, level of virulence in some animal species [10], and it occasionally gives rise to necrosis at the site of inoculation and, rarely, death of the recipient animal [20]. It is because of concern about their safety that live spore vaccines are generally considered as being unsuitable for human use in Western countries. However, in the USSR a live spore vaccine formulated from a Sterne strain derivative, STI, is produced for human use [10,58]. It is claimed to be effective and has even been administered by the aerosol route [59] with success. There appears to be no information that allows a comparison of the relative effectiveness in humans of the Russian live spore vaccine and the protective antigen vaccine used in the West [14]. It is, however, apparent that, notwithstanding problems of safety, live spore vaccines are considered to have greater effectiveness than human protective antigen (cell-free) vaccines [20,60,61], and this has raised interest in the role of antigen presentation in stimulating an effective, protective immune response [33,61].

## C. Protective Antigen Vaccines

The feasibility of developing effective nonliving anthrax vaccines was demonstrated as long ago as 1904 [62], when it was shown that sterile edema fluid from dying animals was immunogenic. Investigation of this phenomenon over the next half century [18] culminated in the understanding of the toxigenic basis of the disease and recognition of the so-called protective antigen component, as described above. Early attempts at producing immunizing anthrax antigens in vitro had limited success, mainly because the quantities of antigen produced were very low and the complex serum-containing culture media used made purification difficult [18,63,64]. Boor and Tresselt [65,66] developed the first effective protective antigen preparation from in vitro culture of a *B. anthracis* strain of intermediate virulence using precipitation techniques to purify the active component from culture supernatants. However, the growth medium still contained serum proteins that impeded the purification process and contaminated the vaccine.

Studies on the nutritional requirements of *B. anthracis* [67] enabled workers in both the United Kingdom [68] and the United States [69] to produce protective antigen successfully in vitro in nonproteinaceous media and hence prepare vaccines of markedly improved purity from toxigenic noncapsulated *B. anthracis*. The PA is readily separated from the other components of the toxin [46] and is immobilized by alum precipitation [70,71] or adsorbed onto aluminum hydroxide [72]. Such vaccines can produce some protective activity in experimental animals [18,60,61] and may be effective in humans [73]. The preparation of an adsorbed, protective antigen vaccine has been described by Soviet workers [74], but its use, other than in animal experiments, appears not to have been described [18].

### D. Efficacy of Protective Antigen Vaccines

There is a considerable body of information on the efficacy of various protective antigen preparations [18,20,60,61], although it is difficult to compare directly different preparations, since the published data often do not readily allow quantitative comparison of the amounts of antigen present in different vaccines nor the levels of protection induced [18]. Although Reimann [75] expressed concern over the reliability of adsorbed protective antigen vaccines, Brachman et al. [73] concluded, on the basis of field trials, that the U.S. vaccine was effective. Although a similar study has not been carried out on the U.K. vaccine, analysis of data from the Public Health Laboratory Service Communicable Disease Surveillance Centre (Table II) shows a clear decline in the incidence of human cases of anthrax in England and Wales corresponding to the introduction of vaccination against anthrax for workers at risk.

Despite their effectiveness in protecting against industrially acquired anthrax, the present human vaccines do have deficiencies. For example, three doses at short intervals are required initially and there is a requirement for repeated booster doses; also, the vaccine gives rise to a variable degree of reactogenicity, which may be rather more than localized. Of perhaps more

**TABLE II. Decline in Reported Cases of Human Anthrax in England and Wales Following the Introduction of Vaccination of Workers at Risk**

| Period | Reported cases | Deaths |
|---|---|---|
| 1961–1965 | 55 | 4 |
| 1966–1970 | 54 | 3 |
| 1971–1975 | 14 | 5 |
| 1976–1980 | 13 | 4 |
| 1981–1985 | 5 | 0 |
| 1986 on | 0 | 0 |

Data from references 18 and 19.

obvious concern are reports [60,61,76–78] that the protective antigen vaccines are unable to give protection, in guinea pigs at least, against certain virulent strains of *B. anthracis*. For the above reasons, there is currently interest in developing a human vaccine that lacks side effects, requires fewer doses to achieve and maintain long-term immunity, and is effective against all virulent challenges.

## E. Recent Developments

Emphasis has been placed on the elucidation of the roles of antigens of *B. anthracis* in eliciting specific, protective immunity; to this end, the immunogenicity of known virulence components and other antigens has been thoroughly assessed.

Somatic antigens such as the poly-D-glutamic acid capsule, outer-membrane proteins, and polysaccharides are generally poor immunogens [20,42,79] and would appear to play no significant role in protecting against virulent challenge. This was confirmed by Ivins et al. [31] using strains of *B. anthracis,* lacking either or both of the plasmids pX01 and pX02, as live vaccines. Nontoxigenic, encapsulated ($pX01^-$, $pX02^+$) or nontoxigenic noncapsulated ($pX01^-$, $pX02^-$) strains did not protect experimental animals against toxin or virulent spore challenge, whereas toxigenic, nonencapsulated ($pX01^+$, $pX02^-$) strains afforded solid protection against a variety of challenge regimens.

This finding concentrated attention on assessing the relationship between protective immunity and antibodies to the various known components of the anthrax toxin (viz. PA, LF, and EF). It has been shown that in some instances small quantities of EF and LF proteins in existing human vaccines give rise to significant serological responses, in addition to that of PA [20,31,60,61], suggesting that such antibodies may be necessary for protection. The serological responses to EF and LF are, however, variable, yet all batches of vaccine conform to the requirements of animal efficacy tests. Of the three toxin components, PA is generally regarded as being by far the most significant immunogen [18,20,61,78]. Using a recombinant *Bacillus subtilis* containing the PA gene of *B. anthracis* alone, it was clearly shown [20,32,80] that PA was most likely the sole and sufficient protective immunogen because the recombinant afforded substantial protection against anthrax to experimental animals.

Berendt et al. [81] demonstrated clearly that the Sterne spore vaccine was significantly more protective to guinea pigs than a human PA vaccine and that EF and LF alone were poorly protective. Surprisingly, the levels of PA antibodies were higher in animals receiving PA vaccine than in the better protected, spore-immunized animals. The view that effective protection against all virulent challenge was more than simply a function of substantial levels of circulating antibodies to PA was confirmed by the work of Turnbull et al. [60] and Little and Knudson [78]. Together, these studies [60,78,80] show clearly that the Sterne live spore vaccine is more protective than human PA vaccines, that all effective anthrax vaccines induce antibodies to PA, and that EF and LF alone are poor immunogens. They also confirm that although antibodies to PA are undoubtedly protective, there is no absolute correlation between anti-PA titers and levels of protection. A more recent analysis of antibodies to *B. anthracis* toxin components and their relevance to protective immunity [61] clearly showed that protection against a "vaccine-resistant" strain of *B. anthracis* (Ames) was possible in the absence of detectable anti-LF antibodies and that PA is, most probably, the essential anthrax-derived antigen for protection.

What these recent studies suggest is that it may be the quality of the PA antibodies that is critical to protection, rather than the overall quantity. Present assays for PA antibody reflect polyclonal responses to the various antigenic domains on the PA molecule, not necessarily those against the specific protective immunogenic epitopes responsible, for example, for binding to the host cell, EF, or LF. It may be that purified, adsorbed PA differs in its three-dimensional antigenic structure from that produced in vivo [20] and hence that correct presentation of the purified PA is critical to the induction of solid immunity [61]. A further possibility is that a critical and sustained level of PA is required in vivo for maximum immunity. This could, of course, be achieved by a live vaccine but possibly not by a nonliving one. It is also possible that cellular immunity contributes to optimal immune protection and that live vaccines stimulate this better than the PA vaccines alone. Finally, it may be that *B. anthracis* antigens, other than toxin components, supplement the response to PA [20], a suggestion partly supported by the augmenting effects of *Corynebacterium ovis,* Freund's complete adjuvant, or pertussis vaccine reported by Turnbull et al. [61].

## V. FUTURE DEVELOPMENTS

### A. Analysis of Functional Epitopes

Functional epitope analyses are particularly suited to investigating specific functional regions of the toxin components. Hence it may be possible to identify the specific protective epitopes of PA. Also, it will be possible to investigate the interactions of the toxin components with each other and with host cells at the molecular level. As an example of this approach, Leppla et al. [48] (see also Ivins and Welkos [20]) have analyzed the functional properties of purified proteolytic fragments of PA and have located a specific binding site for EF and LF on a 65 kDa fragment. The immunogenic properties of intact PA and fragments thereof have been compared and the data suggest that protective antigenic regions are associated with the same 65 kDa fragment of PA. It is possible that by such approaches, immunogenic peptides that would allow the development of synthetic peptide vaccines might be identified. While such vaccines could offer advantages [82], including the possibility of minimal side effects, there might be difficulties in providing effective antibody responses in the host. This latter point is borne out by the suggestion already made [20,61] that the presentation of immunogenic epitopes of PA to the host immune system is critical to the induction of good immunity.

An alternative strategy has been developed from the observation [83] that mutants of *Salmonella typhimurium* that require aromatic metabolites not available in host tissues (*aro* mutants) may act as effective vaccines. Transposon Tn 916 mutagenesis of a noncapsulated, toxigenic *B. anthracis* generated *aro* mutants [84] that showed potential as live vaccines by protecting guinea pigs against challenge by virulent *B. anthracis* spores [20].

The techniques of site-directed mutagenesis offer the possibility of altering specific DNA sequences in the toxin genes of *B. anthracis*. In this way, it would be feasible to produce modified gene products having conserved, or even enhanced, immunogenic properties while being rendered nontoxic. This offers the potential for nonliving vaccines containing cocktails of PA with modified EF and LF possibly combined with surface antigens of *B. anthracis*. It also could allow the construction of modified live vaccine strains containing the DNA sequences for nontoxic proteins [20].

### B. Adjuvants

There are clear implications that the protective immunogenicity of PA vaccines can be potentiated by adjuvants. Present human vaccines, of course, comprise PA adsorbed onto insoluble aluminum salts, but their effectiveness is extended when supplemented by *Corynebacterium ovis,* Freund's complete adjuvant, or the killed whole-cell pertussis vaccine [61]. The role of such immunopotentiating agents in enhancing the prophylactic value of vaccines is well established [9], but some frequently provoke toxic, allergic, ulcerative, or inflammatory reactions that preclude their use in human vaccines.

Because of the effectiveness of biological immunopotentiators, attempts have been made to identify alternative, but innocuous, adjuvant substances of biological origin. The identification of N-acetyl muramyl-L-alanyl-D-isoglutamine (MDP) [85] as the significant immunopotentiating moiety of bacterial peptidoglycans has led to the development of synthetic muramyl dipeptides [86] capable of inducing delayed hypersensitivity without undesirable local reactions. Dimethylglycine (DMG) also has potential as a novel human immunomodulator [20]. The fact that a commercial company, Syntex, produces a synthetic MDP that is incorporated into several investigational new drugs (IND) [9] illustrates the perceived potential for this type of immunopotentiator.

Despite the efficacy of avirulent live spore vaccines, their immunogenicity can be enhanced by addition of glycerol and saponin to the spore suspension [51], although the mechanism(s) whereby either, or both, of these substances exerts this effect is obscure [18,87].

## VI. VACCINE ASSESSMENT

Although there are many strategies that might be used to develop novel anthrax vaccines, it is a fundamental requirement that there should be a reliable basis on which putative vaccines can be assessed and compared. In a previous review [18], an attempt was made to compare various PA vaccines, but the published data did not allow an adequate comparison of the amounts of antigen(s) in the preparations nor quantitative comparison of efficacy because of the various methods used to quantitate antigen(s) and the differing animal models.

### A. Experimental Animal Models

Evaluation of any vaccine requires selection of an appropriate animal model [18]. For anthrax vaccines the main criterion of efficacy is survival following a toxin or virulent spore challenge, although other criteria such as serum antibody levels, toxemia, bacteremia, or time to death have been used.

Guinea pig, rabbit, and rat protection tests have all been used to assess the efficacy of anthrax vaccines with varying success [18,20]. Protection studies have also been carried out using domestic animals [18,20]. Despite a large body of data that has accumulated through years of study, it is still uncertain as to which animal model is most appropriate for extrapolation to humans. There are several possible reasons for this. First, anthrax is still poorly understood and a precise mechanism of death has yet to be established (see above). Second, animal species differ in their natural resistance to the disease [18,20]. Thus, rats are resistant to establishment of infection but sensitive to toxin, whereas guinea pigs are relatively susceptible to spore infection but resistant to toxin administered intravenously [16,20,41]. This implies that resistance to infection and the effects of toxin may involve separate mechanisms of host defense [16,20].

The influence of host genetically determined factors on resistance to microbial infections is not doubted [88], but the precise mechanisms for this are unclear. Attempts are being made to study this aspect of host resistance, and a mouse model has been developed that has enabled genetic studies on innate differences in susceptibility to infection by a virulent strain of *B. anthracis* and the avirulent Sterne strain. Mouse resistance to the Sterne strain appears to be controlled, in the main, by a single, major gene [20,89].

The extent to which host genetic elements influence response to infection must affect the relevance of experimental animal models, particularly when the apparent high resistance of humans to infection [18] is taken into account. Because of this it is probably necessary to continue to assess anthrax vaccine efficacy using several animal models.

### B. Other Assays

Anthrax vaccines can also be assessed in terms of several other in vivo assays. Skin tests in guinea pigs and rabbits were used in early attempts [88,90] to quantitate both immunizing antigen(s) and antibody. A double immunodiffusion method superseded the in vivo assays for titrating protective antigen and is still in use although it is not an assay of great inherent sensitivity.

The use of enzyme-linked immunosorbent assays (ELISAs) to assess serological responses to PA, EF, and LF is now well established [20,60,61]. Although such assays have clearly shown that the relationship between protection and antibodies to PA is not a straightforward one [60,61] they clearly have a role in allowing quantitative comparisons to be made between the immune responses induced by different batches of vaccine. Furthermore, the ability to prepare anthrax toxin components to high purity [91] should enable the development of ELISAs using immobilized homologous antibodies for more precise quantitation of the antigenic components of vaccines.

## VII. CONCLUDING REMARKS

As with other infectious diseases, the development of vaccines against anthrax has mirrored developments in the understanding of the pathogenic mechanisms of the disease. Although the current animal and human vaccines clearly can be seen to be effective, they all have limitations, and improvements are actively being sought. Live attenuated spore vaccines are effective and confer good long-term immunity in animals, but it is unlikely, from present trends, that such vaccines would be generally accepted for human use; future, improved human vaccines will, most probably, be based on the current PA vaccines. The modern tools of biochemistry and molecular biology are uncovering new aspects of the disease and, undoubtedly, there will be concomitant developments toward more effective and safe vaccines.

## ACKNOWLEDGMENTS

The authors thank Ann Bennett for skilled assistance in preparation of the manuscript.

## REFERENCES

1. Bible, Exodus, Ch. 9.
2. Dirckx JM: Virgil on anthrax. Am J Dermatopathol 3:191, 1981.
3. Klemm DM, Klemm WR: A history of anthrax. J Am Vet Med Res 135:458, 1959.
4. Koch R: The aetiology of anthrax based on the ontogeny of the anthrax bacillus. Beitr Biol Pflanz 2:277, 1877.
5. Bell JH: On anthrax and anthracaemia in woolsorters, heifers and sheep. Br Med J 2:656, 1880.
6. Eurich FW: The history of anthrax in the wool industry of Bradford and its control. Lancet 57(1):105, 1926.
7. Tigertt WD: William Smith Greenfield, M.D., F.R.C.P., Professor Superintendent, The Brown Animal Sanatory Institution (1878–81) J Hyg (Camb) 85:415, 1980.
8. Pasteur L: De l'attenuation des virus et de leur retour a la virulence. C R Acad Sci 92:429, 1881.
9. Hambleton P, Prior SD, Robinson A: Approaches to rational design of bacterial vaccines. Prog Drug Res 32:378, 1988.
10. Turnbull PCB: Anthrax. In Smith GR, Eason CR (eds): "Bacterial Diseases, Vol. 3. Topley and Wilson's Principles of Bacteriology, Virology and Immunity." London, Arnold (in press).
11. La Force FM: Woolsorters' disease in England. NY Acad Med 54:956, 1978.
12. FAO/OIE/Who: Animal Health Year Book, 1985.
13. Whitford HW: A guide to the diagnosis, treatment, and prevention of anthrax. Document (unpublished) WHO/Zoon/87.163 cited in Wkly Epidemiol Rec 31, 1987.
14. Hambleton P, Melling J: Exploitation of the bacterial envelope: Rational design of vaccines. Med Microbiol 3:231, 1983.
15. Brachman PS: Anthrax. Ann NY Acad Sci 174:577, 1970.
16. Lincoln RE, Fish DC: Anthrax toxin. In Montie TC, Kadis S, Ajl SI (eds): "Microbial Toxins." New York: Academic Press, 1970, Vol. III, p 361.
17. Wright GG: Anthrax toxin. In Schlessinger D (ed): "Microbiology—1975." Washington, DC: American Society of Microbiology, 1975, p 292.
18. Hambleton P, Carman JA, Melling J: Anthrax: The disease in relation to vaccines. Vaccine 2:125, 1984.

19. Turnbull PCB: Thoroughly modern anthrax. Bureau Hyg Trop Dis 61:R1, 1986.
20. Ivins BE, Welkos SL: Recent advances in the development of an improved human anthrax vaccine. Eur J Epidemiol 4:12, 1988.
21. Manchee RJ, Broster MG, Melling J, Henstridge RM, Stagg AJ: *Bacillus anthracis* on Gruinard Island. Nature 294:254, 1981.
22. Anonymous: Vaccine against anthrax. Br Med J II:717, 1985.
23. Christie AB: "Infectious Diseases: Epidemiology and Clinical Practice," IVth edition. Edinburgh: Churchill Livingstone, 1987, p 983.
24. Turner M: Anthrax in humans in Zimbabwe. Contrib Afr J Med 26:160, 1980.
25. Albrink WS, Brooks SM, Biron RE, Kopel M: Human inhalation anthrax: A report on three cases. Am J Pathol 36:457, 1960.
26. Hughes MH: Anthrax. Br Med J 1:488, 1973.
27. Taafe A: Anthrax—A review. J Ir Med Assoc 72:437, 1979.
28. Smith H: Studies on organisms grown *in vivo* to reveal the bases of microbial pathogenicity. Ann NY Acad Sci 88:1213, 1960.
29. Green BD, Battisti L, Koehler TM, Thorne CB, Ivins BE: Demonstration of a capsule plasmid in *Bacillus anthracis*. Infect Immun 49:291, 1985.
30. Uchida I, Sekizaki T, Hashimoto K, Terikado N: Association of the encapsulation of *Bacillus anthracis* with a 60 megadalton plasmid. J Gen Microbiol 131:363, 1985.
31. Ivins BE, Ezzell JW Jr, Jemski J, Hedlund K, Ristroph JD, Leppla SH: Immunization studies with attenuated strains of *Bacillus anthracis*. Infect Immun 52:454, 1986.
32. Ivins BE, Welkos SL: Cloning and expression of the *Bacillus anthracis* protective antigen gene in *Bacillus subtilis*. Infect Immun 54:537, 1986.
33. Robertson DL, Leppla SH: Molecular cloning and expression in *Escherichia coli* of the lethal factor gene of *Bacillus anthracis*. Gene 44:71, 1986.
34. Smith H, Stoner HB: Anthrax toxic complex. Fed Proc 26:1554, 1967.
35. Bail O, Weil E: Beitrage zum studien der Mitzbrandinfektion. Arch Hyg 73:218, 1911.
36. Cromartie WJ, Bloom WL, Watson DW: Studies on infection with *Bacillus anthracis*, 1. A histophathological study of skin lesions produced by *B. anthracis* in susceptible and resistant species. J Infect Dis 80:1, 1947.
37. Ivanovics G, Bruckner V: Chemiche und immunologische studienuber den mechanismus der milzbrandinfektion und immunitat die chemische struktur der kapselsubstanz des milzbrandbazillus und der serolisch identischen spezifichen substanz des bacillus mesentericus. Z Immunitaetsforsch Exp Ther 90:304, 1937.
38. Keppie J, Harris-Smith PW, Smith H: The chemical basis of the virulence of *Bacillus anthracis*. IX. Its agressins and their mode of action. Br J Exp Pathol 44:446, 1963.
39. Vaughan VC, Novy FG: "Cellular Toxins," 4th edition. New York: Lea Bros, 1902.
40. Smith H, Keppie J: Observations on experimental anthrax: Demonstration of a special lethal factor produced *in vivo* by *Bacillus anthracis*. Nature 173:689, 1954.
41. Keppie J, Smith H, Harris-Smith PW: The chemical basis of virulence *Bacillus anthracis* III. The role of the terminal bacteraemia in death of guinea pigs from anthrax. Br J Exp Pathol 36:315, 1955.
42. Smith H, Gallop RC: The chemical basis of the virulence of *Bacillus anthracis*. VI. An extracellular immunising agressin isolated from exudates of infected guinea pigs. Br J Exp Pathol 37:144, 1956.
43. Smith H, Tempest DW, Stanley JL, Harris-Smith PW, Gallop RC: The chemical basis of the virulence of *Bacillus anthracis* VII. Two components of the anthrax toxin: Their relationship to known immunising agressins. Br J Exp Pathol 37:263, 1956.
44. Sargeant K, Smith H: Purification of factor II of the anthrax toxin produced *in vivo* using cellulose ion-exchange columns. Biochem J 69:55, 1958.

45. Harris-Smith PW, Smith H, Keppie J: Production *in vitro* of the toxin of *Bacillus anthracis* previously recognised *in vivo*. J Gen Microbiol 19:91, 1958.
46. Thorne CB, Molnar DM, Strange RE: Production of toxin *in vitro* by *Bacillus anthracis* and its separation into two components. J Bacteriol 79:450, 1960.
47. Ezzell JWE Jr, Ivins BE, Leppla SH: Immunoelectrophoretic analysis, toxicity and kinetics of *in vitro* production of the protective antigen and lethal factor components of *Bacillus anthracis* toxin. J Bacteriol 45:761, 1984.
48. Leppla SH, Friedlander AM, Cora EM: Proteolytic activation of anthrax toxin bound to cellular receptors. In Fehrenbach FJ, Alouf JE, Falmagne P, Goebel W, Jeljaszewicz J, Jurgens D, Rappuoli R (eds): ''Bacterial Protein Toxins.'' Zentralbl Bakterial Suppl 17. Stuttgart: Gustav Fischer, 1988, p 113.
49. Smith H, Keppie J, Stanley JL, Harris-Smith PW: The chemical basis of the virulence of *Bacillus anthracis*. IV. Secondary shock as a major factor in the death of guinea pigs from anthrax. Br J Exp Pathol 36:323, 1955.
50. Mikesell P, Ivins BE, Ristroph JD, Dreier TM: Evidence for plasmid-mediated toxin production in *Bacillus anthracis*. Infect Immun 39:371, 1983.
51. Cameron J: The assay of the potency of anthrax spore vaccine. Symp Ser Immunobiol Stand 10:83, 1969.
52. Sterne M: Variation in *Bacillus anthracis*. Onderstepoort J Vet Sci Anim Ind 8:271, 1937.
53. Sterne M: The use of anthrax vaccines prepared from virulent (uncapsulated) variants of *Bacillus anthracis*. Onderstepoort J Vet Sci Anim Ind 13:307, 1939.
54. Sterne M: The immunisation of laboratory animals against anthrax. Onderstepoort J Vet Sci Anim Ind 13:313, 1939.
55. Sterne M, Nicol J, Lambrechts MC: The effect of large-scale active immunisation against anthrax. J S Afr Vet Med Assoc 13:53, 1942.
56. Blancou JM: Study of a mixed vaccine against black-leg and anthrax. Rev Elev Med Vet Pays Trop 27:183, 1974.
57. Sterne M: Distribution and economic importance of anthrax. Fed Proc 26:1493, 1967.
58. Fish DC, Mahlandt BG, Dobbs JP, Lincoln RE: Purification and properties of *in vitro* produced anthrax toxin components. J Bacteriol 95:907, 1968.
59. Aleksandrov NI, Gefen NE, Garin NS, Gapochko KG, Sergev VM, Smirnov NS, Tamarin AL, Shliakov EN: Experience in massive aerogenic vaccination against anthrax. Voen Med Zh 8:27, 1959.
60. Turnbull PCB, Broster MG, Carman JA, Manchee RJ, Melling J: Development of antibodies to protective antigen and lethal factor components of anthrax toxin in humans and guinea pigs and their relevance to protective immunity. Infect Immun 52:356, 1986.
61. Turnbull PCB, Leppla SH, Broster MG, Quinn CP, Melling J: Antibodies to anthrax toxin in human and guinea pigs and their relevance to protective immunity. Med Microbiol Immunol 177:293, 1988.
62. Bail O: Untersuchungen uber naturliche und kunstliche Milzbrandimmunitat. Zentralbl Bakteriol Parasitenk Infektionskr 47:270, 1904.
63. Gladstone GP: Immunity to anthrax. Protective antigen present in cell-free culture filtrates. Br J Exp Pathol 27:394, 1946.
64. Heckly RJ, Goldwasser E: Studies on infection with *Bacillus anthracis* VIII. The production of an immunising antigen *in vitro*. J Infect Dis 84:91, 1949.
65. Boor AK: An antigen prepared *in vitro* effective for immunisation against anthrax. 1. Preparation and evaluation of the crude protective antigen. J Infect Dis 96:194, 1954.
66. Boor AK, Tresselt HB: An antigen prepared *in vitro* effective for immunisation against anthrax. 11. Concentration of the protective antigen by chemical fractionation. J Infect Dis 96:203, 1954.

67. Brewer CR, McCullough WG, Mills RC, Roessler WG, Herbst EJ, Howe AF: Studies on the nutritional requirements of *Bacillus anthracis*. Arch Biochem 10:65, 1946.
68. Belton FC, Strange RE: Studies on a protective antigen produced *in vitro* from *Bacillus anthracis*: Medium and methods of production. Br J Exp Pathol 35:144, 1954.
69. Wright GG, Hedberg MA, Slein JB: Studies on immunity in anthrax III. Elaboration of protective antigen in a chemically-defined non-protein medium. J Immunol 72:263, 1954.
70. Thorne CB, Belton FC: An agar-diffusion method for titrating *Bacillus anthracis* immunising antigen and its application to a study of antigen production. J Gen Microbiol 17:505, 1957.
71. Wright GG, Green T, Kanode RGJ: Studies on immunity in anthrax V. Immunising activity of alum-precipitated protective antigen. J Immunol 73:387, 1954.
72. Puziss M, Wright GG: Studies on immunity in anthrax X. Gel-adsorbed protective antigen for immunisation in man. J Bacteriol 85:230, 1963.
73. Brachman PS, Gold H, Plotkin SA, Fekety FR, Werrin M, Ingraham NR: Field evaluation of a human anthrax vaccine. Am J Public Health 52:632, 1962.
74. Aleksandrov NI, Gefen NE, Budak AP, Ezepchuck Luv V, Filippenko AI, Runova VF: Utilisation of effective chemical vaccines against certain zoonoses I. Production of chemical depot anthrax vaccine and determination of its effectiveness in animal experiments. Zh Mikrobiol Epidemiol Immunobiol 32:42, 1960.
75. Reimann HA: Infectious diseases. Annual review of significant publications. Postgrad Med 42:247, 1966.
76. Auerbach S, Wright GG: Studies on immunity in anthrax. VI. Immunising activity of protective antigen against various strains of *Bacillus anthracis*. J Immunol 75:129, 1955.
77. Ward MK, McGann VG, Hogge AL Jr, Huff ML, Kanode RG Jr, Roberts EO: Studies on anthrax infections in immunised guinea pigs. J Infect Dis 115:59, 1965.
78. Little SF, Knudson GB: Comparative efficacy of *Bacillus anthracis* live vaccines and protective antigen vaccine in the guinea pig. Infect Immun 52:509, 1986.
79. Ivanovics G: Immunity to anthrax. Z Immunitat Allerg Klin Immunol 94:436, 1983.
80. Vodkin M, Leppla SH: Cloning of the protective antigen gene of *Bacillus anthracis*. Cell 34:693, 1983.
81. Berendt R, Jemski J, Johnson-Winegar A, Ivins BE: The use of toxin components for the immunoprophylaxis of inhalation anthrax. Abstr Ann Meeting Am Soc Microbiol E61:85, 1985.
82. Dorner F, McDonel JL: Bacterial toxin vaccines. Vaccine 3:94, 1985.
83. Hoiseth SK, Stocker BAD: Aromatic dependent *Salmonella typhimurium* are non-virulent and effective as live vaccines. Nature 291:238, 1981.
84. Ivins BE, Welkos SL, Knudson GB, Le Blanc DJ: Transposon Tn 916 mutagenesis in *Bacillus anthracis*. Infect Immun 56:176, 1988.
85. Stewart-Tull DES: The immunological activities of bacterial peptidoglycans. Annu Rev Microbiol 34:311, 1980.
86. Takada H, Kotani S: Immunopharmacological activities of synthetic muramyl-peptides. In Stewart-Tull DES, Davies M (eds): ''Immunology of the Bacterial Cell Envelope.'' Chichester: John Wiley and Sons, 1985, p 119.
87. Cameron J: Letter. Vaccine 2:237, 1984.
88. Smith M, Keppie J, Stanley JL: The chemical basis of the virulence of *Bacillus anthracis* V. The specific toxin produced by *B. anthracis in vivo*. Br J Exp Pathol 36:340, 1955.
89. Welkos SL, Keener TJ, Gibbs PH: Difference in susceptibility of inbred mice to *Bacillus anthracis*. Infect Immun 51:795, 1986.
90. Belton FC, Henderson DW: A method for assaying anthrax immunising antigen and antibody. Br J Exp Pathol 37:156, 1956.

91. Quinn CP, Shone CC, Turnbull PCB, Melling J: Purification of anthrax toxin components by high performance anion exchange, gel-filtration and hydrophobic interaction chromatography. Biochem J 252:753, 1988.

Bacterial Vaccines, pages 123–145


# Production and Control of *Neisseria meningitidis* Vaccines

**Carl E. Frasch**

Division of Bacterial Products, Center for Biologics Evaluation and Research, Bethesda, Maryland 20892



## I. INTRODUCTION

*Neisseria meningitidis* regularly causes severe outbreaks and epidemics throughout the world. The meningococcal disease problem is particularly

**TABLE I. Structures of *N. meningitidis* Group-Specific Capsular Polysaccharides**

| Serogroup | Repeating unit of polysaccharide | Reference |
|---|---|---|
| A | N-acetyl 3-O-acetylmannosamine phosphate α-(1-6) | 2 |
| B | N-acetylneuraminic acid α-(2-8) | 3 |
| C | N-acetyl-7,8-O-acetylneuraminic acid α-(2-9) | 3 |
| H | Galactose (1-2) glycerolphosphate α-(-4) | 4 |
| I | Not determined | |
| K | Not determined | |
| L | N-acetylglucosamine β-(1-3) N-acetylglucosamine α-(1-3) N-acetylglucosamine phosphate β-(-3) | 5 |
| X | N-acetylglucosamine phosphate α-(1-4) | 6 |
| Y | N-acetylneuraminic acid α-(1-4) glucose α-(2-6) | 7 |
| Z | N-acetylgalactosamine α-(1-1) glycerol 3- phosphate (-4) | 8 |
| 29E (Z′) | 3-D-manno-octulosonic acid α-(1-7) N-acetylgalactosamine β-(2-3) | 9 |
| W135 | N-acetylneuraminic acid α-(1-4) galactose α-(2-6) | 7 |

severe in the sub-Sahara area of Africa, described by Lapeyssonie as the "meningitis belt," where the average incidence of meningococcal disease is approximately 40–50/100,000 [1]. In developed countries the incidence of meningococcal disease generally ranges from 1 to 5/100,000.

*N. meningitidis* is divided into 12 different serogroups based on the chemical and immunological specificity of their capsular polysaccharides (PSs). The structures of most of the PSs are known (Table I). Approximately 90% of meningococcal disease worldwide is caused by serogroups A, B, and C. Group A is uncommon outside of Asia and Africa, where it is the major cause of bacterial meningitis. Groups B and C are the predominant causes of meningococcal disease in Europe and the Americas. Approximately 10 to 15% of the disease in the United States is caused by groups Y and W135. The other seven serogroups rarely cause disease.

Immunization against meningococcal disease became a reality with the accomplishments of Gotschlich et al. [10]. The PS vaccines developed by these investigators in the late 1960s and early 1970s were the first chemically pure bacterial vaccines, permitting the use of physical and chemical methods for the characterization and standardization.

## II. PRODUCTION OF POLYSACCHARIDE VACCINES

### A. Background

Following successful field trials [11–14], PS vaccines for the prevention of *N. meningitidis* groups A and C were licensed in 1975 in the United States [15] and in Europe by Institute Merieux [16]. The tetravalent ACYW135 vaccine was licensed in the United States in 1981 [17] based on demonstration of a satisfactory bactericidal antibody response to each of the four PSs. A detailed description of the steps in the production of the meningococcal PS vaccines was published by the World Health Organization in their Technical Report Series [18,19]. Additional descriptions were published by Gotschlich [20,21].

### B. Production Phases

A flow diagram for purification of meningococcal PS is shown in Figure 1. The production method shown and possible alternative steps to be described do not necessarily represent methods used by any one biologics manufacturer, but *N. meningitidis* vaccines produced by these methods should be able to meet the U.S. and WHO release requirements.

**1. Growth and inactivation.** There are no requirements as to which meningococcal A, C, Y, and W135 strains are to be used, but the following have been used: A—A1, A4, and M1027; C—C11 (60E) and C2241; Y—6306Y and IM2261; W135—6308, S4383, and IM2263. The strains should be verified as to serogroup and purity, then retained in a three-level seed-lot system. The master seed should be saved as a small number of freeze-dried ampoules. A suitable menstruum in which to freeze-dry the organisms is 5% monosodium glutamate and 5% bovine serum albumin with 2% mannitol or inositol [22]. One of the master-seed ampoules should be used to produce a number of ampoules of a working master. The two masters should be stored in separate locations. One of the working-master seeds is grown, strain verified, then used to produce a large number of liquid working seed stock, which is stored at −70°C.

The fluid production medium should be free of higher-molecular-weight substances, including human blood group substances, that could precipitate upon later addition of the hexadecyl trimethylammonium bromide (Cetavlon). The medium should therefore contain only dialyzable low-molecular-weight components. The Frantz medium supplemented with 0.2% dialyzed yeast extract has been used [21], as has a modification of this medium containing sodium glycerol phosphate in place of glucose and inorganic phosphate [23]. Studies conducted at the Institute Merieux [16] indicate that optimal production of high-molecular-weight PS is achieved by growth with only moderate aeration (about 6 liters/min by sparge) for 10 to 12 h.

**2. Isolation of polysaccharide.** The fermentor cultures are checked for purity, then brought to 55°C for 10 min to inactivate the meningococci. At this point, two alternative methods may be used to obtain the initial crude PS-containing paste. The negatively charged PS may be precipitated along with the cells by the addition of 10% Cetavlon to the culture to a final concentration of 0.1%. This results in extensive cell lysis and precipitation of much of the nucleic acid. Alternatively, the cells may first be removed from

Grow organisms in modified Frantz medium 10 to 12 hrs

↓ Heat kill, 55 C 10 min

Remove inactivated cells by centrifugation

↓ Add Cetavlon to 0.1%

Precipitate PS from culture broth

↓ Add calcium chloride to 1M

Dissolve PS then centrifuge to remove cellular debris

↓ Add ethyl alcohol to 25%

Remove precipitated nucleic acids by centrifugation

↓ Add ethyl alcohol to 80%

Precipitate crude PS and remove alcohol

↓ 45 % phenol extractions

Extract protein contaminants into the phenol

↓

Dialyze PS against 0.1 M calcium chloride

↓ Ultracentrifuge 4 hrs

Pellet lipopolysaccharide and discard

↓ Add ethyl alcohol to 75%

Precipitate purified bulk PS

↓ Ethyl alcohol and acetone washes

Freeze–dry PS bulk powder and store at − 50 C

**Fig. 1.** Production of meningococcal polysaccharides.

the broth by continuous-flow centrifugation through either a Sharples or a Westfalia centrifuge. Much of the PS is released into the medium during growth and by shear forces during the centrifugation step. Cetavlon is then added to the culture supernatant.

After allowing the PS-Cetavlon precipitation to occur for 30 to 60 min at 4°C, the Cetavlon precipitate may be recovered by centrifugation in a continuous-flow centrifuge. The gray Cetavlon-PS–containing paste may be immediately processed further or frozen at −50°C or lower for later purification.

The Cetavlon paste should be smoothly suspended in 100 ml of cold (4°C) sterile distilled water per 50 g paste. An equal volume of cold 2 M calcium chloride is added to the homogenous suspension to dissolve the PS. Nucleic acids, lipopolysaccharide (LPS), and some protein will also be dissolved. Cell debris is removed by centrifugation. The nucleic acid is then removed by fractional ethyl alcohol precipitation, and the amount that will be present depends upon how the initial Cetavlon precipitation was done. Ethyl alcohol is added slowly with constant stirring to a final concentration of 25%. Much of the DNA can be removed immediately by spooling onto a glass rod. After standing 1 hr, the liquid is centrifuged at approximately 15,000*g* for 30 min to pellet the nucleic acids. It may be necessary to decant the liquid through cheesecloth if lipidlike material is found floating on the surface following centrifugation.

The PS is precipitated from the clear liquid by adjusting the ethyl alcohol concentration to 80% (4 volumes counting that added earlier). The mixture is left at 4°C for 12 to 18 h to permit complete precipitation of the PS. The PS is recovered by centrifugation at 10,000*g* for 20 min. The residual Cetavlon and calcium chloride are removed by washing the pellet three times with absolute ethyl alcohol. It is important to use absolute ethyl alcohol at this step because loss of PS occurs if 95% alcohol is used. The pellet is next washed three times with acetone, then dried under vacuum at 4°C. This completes the initial purification steps to obtain the crude PS. However, at this stage the PS contains far too much protein and endotoxin (LPS).

**3. Final purification and formulation.** The contaminating protein is removed from the crude PS by extraction with cold phenol. First, saturated sodium acetate is prepared and the pH is adjusted to 7.4. From this, a 10% saturated solution in water is made and 180 ml of the solution is used to dissolve 454 g (1 lb) of phenol crystals. A 10 mg/ml solution of the PS in 10% saturated sodium acetate is then prepared, and an equal volume of the phenol solution is added; this is shaken vigorously for 1 min. Phase separation is achieved by centrifugation at 10,000*g* for 30 min. The aqueous phase is recovered and is reextracted twice more with phenol. To minimize loss of PS, each of the three phenol phases should be extracted with one-half volume of distilled water and these extracts pooled with the original aqueous phases.

The pooled aqueous phases are extensively dialyzed against distilled water to remove the phenol. The dialysis should be continued until all traces of phenol odor are gone from the sample. At this point it may be necessary to reduce the volume either by ultrafiltration using a 30,000–50,000 molecular weight cutoff membrane or by addition of sodium acetate and precipitation in 80% ethyl alcohol (final concentration).

The endotoxic LPS must be removed from the PS. This is accomplished by ultracentrifugation. An 8 to 10 mg/ml solution of the PS in 0.1 M calcium chloride is prepared, or the PS solution is dialyzed against 0.1 M calcium chloride. The solution is then ultracentrifuged at 100,000*g* for 3 to 4 h to obtain a small clear LPS pellet. Removal of LPS can be made more complete with little loss of PS by the addition of ethyl alcohol to 25% prior to centrifugation.

The purified PS is precipitated by addition of ethyl alcohol to a final concentration of 75%. The precipitate is recovered as before by centrifugation. The precipitate is washed twice with absolute ethyl alcohol and twice with acetone, then dried at 4°C under vacuum over anhydrous calcium chloride. The purified bulk PS is stored frozen at −50°C or colder. Polysaccharide stored under these conditions is stable for several years; some have been stored for 10 years without loss of potency.

The final vaccine is formulated to contain 50 μg/0.5 ml dose of each polysaccharide and 2.5 to 5.0 mg lactose as a stabilizer per dose. There are four licensed vaccine formulations, A, C, AC, and ACYW135. For formulation of final bulk vaccine, the PS and lactose are dissolved in distilled water at 5× concentration and filled into final containers. The vaccine is then freeze-dried.

The initial meningococcal group A PS vaccines were unstable and had to be stored at −20°C. Studies at the National Institute of Public Health, Bilthoven, The Netherlands, showed that addition of lactose to the vaccine provided enhanced stability [24]. In these studies, lyophilization in sodium chloride, mannitol, and lactose were compared. Vaccines dried with sodium chloride were not stable. Both lactose and mannitol improved the stability, but lactose was found to be superior.

## III. PHYSICAL AND CHEMICAL REQUIREMENTS

### A. Bulk Polysaccharides

The bulk PSs purified from the four different meningococcal serogroups must be analyzed for purity and molecular size prior to formulation into the final product, because most of these determinations cannot be done after mixing the PSs. The WHO and U.S. Food and Drug Administration requirements are nearly the same and are shown in Table II.

**1. Chemical determinations.** The chemical determinations are done based on the dry weights of the PS. The moisture content may be determined by the thermogravimetric method at 50°C [25]. Alternatively, the moisture content may be determined by the Karl Fischer method [25] or by drying over $P_2O_5$ at 37°C until a constant weight is obtained.

The protein content is determined by the method of Lowry et al. [26], and nucleic acid content by spectrometry at an absorbance of 260 nm. The amount of nucleic acid is calculated assuming an absorbance of 1 = 50 μg

**TABLE II. Chemical and Physical Release Requirements for Polysaccharides Contained in Different Meningococcal Vaccine Formulations**

| | U.S. FDA | | | | WHO | | | |
|---|---|---|---|---|---|---|---|---|
| Assay | A | C | Y | W135 | A | C | Y | W135 |
| Phosphorus (> %) | 8.0 | — | — | — | 8.0 | — | — | — |
| Neuraminic acid (> %) | — | 80.0 | 56.0 | 56.0 | — | 80.0 | 56.0 | 56.0 |
| O-acetyl (> mmol/g) | 2.0 | 2.0 | 0.3 | 0.3 | 2.0 | 1.5 | 0.3 | 0.3 |
| Protein (< %) | 1.0 | 1.0 | 1.0 | 1.0 | 1.0 | 1.0 | 5.0 | 5.0 |
| Nucleic acids (< %) | 1.0 | 1.0 | 1.0 | 1.0 | 1.0 | 1.0 | 2.0 | 2.0 |
| Molecular size (< $K_d$)[a] | 0.70 | 0.70 | 0.57 | 0.68 | 65% | 75% | 80% | 80% |
| Pyrogenicity (rabbit)[b] | 0.025 each polysaccharide | | | | 0.025 each polysaccharide | | | |

[a]The molecular sizing (potency) assay is different for U.S. FDA an WHO. U.S. FDA: the maximum $K_d$ of the major peak on Sepharose CL-2B. WHO: the percent of total polysaccharide recovered before a $K_d$ of 0.50 is reached using Sepharose CL- 4B.
[b]The micrograms of each polysaccharide in the vaccine that must pass the pyrogen test. Thus the bivalent and tetravalent vaccine must pass at 0.05 and 0.1 μg/kg of rabbit, respectively.

(1 cm wide cell). The goal should be to obtain purity levels of under 1% protein and nucleic acid.

Purity should also be assessed by direct measurements of the PSs. As seen in Table III, purity of the group A PS can be estimated by phosphorus determination. The recommended procedure is that of Chen et al. [27]. Of particular importance is to use a phosphorus standard that has been standardized by independent methods such as flame photometry. We have found that differences in assay results between laboratories could be traced to differences in phosphorus values assigned to the reference standard. The purity of the other three PSs is determined by assay for sialic acid (N-acetyl nueraminic acid) content using the method of Svennerholm [28]. The reference standard is N-acetylneuraminic acid.

Each of the strains used for vaccine production elaborate PSs containing O-acetyl groups. However, approximately 15% of naturally occurring group C meningococcal strains do not elaborate capsular PS containing O-acetyl groups [29]. Immunogenicity studies comparing O-acetyl–minus and O-acetyl–containing group C PS have shown that comparable antibody titers were induced by both PSs [30]. Unpublished studies by Tiesjema of the National Institute of Health, Bilthoven, The Netherlands, indicate that there are group Y and W135 strains that do not elaborate O-acetyl–containing PS, but the relative immunogenicity of the PSs have not been studied. Because the O-acetyl groups are not stable above pH 8.5, measurement of O-acetyl content can provide an indication of harsh treatment. The O-acetyl content should be determined by the method of Hestrin, using acetylcholine as the reference preparation [31], or alternatively by NMR spectroscopy.

**TABLE III. Chemical Analysis of Meningococcal Polysaccharides—Determination of Purity by Comparison With Theoretical Percent Dry Weight of Calcium Salt**

| | Percent of dry weight of polysaccharide | | | |
|---|---|---|---|---|
| Constituent | A | C | Y | W135 |
| Phosphorus | 9.1 | 0 | 0 | 0 |
| Hexosamine | 88.0[a] | 0 | 0 | 0 |
| Neuraminic acid | 0 | 90.0 | 63.0 | 63.0 |

[a]The group A polysaccharide is generally not assayed for hexosamine content.

**2. Molecular sizing for potency.** The PSs must be of high molecular size to stimulate an optimal immune response. In 1944, Kabat et al. [32] reported clinical studies on a group A meningococcal PS vaccine, finding it essentially nonimmunogenic in human adults. In later studies by Liu et al. [3], it was determined that group A PS prepared by methods similar to those described by Kabat probably had a molecular weight of less than 20,000, thus providing an explanation for the lack of immunogenicity observed by Kabat. Gold et al. [33] immunized infants with two lots of group A vaccine differing significantly in molecular size. The larger PS induced twofold higher peak antibody levels than the lower-molecular-weight PS. In a similar study in Finland, Kayhty et al. [34] compared the antibody responses of children to a high-molecular-weight PS and four preparations of lower-molecular-weight PS. Again the antibody response was best to the larger PS. Finally, the critical importance of high-molecular-weight vaccine was demonstrated in group A meningococcal PS efficacy trials in Africa where only high-molecular-weight vaccine provided significant protection [35].

Molecular size determination is the potency assay required for meningococcal PS vaccines. The method for molecular size determination has been described in detail by Wong et al. [35]. The molecular size of each PS lot is estimated by gel filtration using Sepharose CL-2B or CL-4B (Pharmacia Fine Chemicals, Uppsala, Sweden) by determining its distribution coefficient ($K_d$). The PS is dissolved in distilled water to 1% of the column volume and applied to the column. It is eluted using 0.2 M ammonium acetate buffer pH 7.0. The void volume ($V_o$) of the column is determined using blue dextran (Pharmacia) or inactivated viral particles for Sepharose CL-2B. The included volume ($V_i$) within the gel beads may be determined using tritiated water. The elution volume ($V_e$) of the major PS peak is determined when CL-2B is used, as for the U.S. FDA release requirements. For the WHO requirements, the percent of the PS eluting before a $K_d$ of 0.5 is determined. The $K_d$ is calculated as $K_d = (V_e - V_o/V_i$.

For molecular size estimation, elution of the PS from the column may be detected using a refractometer or by chemical assay. For the group A PS, use of the Park and Johnson method for reducing sugars is more sensitive than phosphorus determination [36].

Each batch (lot) of bulk PS should be examined for endotoxin content to be certain that the final product will pass the rabbit pyrogenicity requirements. The endotoxic LPS may be accurately quantitated by the polyacrylamide gel silver stain method [37]. Alternatively it may be estimated by limulus amebocyte lysate (LAL) gelation.

### B. Final Product

The final vaccine is formulated to contain 50 ± 10 μg of each PS to be included in the vaccine and 2.5 to 5 mg of lactose as a stabilizer per dose. The presence (identity test) and amount of each PS to be included in the vaccine should be determined. This can be done serologically by either rocket immunoelectrophoresis [38] or by rate nephelometry [39]. The reference standard for each of the PSs should be bulk material that has passed all requirements shown in Table II.

It is important to determine the molecular size of the PS after completion of the filling and lyophilization process, because decomposition of the group A PS during the filling process has been found to occur [35]. The group A PS is the least stable of the four PSs because of the presence of phosphodiester linkages and may serve as an indicator of possible breakdown during filling, freeze-drying, or storage in the bivalent or tetravalent vaccines.

The moisture content in multidose final containers should be determined on one vial per 1,000 filled up to a maximum of ten, but on no less than five vials. The WHO requirements are an average moisture by gravimetric analysis of not more than 2.5%, with no vial over 3.0%. The U.S. FDA requirements are an average of less than 1%. Moisture determination on single-dose vials is best done by the coulometric Karl Fischer method [25]. The Karl Fischer method will detect more water, necessitating a higher limit.

The vaccine should be tested for endotoxin content by rabbit pyrogenicity, or LAL if the LAL test has been standardized to the rabbit test. The vaccine should pass the rabbit pyrogenicity test at a test dose of 0.025 μg for each PS included in the vaccine per kilogram body weight of rabbit. Vaccines must also pass the general safety and sterility test requirements.

## IV. CLINICAL EVALUATION OF POLYSACCHARIDE VACCINES

Manufacturers of meningococcal vaccines in the United States were required to clinically evaluate at least the first two consistency lots of each meningococcal PS vaccine formulation. These clinical studies were done in groups of at least 25 adults using the recommended dose of 50 μg per PS.

Clinical potency was determined because there is no satisfactory animal model. The bactericidal antibody response was determined on sera obtained immediately before immunization and 3 to 4 weeks later. At least 90% of the individuals had to show a fourfold or greater rise in bactericidal titer to each serogroup in the vaccine.

The bactericidal method used for determination of clinical potency has been described by Wong et al. [35]. Since different strains of the same serogroup show varying sensitivity to serum killing, the U.S. and WHO meningococcal vaccine requirements specify that the following strains be used in the bactericidal assays: A1, C11, S-1975, and S-4383 for groups A, C, Y, and W135 respectively.

The Immunization Practices Advisory Committee of the U.S. Public Health Service has made recommendations for the use of the meningococcal PS vaccines. The recommended immunization scheme is a single 0.5 ml intramuscular injection containing 50 μg per PS. The vaccine is presently available in the United States from Connaught Laboratories and in Europe from Smith Kline Biologicals and Institute Merieux.

Routine vaccination of the civilian population in industrialized countries is not currently recommended, because the risk of infection is low, a vaccine against group B is not available, and much of the endemic disease occurs in young children. Vaccination is advised to control outbreaks due to meningococcal serogroups covered by the vaccine. Routine vaccination is recommended for travelers to countries recognized as having hyperendemic or epidemic meningococcal disease, as in the "meningitis belt" of Africa.

Recent studies have been conducted examining the antibody response of young children and infants to the ACYW135 tetravalent vaccine [30,40]. Lepow et al. [40] found that a greater percentage of children between 2 and 5 years of age responded to the Y and W135 PSs than to the A and C PSs. An 80% or better response rate was found to all four PSs in children 9 years of age or older. One problem was the rapid falloff in antibody levels in children under 9 years of age, in which they found a 70 to 80% drop 1 year later.

Peltola et al. [30] examined the immune response of infants aged 6 to 24 months to the tetravalent vaccine at 30 μg per PS. Children under 1 year of age at time of immunization received a second injection 3 months later. Over 90% of the children responded with bactericidal antibodies to group A. The response to the other PSs was more age dependent, with 80 to 100% of the children aged 18–23 months responding and 40 to 50% of the younger children responding to the PSs. Children under 1 year of age had a rapid decline in bactericidal antibodies and 1 year later were not different from unvaccinated children of the same age. Peltola et al. recommended that children under 1 year of age receive two injections 3 months apart. They felt that routine immunization would be justified if an effective group B vaccine component could be added.

Vaccination of individuals with a deficiency in one of the terminal complement components or with properdin deficiency may be effective. Patients with deficiencies in C5, C6, C7, or C8 have a prolonged susceptibility to meningococcal disease and often develop the disease later in life ;median age at first episode is 17 years [41]. Their low mortality rate (4%) from such infections suggests that presence of antibodies that can promote opsonophagocytosis is important in such individuals. Therefore it follows that vaccination can provide some protection.

Normal properdin function appears to be important in resistance to meningococcal disease. Properdin promotes bactericidal killing through activation of the alternative pathway by stabilizing the C3 convertase C3b,Bb [42]. Many individuals with meningococcal disease associated with properdin deficiency are young children, who would not be expected to have developed protective antibodies [42]. Vaccination of properdin-deficient individuals with the meningococcal PS vaccines has been shown to induce antibodies that cause meningococcal killing through the classical complement pathway [43].

## V. CONJUGATE VACCINES

For meningococcal PS vaccines to provide long-term protection in young children, they will need to be conjugated to protein carriers. The duration of protection in young children is relatively short, and the antibody response is not boostable as is the case with T-cell independent antigens.

Meningococcal group A, B, and C PSs, as well as oligosaccharides from these PSs, have been covalently attached to tetanus toxoid [39,40]. Jennings and Lugowski [44] produced conjugates of the A, B, and C PSs by periodate oxidation and direct attachment to tetanus toxoid through the lysines. In addition, they prepared oligosaccharides from the group C PS and linked them through a spacer molecule to bovine serum albumin. These conjugates were used to hyperimmunize mice. All except the B conjugate were highly immunogenic. The group B conjugate failed to stimulate antibodies.

Beuvery et al. [45,46] produced group A and group C PS conjugates by conjugation of the PSs to tetanus toxoid. These conjugates were compared to the native PSs for their immunogenicity in mice. The native PSs were essentially nonimmunogenic when administered in two doses 10 weeks apart. In contrast, a single injection of conjugate elicited high antibody levels. A second injection at week 10 elicited a booster response, typical of a T-cell–dependent response. Adsorption of the conjugates to aluminum phosphate significantly increased the primary immune response, but failed to have an effect on the booster response [45,46]. Thus, conjugation of the A and C PSs to a protein carrier converted the PSs from T-cell–independent to T-cell–dependent antigens.

Meningococcal PS-protein conjugates have not been evaluated to date in humans. However, extensive human studies with *Haemophilus influenzae* type b PS conjugates indicate that conjugates can be safe and much more immunogenic in young children than the native PS [47]. A second injection of a *Haemophilus* b conjugate or of the native PS following primary immunization with the conjugate elicits a strong booster response. Whereas the *Haemophilus* b PS is not effective in children under 18 to 24 months of age, a dipththeria toxoid conjugate proved to be highly effective in infants at 6 months of age [48]. Meningococcal conjugates will likely also prove to be highly immunogenic in young infants.

## VI. GROUP B MENINGOCOCCAL VACCINES

### A. Polysaccharide

Although the group B PS, a homopolymer of alpha 2-8 linked sialic acid, is the logical choice for production of a group B meningococcal vaccine, only limited success has been achieved in animal studies [49,50]. In humans, the native PS induces only a transient IgM response [51], and a number of possible explanations have been given for the poor immunogenicity, including sensitivity to neuraminidases and similarity to sialic acid moieties on human tissues.

Moreno et al. have attempted to improve the immunogenicity of the group B PS by noncovalently complexing it to outer-membrane proteins (OMPs) [48] and by adsorption of the complexed vaccine to aluminum hydroxide [52]. These vaccines were used to immunize mice in a comparison with the pure PS. While the PS itself was nonimmunogenic, both of the other formulations induced transient increases in IgM antibodies peaking on day 7. The aluminum hydroxide–adsorbed vaccine induced higher responses in the mice than unadsorbed vaccine. However, these transient IgM responses would not be expected to provide useful protection. Furthermore, a similar aluminum hydroxide serotype protein group B PS vaccine failed to stimulate measurable increases in group B PS antibodies in human studies [53].

Another approach to preparing immunogenic group B PS vaccines is to chemically modify the PS. Jennings et al. made a number of chemical modifications to the PS [49]. The only modification that did not abrogate the ability of group B–specific antibodies to recognize the PS was substitution of the N-acetyl groups with N-propionyl groups.

The N-propionylated PS and its tetanus toxoid conjugate were used to immunize mice using Freund's complete adjuvant in a series of three injections [49]. The altered PS alone failed to induce group B PS-reactive antibodies. In contrast, two injections of the conjugate induced antibody levels severalfold above background, and a third injection gave a good booster response. Interestingly, the antibody response was mostly IgG. These antibodies were shown to be bactericidal against group B strains of different serotypes [54], but not for group A, C, or W135 strains. Quantitative precipitin experiments indicated that about half the antibodies induced to the N-propionylated PS were specific for the altered PS [55]. Unexpectedly, only absorption of the sera with the N-propionylated PS removed the group B bactericidal antibodies, even though the native PS adsorbed all group B PS radioimmunoassay reactive antibodies. Thus, the N-propionylated group B PS mimics a bactericidal epitope found on the group B organism. Additional studies on the nature of this epitope are important, and limited adult human studies are warranted.

### B. Outer-Membrane Proteins

A promising approach to development of an effective group B meningococcal vaccine is the use of LPS-depleted outer membranes. There are a number of proteins in the outer membrane that are antigenically heterogeneous among different group B meningococcal strains. *N. meningitidis* serogroups B, C, Y, and W135 are divided into serotypes based on the immunological specificity of the $M_r$ 38,000 to 41,000 class 2 and 3 proteins and into subtypes based on the specificity of the $M_r$ 45,000 class 1 proteins [55]. Antibodies to both the class 1 and class 2/3 proteins are bactericidal and therefore probably protective [56]. However, one potential problem with outer-membrane protein vaccines is that such vaccines will not induce protection against heterologous serotypes or subtypes.

Relatively few serotypes and subtypes are associated with the majority of group B meningococcal disease occurring in a given geographic region at a given time, but as with the influenza virus, the antigenic types causing disease vary with geographic area and year. We have, however, shown that the vaccine production method described below can be used with little or no modification for most serotypes [57]. A serotype protein vaccine would need to contain proteins from at least two or three different serotypes.

### C. Preparation of an Outer-Membrane Vaccine (Fig. 2)

**1. Recovery of outer membranes.** The meningococcal strains are grown in a dialysate or semisynthetic medium to eliminate possible contamination from media components. We obtain optimal growth of the organisms using a dialysate of tryptic soy broth (TSB, Difco) and 1% yeast extract. The dialysate is prepared by dissolving the TSB and yeast extract in water, heating briefly to 60–70°C, and collecting the ultrafiltrate from a 10,000

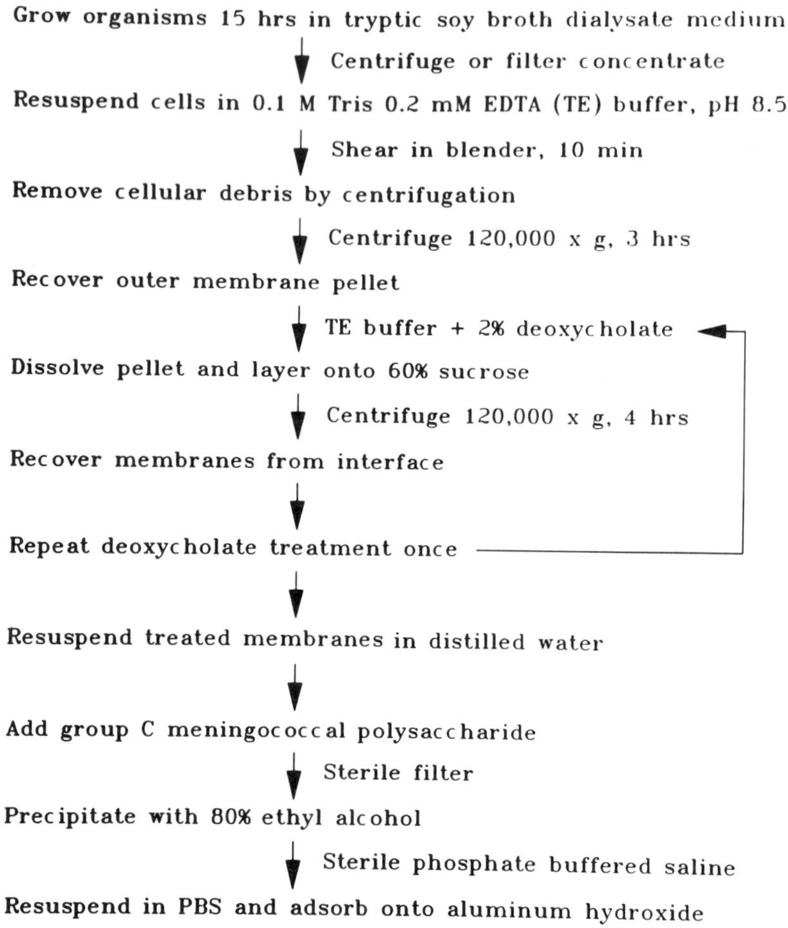


**Fig. 2.** Production of meningococcal outer-membrane protein vaccines.

molecular weight cutoff ultrafiltration membrane. The dialysate medium is inoculated from seed cultures and incubated at 35°C for 12 to 14 h. The cells are killed by heating to 55°C for 10 min and are then harvested by centrifugation.

The cells contain large amounts of loosely bound outer-membrane material, which are recovered as small vesicles by shearing. The cell pellet is resuspended in 0.1 M Tris-HCl, 0.2 mM EDTA (TE) buffer, pH 8.5, and sheared in a sealed blender or Omnimixer (Sorvall) for 10 min. The largely intact cells are removed by centrifugation at 10,000*g* for 20 min. A second centrifugation at 10,000$g$ is required to remove the remaining cells. The supernatant is treated with RNase and DNase for 2 h at 37°C. The outer-membrane vesicles (OMV) are recovered by centrifugation at 120,000$g$ for 3 h. The OMV pellet should appear clear with a straw yellow color.

**2. Removal of LPS.** The OMV contains approximately equal amounts of protein and LPS. The LPS must be removed by treatment with detergent. If all of the LPS is removed, the vesicle structure is destroyed and the resulting material has poor immunogenicity. Although a number of different nonionic and zwitterionic detergents have been used, we have found that sodium deoxycholate, a normal bile metabolite, works well. The remaining protein-bound LPS is much less toxic than free LPS [58].

To remove the LPS, the pelleted OMV are resuspended in TE buffer containing 2% sodium deoxycholate. After solubilization for 10 min at room temperature, the LPS-depleted OMV may be purified by one of two methods, gel filtration or ultracentrifugation. For gel filtration the solubilized material is separated on a Sephacryl S-400 or sepharose CL-6B column using 1% deoxycholate in TE buffer as the eluant. The OMV will elute at the void volume of either column and is recovered by precipitation with 80% ethyl alcohol and redissolved in distilled water. If the OMV are purified by ultracentrifugation they should be pelleted onto a 60% sucrose bed to allow for rapid resolubilization. The sucrose bed is prepared by dissolving ultrapure sucrose in TE buffer to a density of 1.29 g/ml at room temperature. The solubilized OMV is layered over the sucrose bed and centrifuged at 120,000$g$ for 4 h; the clear liquid above the white band is aspirated and discarded. Most of the sucrose bed is aspirated from under the band and discarded. The tube is refilled again with TE buffer containing 2% deoxycholate, leaving sufficient space to replace the 60% sucrose bed. The OMV are pelleted as before. Finally, the OMV band is taken up in distilled water.

**3. Vaccine formulation.** The concentration of protein in the LPS-depleted OMV is determined by the Lowry procedure [26]; if purified by the sucrose-bed ultracentrifugation method, a modified Lowry procedure [59 is used. Based on the protein assay results, an equal amount by weight of group C meningococcal PS is added. Addition of the PS improves the solubility of the final vaccine. The vaccine is prefiltered, then sterile filtered through a combination of depth prefilters and a 0.45 μm membrane filter.

The sterilized protein PS combination is precipitated with 80% ethyl alcohol after addition of 10 ml per liter of sterile 10% saturated sodium acetate. The 95% ethyl alcohol to be used must be prefiltered through a 0.22 μm solvent resistant membrane filter such as the Millipore Durapore filter, since the alcohol may contain viable bacillus spores. The pellet is recovered by centrifugation and dissolved to 200 μg/ml of protein in sterile phosphate-buffered saline, pH 7.2–7.4. The final vaccine is prepared by adsorption onto

**TABLE IV. Preliminary Chemical and Physical Release Specifications for Meningococcal Outer-Membrane Protein Vaccines**

| Vaccine intermediate | Assay | Specification |
|---|---|---|
| LPS-depleted OMV | Protein | For calculation |
| | LPS (By KDO) | nmt 50 μg/mg protein |
| | SDS-PAGE | Purity of OMV |
| | Identity | Serological identity |
| | Molecular size[a] | $K_d$ = 0.0 on Sepharose CL-6B or Sephacryl S-400 |
| Combined protein-PS | PS identity | Serological identity |
| | Sialic acid/protein | 0.8–1.2:1 |
| Adsorbed vaccine | Percent adsorption | nlt 80% |
| | Pyrogenicity (rabbit) | Pass: 1 μg/kg |

Abbreviations: OMV, outer-membrane vesicle; LPS, lipopolysaccharide; PS, polysaccharide; nmt, not more than; nlt, not less than; KDO, ketodeoxyoctonate.
[a]The molecular size is measured in the presence of 1% deoxycholate in TE buffer.

aluminum hydroxide. The aluminum hydroxide is prepared by slow addition of sterile 5% (wt/vol) sodium carbonate to sterile 3% (wt/vol) aluminum chloride as previously described [60]. An equal volume of 4 mg/ml aluminum hydroxide is mixed with the vaccine. A 0.5 ml dose of vaccine will contain 50 μg protein and 50 μg group C PS.

**4. Physical and chemical analysis.** Although no specifications have been agreed upon for outer-membrane protein vaccines, possible specifications are presented in Table IV. The OMV should retain their membrane structure after detergent treatment as determined by electron microscopy, but at the same time should have an LPS/protein ratio of less than 0.05. The LPS should be measured by a chemical assay or by SDS-PAGE because membrane-bound LPS is not accurately estimated by the LAL assay [58]. The chemical assay can be for either KDO or glucosamine. Meningococcal LPS contains about 4 to 5% measurable KDO as estimated by the Osborn method [61]. Alternatively, the LPS can be accurately estimated using silver stain following SDS-PAGE [37].

All chemical assays need to be done on the LPS-depleted OMV and combined OMV+PS prior to adsorption onto aluminum hydroxide. The ratio of PS to protein in the mixture is formulated to be 1:1. The PS content is estimated by sialic acid determination using the Svennerholm method [28]. The final vaccine should be examined for adsorption to the adjuvant by centrifuging a sample of the vaccine and examining the supernatant for protein content. This is then compared to one-half the protein content measured on the combined vaccine prior to adsorption. General safety, sterility, and pyrogenicity tests are done on the final containers.

### D. Clinical Studies on Protein Vaccines

A number of clinical studies have been conducted with OMP vaccines; all indicate that these vaccines are safe and immunogenic. Most individuals will experience local reactions of redness or tenderness at the injection site, but systemic reactions are minimal [53].

Rosenqvist et al. [62] examined the antibody responses of 57 adults to a vaccine containing OMPs of sterotypes 2b:P1.2 and 15:P1.16 noncovalently complexed to meningococcal PSs A, C, Y, and W135. Volunteers received one dose of the vaccine containing 52 μg of protein of each serotype, 40 μg PS, and 3.5 μg LPS. Overall, 70% of the vaccines had immune responses to both serotypes judged by ELISA and bactericidal assay. The vaccinee's responses were equivalent to those of group B meningococcal disease patients (mostly B:15:P1.15), and were predominantly IgG, which we also found in similar studies [53]. The majority of vaccines had substantial increases in bactericidal antibodies to both serotypes included in the vaccine. Before vaccination, 60% of the individuals had no detectable bactericidal activity against either serotype, and after immunization 21% and 24% still lacked bactericidal antibodies. A second immunization or use of adjuvants would likely increase the rate of response.

An important observation of Rosenqvist et al. [62] was the excellent correlation ($R = 0.8$) between the IgG levels measured by ELISA and bactericidal titers measured using human complement. Earlier studies using baby rabbit serum as a source of complement failed to show a correlation between the two assays.

The antigen specificity of the adult response to a serotype 2a OMP vaccine was examined by immunoblotting [63,64]. Most individuals responded to some of the high molecular proteins and to the class 1 and 5 proteins. It was particularly noteworthy that those who responded with high bactericidal titers had high antibody levels to the P1.2 class 1 protein. The class 1 protein is therefore an important antigen for inclusion in an OMP vaccine. In immunoblots, antibody binding to the class 2/3 proteins is dependent upon renaturation in a zwitterionic detergent [65]. Examination of the IgG response to the serotype 2a protein in the presence of the detergent Empigen BB resulted in more postvaccination antibodies being bound to the class 2 protein [64]. However, the response to the class 1 protein remained highest, the subclass response being primarily IgG1 and IgG3.

The major target population for a group B vaccine is young children. We therefore compared the antibody responses of adults and children to vaccines containing 2a OMP and group B meningococcal polysaccharide. The antibody responses were measured by ELISA using purified serotype 2a outer membrane vesicles and by bactericidal assay using a group C:2a strain and baby rabbit serum as a complement source. In both assays young children (<6 years old) had lower antibody responses than older children and adults. Over 50% of children over 2 years of age had postimmunization bactericidal titers of 1:8 or greater, compared to only 20% for those under 2 years of age. By comparison, 70 to 80% of adults had a fourfold rise in bactericidal titer after immunization [66].

In an effort to improve the immune response of young children to serotype protein vaccines, we have evaluated the effect of aluminum hydroxide on serotype 2b protein vaccines. Initial studies in animals indicated that adsorption of the vaccine significantly improved the immune response [60]. The experiments have now been extended to studies in adult volunteers [53]. Groups of adults were immunized with a serotype 2b-group B PS vaccine with and without aluminum hydroxide adjuvant. The adjuvant improved the antibody response as measured by ELISA and bactericidal assays. Only those who received the first immunization with adjuvant had a booster effect on reimmunization. These studies indicate that aluminum hydroxide or other adjuvants should be used to increase the percentage of children who respond to OMP vaccines.

Trials to estimate the effectiveness of OMP vaccines for preventing group B meningococcal disease have been conducted in South Africa and are in progress in Chile, Cuba, and Norway. A randomized serotype 2a vaccine trial was carried out in Cape Town, South Africa, in 1981 [67]. The vaccine was administered to 2,200 children between 4 months and 6 years of age (mean = 2 years). Another 2,200 received a group A/C PS vaccine. The trial was conducted during a group B serotype 2b epidemic and provided some evidence for protection against group B serotype 2 disease, but insufficient numbers of children were immunized to allow for a statistical estimate of efficacy. Interim results of an ongoing vaccine trial against group B serotype 4 (B:4:P1.15) disease in Cuba were reported at the Sixth International Pathogenic Neisseria Conference (Atlanta, Georgia, October, 1988), and after 1 year of observation a preliminary analysis indicated that there was protection against group B meningococcal disease.

One problem with the serotype protein approach to prevention of group B meningococcal disease is the restricted range of protection induced by the serotype and subtype proteins. To achieve broad protection, a vaccine would need to contain outer-membrane proteins from three or four serotypes. Also, the serotypes associated with significant levels of disease vary from one geographic region to another and vary over a period of years, necessitating continued epidemiological surveillance of disease-associated serotypes to produce the optimal vaccine for a given time and location.

Although serotype protein vaccines offer the best immediate approach, there are alternatives. An ideal meningococcal vaccine would be immunogenic in all age groups and protect against all group B strains and possibly all meningococcci. However, such a vaccine can only be developed based on a much better understanding of the basic mechanisms by which only some meningococcal strains are able to gain entrance into the host, evade host defenses, and cause invasive disease. Alternative antigens that need to be examined include the iron-inducible outer-membrane proteins of $M_r$ ~37,000 and 70,000–110,000 found in all meningococci. Another alternative vaccine is LPS-derived oligosaccharides covalently bound to a protein carrier.

## VII. CONCLUSIONS

Meningococcal disease remains a problem in many countries, not because of the normal endemic levels, but because outbreaks and severe epidemics periodically occur. The group A PS is particularly effective in the control of epidemic disease. In Africa, ongoing group A epidemics have been rapidly halted by mass vaccination. PS vaccines have been licensed for serogroups A, C, Y, and W135, and of these only groups A and C cause significant disease. Group C meningococcal disease occurs more often than group A in young children, but the group C PS is not effective in children under 2 years of age. This problem will probably be alleviated by covalent attachment of the PS to a protein carrier.

This chapter has described the production of both meningococcal polysaccharide and outer-membrane protein vaccines. Although most national control agencies will not produce meningococcal vaccines, they need guidelines for proper control of these vaccines. Guidelines are provided by WHO in their Technical Report Series, by the U.S. Food and Drug Administration in "Guidelines for the Production of Meningococcal Polysaccharide Vaccines," in Tables II and IV above. In the United States, vaccine manufactures are requested to submit with each lot or batch of vaccine a release protocol providing their test results for each required analysis. The control agency may wish to test final container material and to request samples of individual bulk polysaccharides for analysis and samples of LPS-depleted OMV if such vaccines are approved for use.

## REFERENCES

1. Lapeysonnie L: La meningite cerebro-spinale en Afrique. Bull WHO 28(Suppl 1):3, 1963.
2. Liu TY, Gotschlich EC, Jonessen EK, Wysocki JR: Studies on the meningococcal

polysaccharides. I. Composition and chemical properties of the group A polysaccharide. J Biol Chem 246:2849, 1971.
3. Liu TY, Gotschlich EC, Dunne FT, Jonssen EK: Studies on the meningococcal polysaccharides. II. Composition and properties of the group B and group C polysaccharides. J Biol Chem 246:4703, 1971.
4. Van der Kaaden A, Van Dooren-Van Wakeren JIM, Kamerling JP, Vliegenthart FG, Tiejsjema RH: Structure of the capsular antigen of *Neisseria meningitidis* serogroup H. Eur J Biochem 141:513, 1984.
5. Jennings HJ, Lugoski CW, Ashton FA, Ryan JA: The structure of the capsular polysaccharide obtained from a new serogroup (L) of *Neisseria meningitidis*. Carbohydr Res 112:105, 1983.
6. Bundle DR, Jennings HJ, Kenney CP: Studies on the group-specific polysaccharide of *Neisseria meningitidis* serogroup X and an improved procedure for its isolation. J Biol Chem 249:4797, 1974.
7. Bhattacharjee AB, Jennings HJ, Kenney CP, Martin A, Smith ICP: Structural determination of the polysaccharide antigens of *Neisseria meningitidis* serogroups Y, W-135, and Bo. Can J Biochem 54:1, 1976.
8. Jennings HJ, Rosell K-G, Kenny CP: Structural elucidation of the capsular polysaccharide antigen of *Neisseria meningitidis* serogroup Z using $^{13}C$ nuclear magnetic resonance. Can J Chem 57:2902 1979.
9. Bhattacharjee AK, Jennings HJ, Kenny CP: Structural elucidation of the 3-deoxy-D-manno-octulosonic acid containing meningococcal 29-e capsular polysaccharide antigen using carbon-13 nuclear magnetic resonance. Biochemistry 17:645, 1978.
10. Gotschlich EC, Goldschneider I, Artenstein MS: Human immunity to the meningococcus IV. Immunogenicity of group A and group C polysaccharides in human volunteers. J Exp Med 129:1367, 1969.
11. Peltola H, Makela PH, Kayhty H, Jousimies H, Herva E, Hallstrom K, Sivonen A, Renkonen O-V, Petty O, Karanko V, Ahonen P, Sarna S: Clinical efficacy of meningococcus group A capsular polysaccharide vaccine in children three months to five years of age. N Engl J Med 297:686, 1977.
12. Whadan MH, Rizk F, El-Akkad AM, El Ghoroury AA, Hablas R, Girgis NI, Amer A, Boctar W, Sippel JE, Gotschlich EC, Triau R: A controlled field trial of serogroup A meningococcal polysaccharide vaccine. Bull WHO 48:667, 1973.
13. Artenstein MS, Gold R, Zimmerly JG, Wyle FA, Schneider H, Harkins C: Prevention of meningococcal disease by group C polysaccharide vaccine. N Engl J Med 282:417, 1970.
14. Sanborn WA: Development of meningococcal vaccines. In NA Vedros (ed): "Evolution of Meningococcal Disease." Boca Raton, FL: CRC Press, 1987, Vol. 2, p. 121.
15. Centers for Disease Control: Recommendations of the Immunization Practices Advisory Committee on Immunization Practices—Meningococcal polysaccharide vaccines. MMWR 27:327, 1978.
16. Ayme G, Donikian R, Mynard MC, Lagrandeur G: Production and controls of serogroup A *Neisseria meningitidis* polysaccharide vaccine. In R Triav (ed): "Table Ronde sur l'Immunoprophylaxie de la Meningite Cerebro-spinale." Lyon: Foundation Merieux, 1973, p. 4.
17. Centers for Disease Control: Recommendations of the Immunization Practices Advisory Committee (ACIP)—Meningococcal vaccines. MWR 34:255, 1985.
18. WHO Expert Committee on Biological Standardization: Requirements for meningococcal polysaccharide vaccine. WHO Technical Report Series No. 658, 1981, p. 174.
19. WHO Expert Committee on Biological Standardization: Requirements for meningococcal polysaccharide vaccine. WHO Tech Rep Ser 594:50, 1976.

20. Gotschlich EC, Liu TY, Artenstein MS: Human immunity to the meningococcus III. Preparation and immunochemical properties of the group A, group B, and group C meningococcal polysaccharides. J Exp Med 129:1349, 1969.
21. Gotschlich EC: Development of polysaccharide vaccines for the prevention of meningococcal diseases. Monogr Allergy 9:245, 1975.
22. Redway KF, Lapage SP: Effects of carbohydrates and related compounds on the long term preservation of freeze-dried bacteria. Cryobiology 11:73, 1974.
23. Frasch CE, McNelis RM, Gotschlich EC: Strain-specific variation in the protein and lipopolysaccharide composition of the group B meningococcal outer membrane. J Bacteriol 127:973, 1976.
24. Tiesjema RH, Beuvery EC, te Pas BJ: Enhanced stability of meningococcal polysaccharide vaccines by using lactose as a menstruum for lyophilization. Bull WHO 55:43, 1977.
25. May JC, Grim E, Wheeler RM, West J: Determination of residual moisture in freeze-dried viral vaccines: Karl Fischer, gravimetric and thermogravimetric methodologies. J Biol Stand 10:249, 1982.
26. Lowry OH, Rosenbrough NJ, Farr AL, Randall RH: Protein measurement with the Folin phenol reagent. J Biol Chem 193:165, 1951.
27. Chen PS Jr, Toribara TY, Warner H: Microdetermination of phosphorus. Anal Chem 28:1756, 1956.
28. Svennerholm L: Quantitative estimation of sialic acids. II. A colorimetric resorcinol-hydrochloric acid method. Biochim Biophys Acta 241:604, 1957.
29. Apicella MA: Identification of a subgroup antigen on the *Neisseria meningitidis* group C capsular polysaccharide. J Infect Dis 129:147, 1974.
30. Peltola H, Safary A, Kayhty H, Karanko V, Andre FE: Evaluation of 2 tetravalent (ACYW135) meningococcal vaccines in infants and small children- a clinical stuty comparing immunogenicity of O-acetyl-negative and O-acetyl-positive group C polysaccharides. Pediatrics 76:91, 1985.
31. Hestrin S: The reaction of acetyl choline and other carboxylic acid derivatives with hydroxylic acid derivatives with hydroxylamine, and its analytical applications. J Biol Chem 180:249, 1949.
32. Kabat EA, Kaiser H, Sikorski H: Preparation of the type-specific polysaccharide of the type I meningococcus and a study of its effectiveness as an antigen in human beings. J Exp Med 80:299, 1945.
33. Gold R, Lepow ML, Goldschneider I, Draper TL, Gotschlich EC: Clinical evaluation of group A and group C meningococcal polysaccharide vaccines in infants. J Clin Invest 56:1536, 1975.
34. Kayhty H, Karanko V, Peltola H, Sarna S, Makela PH: Serum antibodies to capsular polysaccharide vaccine of group A *Neisseria meningitidis* followed for three years in infants and children. J Infect Dis 142:861, 1981.
35. Wong KH, Barrera O, Sutton A, May J, Hochstein DH, Robbins JD, Robbins JB, Parkman PD, Seligmann EB Jr: Standardization and control of meningococcal vaccines, group A and group C polysaccharides. J Biol Stand 5:197, 1977.
36. Helting TB, Kumpe M: A sensitive method for the determination of the molecular size of meningococcal polysaccharide vaccine. J BIol Stand 6:111, 1978.
37. Tsai C-M, Frasch CE: A sensitive silver stain for detecting lipopolysaccharides in polyacrylamide gells. Anal Biochem 119:115, 1982.
38. Lee CJ, Pearson SJ, Robbins JB: Quantitative determination of pneumococcal polysaccharides in the polyvalent vaccine. I. Standardization of an immunoelectrophoric method. J Biol Stand 8:271, 1980.
39. Lee CJ: The quantitative immunochemical determination of pneumococcal and meningo-

coccal capsular polysaccharides by light scattering rate nephelometry. J Biol Stand 11:55, 1983.

40. Lepow ML, Beeler J, Randolph M, Samuelson JS, Hankins WA: Reactogenicity and immunogenicity of a quadravalent combined meningococcal polysaccharide vaccine in children. J Infect Dis 154:1033, 1986.
41. Ross SC, Densen P: Complement deficiency states and infection: Epidemiology, pathogenesis and consequences of neisserial and other infections in an immune deficiency. Medicine 63:243, 1984.
42. Sjoholm AG, Kuijper EJ, Tijssen CC, Jansz A, Bol P, Spanjaard L, Zanen HC: Dysfunctional properdin in a Dutch family with meningococcal disease. N Engl J Med 319:33, 1988.
43. Densen P, Weiler JM, Griffiss JM, Hoffmann LG: Familial properdin deficiency and fatal meningococcemia: Correction of the bactericidal defect by vaccination. N Engl J Med 316:922, 1987.
44. Jennings HJ, Lugowski C: Immunochemistry of groups A, B, and C meningococcal polysaccharide-tetanus toxoid conjugates. J Immunol 127:1011, 1981.
45. Beuvery EC, Kaaden A, Kanhai V, Leussink AB: Physiochemical and immunological characterization of meningococcal group A polysaccharide-tetanus toxid conjugates prepared by two methods. Vaccine 1:31, 1983.
46. Beuvery EC, Miedema F, Van Delft R, Haverkamp K: Preparation and immunochemical characterization of meningococcal group C polysaccharide-tetanus toxoid conjugates as a new generation of vaccines. Infect Immun 40:39, 1983.
47. Granoff DM, Munson RS Jr: Prospects for prevention of Haemophilus influenzae type b disease by immunization. J Infect Dis 153:448, 1986.
48. Eskola J, Peltola H, Takala AK, Kayhty H, Hakulinen M, Karanko V, Kela E, Rekola P, Ronnberg P-R, Samuelson JS, Gordon LK, Makela PH: Efficacy of *Haemophilus influenzae* type b polysaccharide-diphtheria toxoid conjugate vaccine in infancy. N Engl J Med 317:717, 1987.
49. Jennings HJ, Roy R, Gamian A: Induction of meningococcal group B polysaccharide-specific IgG antibodies in mice using an N-propionylated B polysaccharide-tetanus toxoid conjugate vaccine. J Immunol 137:1708, 1986.
50. Moreno C, Liefly MR, Esdaile J: Immunity and protection of mice against *Neisseria meningitidis* group B by vaccination using polysaccharide complexed with outer membrane proteins: A comparison with purified B polysaccharide. Infect Immun 47:527, 1985.
51. Zollinger WD, Mandrell RE, Griffiss JM, Altieri P, Berman S: Complex of meningococcal group B polysaccharide and type 2 outer membrane protein immunogenic in man. J Clin Invest 63:836 1979.
52. Moreno C, Lifely MR, Esdaile J: Effect of aluminum ions on chemical and immunological properties of meningococcal group B polysaccharide. Infect Immun 49:587, 1985.
53. Frasch CE, Zahradnik JM, Wang LY, Mocca LF, Tsai C-M: Antibody response of adults to an aluminum hydroxide adsorbed *Neisseria meningitidis* serotype 2b protein group B-polysaccharide vaccine. J Infect Dis 158:700, 1988.
54. Jennings HJ, Gamian A, Ashton FE: N-propionylated group B meningococcal polysaccharide mimics a unique epitope on group B *Neisseria meningitidis,* J Exp Med 165:1207, 1987.
55. Frasch CE, Zollinger WD, Poolman JT: Serotype antigens of *Neisseria meningitidis* and a proposed scheme for designation of serotypes. Rev Infect Dis 7:504, 1985.
56. Saukkonen K, Abdillahi H, Poolman JT, Leinonen M: Protective efficacy of monoclonal antibodies to class 1 and class 3 outer membrane proteins of *Neisseria meningitidis* B;15:

P1.16 in infant rat infection model: new prospects for vaccine development. Microb Pathogen 3:261, 1987.
57. Frasch CE, Mocca LF, Karpas AB: Appearance of new strains associated with group B meningococcal disease and their use for rapid vaccine development. Antonie Van Leeuwenhoek J Microbiol 53:403, 1987.
58. Tsai C-M, Frasch CE, Rivera E, Hochstein HD: Measurements of lipopolysaccharide (endotoxin) in meningococcal protein and polysaccharide preparations for vaccine usage. J Biol Stand 14:25, 1989.
59. Bensadoun A, Weinstein D: Assay of proteins in the presence of interfering materials. Anal Biochem 70:241, 1976.
60. Wang L-Y, Frasch CE: Development of a *Neisseria meningitidis* group B serotype 2b protein vaccine and evaluation in a mouse model. Infect Immun 46:408, 1984.
61. Osborn MJ: Studies on the gram negative cell wall. I. Evidence for the role of 2-keto-3-deoxy-octonate in the lipopolysaccharide of *Salmonella typhimurium*. Proc Natl Acad Sci USA 50:499, 1963.
62. Rosenqvist E, Harthug S, Froholm LO, Hoiby EA, Bovre K, Zollinger WD: Antibody responses to serogroup B meningococcal outer membrane antigens after vaccination and infection. J Clin Microbiol 26:1543, 1988.
63. Wedege E, Michaelsen TE: Human immunoglobulin G subclass immune response to outer membrane antigens in meningococcal group B vaccine. J Clin Microbiol 25:1349, 1987.
64. Wedege E, Froholm LO: Human antibody response to a group B serotype 2a meningococcal vaccine determined by immunoblotting. Infect Immun 51:571, 1986.
65. Mandrell RE, Zollinger WD: Use of a zwitterionic detergent for the restoration of the antibody-binding capacity of electroblotted meningococcal outer membrane proteins. J Immunol Methods 67:1, 1984.
66. Frasch CE: Immunization against *Neisseria meningitidis*. In Easmon CSF, Jeljaszewicz J (eds): ''Medical Microbiology.'' New York: Academic Press, 1983, Vol. 2, p. 115.
67. Frasch CE, Coetzee G, Zahradnik JM, Feldman HA, Koornhof HJ: Development and evaluation of group B serotype 2 protein vaccines: Report of a group B field trial. Med Trop 43:177, 1983.

**Bacterial Vaccines, pages 147–168**


# Vaccination Against Brucella

**Paul Nicoletti**

Department of Infectious Diseases, College of Veterinary Medicine, University of Florida, Gainesville, Florida 32610

## I. INTRODUCTION

The development of vaccines to prevent brucella infections in animals and humans has been a subject of research for nearly a century. Ideal products do not exist. Some of the goals are that any vaccine be harmless to the recipient, including accidental inoculation; induce long-term immunity from one administration; result in minimal influence on diagnostic methods such as serologic or dermal hypersensitivity tests; and that it be easily produced and stored, inexpensive, and of consistent quality.

In simple terms, vaccination means to administer a preparation to a susceptible host to induce immunity. This product may be a suspension of living or dead cells or fractions of these used with or without adjuvants. The purpose of adjuvants is to delay the diffusion of antigens from the inoculation site so that a progressive stimulus that simulates that of living vaccines is produced or may occur from subinfective exposure to more pathogenic organisms.

Most countries with brucellosis in one or more animal species have used vaccines extensively in those for which a satisfactory vaccine is available. This is especially important in the early phases of organized efforts to reduce and eventually eliminate the disease. Few countries have used vaccines extensively in humans, even among populations at high risk.

The literature concerning brucella vaccines is voluminous; this chapter will largely focus on published results of studies during the past 20 years.

## II. IMMUNOGENESIS

### A. General

Vaccination with brucella antigens results in induction of both humoral and cell-mediated responses. The duration and magnitude of these responses is affected by factors such as virulence or potency of the antigen(s), the dosage and method of introduction, and the age, sex, pregnancy, and immune status of the recipient. Humoral responses are easily quantified by serologic methods. However, cell-mediated responses are much more difficult to measure and equating these with immunity is often questionable.

The effectiveness of vaccines can be evaluated in several ways. These are usually by one or more of the following: 1) studies performed in laboratory animals and data extrapolated to natural hosts; 2) vaccination of natural hosts under controlled conditions and the data compared with unvaccinated controls that are also challenged; and 3) studies of field herds where prevaccination and postvaccination prevalences are compared and/or those of subpopulations that are vaccinated and unvaccinated.

Transposing results of studies from one animal species to another must always be done with caution.

The classic model for evaluating vaccines in livestock is to divide susceptible animals into random groups, vaccinate, and then challenge through the conjunctival route in the initial midgestation with a dose of pathogenic organisms. The challenge dose should infect approximately 95% of the controls with abortions in approximately 50%. As a rule, 70 to 80% of the vaccinates should resist infection. The criteria for infection vary but are usually based upon clinical signs, bacteriological results of tissues and fluids, and serologic tests. The challenge strains are often the "European" *Brucella abortus* 544 and the "American" 2308.

The criteria for evaluation of vaccines in sheep, goats, and pigs are less standardized than those for cattle experiments.

Laboratory animal models, mostly with guinea pigs or mice, are often used to evaluate new products. A guinea pig model is often used to test new lots of long-established vaccines. For example, an intramuscular inoculation of a pathogenic strain is made 45–60 days postvaccination. Lymph node and splenic infections are determined during necropsy approximately 45 days after challenge. In mice, the principal criterion is the severity of infection in the spleen.

### B. Types of Vaccines

Vaccines are usually described as living or dead. Dead vaccines are usually given with an adjuvant. Active research continues to develop a satisfactory vaccine from cell fractions.

**1. Live vaccines.** *B. abortus* strain 19 was isolated as a virulent culture from the milk of a Jersey cow in 1923. It attenuated at room temperature and has not altered in several decades by animal passage. It is stable, low in pathogenicity, and high in immunogenicity. It should be used if smooth cells are at least 95% of the culture.

Research on strain 19 and its application has been more extensive than on most biologics. The original concentration was established at $10^{10}$ per milliliter because maximum viability could best be maintained. The optimal dose has not been established, but the minimal dose for cattle is probably near $5 \times 10^8$ live cells. It is preeminent among vaccines used in the control of bovine brucellosis. The major disadvantage is the production of postvaccinal antibodies. When the dose of the vaccine is similar, the method of administration has little effect upon the degree of protection. The age of the animal at the time of administration has more influence upon the seroagglutination response than either the dose or the method of vaccination.

*B. melitensis* Rev. 1 was developed for immunization of sheep and goats. It was isolated as a nonstreptomycin-dependent revertant from a population of streptomycin-dependent cells derived from a virulent strain. It has been shown to be of low virulence and to produce good immunity. It causes production of serum antibodies that persist for various periods of time.

The stability of Rev. 1 has been proved by animal passage in several animal species. Vaccination has usually been performed in sheep and goats at 4–6 months with a subcutaneous dose of $10^9$ cells. In recent years it has been suggested that a lower dose of approximately $10^5$ to $10^6$ cells be used in adult sheep and goats.

*B. suis* strain 2 [1] is an attenuated smooth strain of biovar 1. It was developed in China, where it has been extensively used in several animal species. Its pathogenicity for guinea pigs is similar to that of strain 19. It does not cause abortion in pregnant animals and persistence in tissues has not been reported.

*B. abortus* strain 104-M [2] was isolated from a cow and developed as a vaccine in the USSR. It is reported to be stable in antigenic structure, virulence, and immunogenicity.

**2. Dead vaccines.** *B. abortus* strain 45/20 is an adjuvant vaccine that is composed of rough whole cells. They do not contain smooth-cell lipopolysaccharide (LPS), which causes postvaccinal agglutinins. Two inoculations spaced 6 to 12 weeks apart with an annual booster are recommended.

*B. melitensis* H38 is a formol-killed smooth strain in an adjuvant and was first developed for immunizing sheep and goats. Results of studies on induced immunity have varied and a booster dose at 3 months is recommended. However, the vaccine often causes local reactions of varying size and results in long-term serologic reactions.

A large number of experiments in laboratory animals and livestock have examined killed whole cells of *Brucella* spp. using differing adjuvants. In general, the results have been disappointing.

Extracts or fractions of brucella cells have been prepared in various ways and examined in several animal species and in humans. Results of some of these studies are given below.

### C. Host Responses to Antigens

Brucella are bacteria that may localize within cells. An effective defense requires cell-mediated responses, and the macrophage is considered to be the main immunizing effector cell. This has been demonstrated by intracellular killing of brucella in vitro following activation by specific antigens. In addition, T lymphocytes are stimulated by antibodies that participate in phagocytosis and lysis of brucella. Antibodies, therefore, play a role in immunity during the extracellular phases of infection by limiting multiplication of the bacteria [3]. However, the exact role of antibodies in protection against natural exposure in most animal species is poorly understood.

Immunologically competent T lymphocytes proliferate into several types of cells and functions. These are cytotoxic (killer) cells, those that are activated by a second exposure (memory cells), those that enhance conversion of B lymphocytes into antibody-producing plasma cells (helper cells), and those that appear to prevent autoimmune reactions (suppressor cells). Macrophage activation occurs when T lymphocytes release lymphokines (interleukins).

The correlates of cell-mediated immunity (CMI) include delayed-type hypersensitivity (DTH) to brucella antigens in vivo and mitogenic responses of peripheral blood leukocytes to brucella antigens in vitro. LPS-free antigens injected intradermally reach maximum intensity at 48–72 h. However, these occur in the apparent absence of protective immunity, and reactions are sometimes used, especially in sheep and goats, as a diagnostic procedure for detecting infection. most brucella vaccines, especially those containing adjuvant, are effective inducers of DTH, but this response may not reflect protective immunity as measured by challenge with a pathogenic strain [4].

The types of soluble brucella antigens used to measure cell-mediated responses significantly affect the results. Most antigens contain complex mixtures of cell components that have been extracted in various ways. They usually contain proteins and may or may not contain LPS.

Peripheral blood lymphocytes from infected and vaccinated cattle have greater mitogenic responses to antigens containing LPS. These responses have been demonstrated as early as 2 weeks after exposure to pathogenic organisms and before antibodies were present [5]. Further evidence that lymphocyte stimulation is independent of antibodies was shown when cells from cattle vaccinated with strain 19 3 or more months previously were not stimulated by LPS or LPS-free antigens even if the serum contained antibodies [6].

After infection with a virulent strain or vaccination with whole smooth brucella cells, antibodies of the immunoglobulin M (IgM) class appear within a few days. These are soon followed by those of the immunoglobulin G (IgG) class. Both $IgG_1$ and $IgG_2$ subclasses are produced. IgA antibodies may be found in mucus or other body fluids. The quantity and duration of antibody responses following vaccination are directly related to the type and dose of vaccine and age at time of vaccination. Residual antibodies are predominantly of the IgM class. Antibodies of the IgG class persist from constant antigenic stimulation such as infection.

Antibodies of the IgA, IgM, $IgG_1$, and $IgG_2$ isotypes will agglutinate brucella antigens, but IgM is the most active at neutral pH. The type of serological test influences the titer. Antigen in which the pH has been lowered below 4.0 (such as the card or rose bengal) will have reduced activity if the antibodies are largely IgM. Treatment of the serum with the acridine dye rivanol (6, 9-diamino-2-ethoxyacridine lactate) selectively precipitates most IgM antibodies. Sulfhydryl reducing agents such as 2-mercaptoethanol dissociates the IgM pentamer. In addition, IgM is more labile to heat of 56°C or higher. These procedures, which inactivate activity of IgM, are very useful for the differentiation of antibodies caused by vaccination or infection of pathogenic strains. The complement fixation test is also very useful for determination of Ig isotypes since fixation is largely due to $IgG_1$.

## III. LABORATORY ANIMALS

Although the results of experiments in laboratory animals cannot always be directly applied to natural hosts, they can provide assistance in screening potential vaccines.

Early studies in laboratory animals were largely in guinea pigs. In recent years, mouse models have become common. This is because the genetic purity is often known and more easily controlled and because a large number of experiments have elucidated various aspects of the broad subject of immunology in the mouse.

The development of live strains 19 and Rev. 1 led to a large number of experiments in laboratory animals, sometimes with conflicting findings. More recently, studies have focused on extracts of brucella cells, with or without adjuvants, and their potential use as vaccines in livestock.

### A. Guinea Pigs

Several decades have elapsed since it was shown that strain 19 conferred an earlier, more complete, and more persistent protection in guinea pigs than vaccines prepared from rough strain 45(20) [7]. A positive correlation was demonstrated between spleen count, spleen weight, and agglutinins following challenge. It was later shown that strain 19 increased protection by 2,000-fold in guinea pigs at 1/15th the cattle dose [8]. Six brucella vaccines were compared for protection against infection in guinea pigs [9]. These were two 45/20 commercial vaccines, two strain 19 preparations, *B. melitensis* H38, and Rev. 1. It was concluded that live vaccines are superior to inactivated products.

Further studies on Rev. 1 vaccine were conducted by several workers. Vaccination with as few as 100 live cells was as effective as $10^8$. Killed 45/20 adjuvant vaccine and a living experimental attenuated strain were poorly immunogenic [10]. Another experiment found that larger doses of Rev. 1 were not necessary and that Rev. 1 was superior to strain 19 as an immunogen [11]. Cutaneous inoculation of Rev. 1 in guinea pigs yielded solid immunity, whereas there was no protection from nonliving Rev. 1 administered orally [12].

In contrast to some other published studies on killed vaccines, ethanol-killed cells and cell walls prepared from *B. abortus, B. melitensis,* and *B. suis* conferred significant protection against challenge [13]. The immunogenicity of both killed smooth and rough strains of *B. abortus* were markedly enhanced by use of water-in-oil emulsion [14]. A cell wall fraction from *B. abortus* strain 99 extracted by boiling sodium dodecylsulfate was used with oil adjuvant in guinea pigs [15]; H38 vaccine was used as a reference. Other guinea pigs received a smooth LPS of *B. melitensis* 16M in adjuvant. Protection was dose dependent in two, while the latter was not immunogenic.

A series of experiments on antigens of strain 45/20 and two adjuvants were conducted [16]. It was found that guinea pigs inoculated with 300 μg of killed whole cells combined in oil emulsion and trehalose dimycolate were protected against splenic infections in a manner similar to that afforded by strain 19. A similar product using muramyl dipeptide was inferior to the other vaccines.

The possible use of whole bacterial cells as immunomodulators was examined in guinea pigs using *Propionibacterium acnes* [17]. The mean splenic counts of strain 2308 (a pathogenic challenge strain) in groups of immunopotentiated guinea pigs was significantly decreased when compared to those given strain 19 alone. The authors concluded that *P. acnes* augmented immune responses presumably by activation of macrophages or a T-cell–mediated cytolytic mechanism or both.

### B. Mice

The mouse model for titration of vaccinal activity was recently reviewed [18]. The authors concluded that a poor response to any product is a correct indicator of low immunogenicity. They hypothesized that colonization of a living vaccine in a lymph node is sufficient to induce a good level of cell-mediated immunity.

Protection in mice is mediated by T lymphocytes [19]. However, there is also evidence that antibodies confer immunity [20].

A review of the efforts to identify antigens of brucella cells that protect against infection was recently published [21]. They have been found to be in either the soluble or insoluble fraction of extracted cells or the cell wall. Soluble fractions contain mainly LPS of smooth cells and proteins. Insoluble fractions are found in the residues of phenol extraction of cell walls. A residue to sodium dodecyl sulfate–boiled cell walls (SDS-1) is a potential product for vaccination [22]. It contains protective antigen(s) and peptidoglycan acts as an adjuvant. When injected into mice using Freund's incomplete adjuvant, the immunity 1 month later was comparable to whole cells of *B. melitensis* H38 vaccine. SDS-1 fractions prepared from rough cells showed varying degrees of protection, but the protection was less than from fractions prepared from smooth cells [21].

Immunity in mice was compared using strain 19 and nonliving vaccines without adjuvant, with complete Freund's adjuvant, and with trehalose dimycolate (TDM) and muramyl dipeptide (MDP) [23]. The protection was evaluated by a challenge and subsequent numbers of the pathogenic strain in the spleen at differing time intervals. The strain 19 group was protected at 1 and 4 weeks postinfection, but by 6 weeks the level was diminished. Mice given nonliving vaccines in mineral oil with TDM and MDP were similar to the strain 19 group. The protection was short lived from cell envelopes, outer-membrane proteins, or whole killed cells in saline, TDM, or MDP. The workers concluded that identifying a nonliving vaccine that will confer effective immunity on a long-term basis remains a problem.

## IV. IMMUNIZATION IN DOMESTIC ANIMALS

### A. General

The control of most infectious disease, including brucellosis, is based upon three major methods: vaccination, limiting exposure of infected and susceptible animals, and hygiene. It is generally agreed that vaccination is the most effective of these procedures. Indeed, vaccination is necessary to control brucellosis in large populations of cattle, sheep, and goats.

Vaccination practically eliminates clinical brucellosis and reduces exposure to animals at risk. In a retrospective study of 79 large dairy herds in California, herds that contained only vaccinated seropositive cows had a herd prevalence level of 0.69%. The presence of one or more unvaccinated infected cows increased the prevalence eightfold, or 5.95% [24]. Another study found the prevalence of infection in vaccinated herds was 1.8% as compared to 9.1% in unvaccinated herds [25]. In subsequent herd tests only 0.5% seropositives were found in vaccinated herds, as compared to 6.9% in others.

The effect of rapid removal of seropositive cattle from vaccinated herds was compared to other herds where they were retained. There were fewer new seropositive cows in the latter and the authors concluded there was little risk in keeping reproductively active infected cows in vaccinated herds [26]. Similar studies have apparently not been done in sheep and goat populations.

Because the two major vaccines used in livestock, strain 19 and Rev. 1, are live smooth-cell products, recommendations have largely been to administer them subcutaneously to sexually immature animals. This was to practically eliminate postvaccinal titers when the animals were tested following sexual maturity. In recent years, many studies have been concerned with expanded use of the vaccines to include the entire populations. These studies have concentrated in three areas: reducing the dose, alternate methods of administration, and improved diagnostic tests. The results have been very successful and some will be presented in the next two sections.

## B. Cattle

**1. Strain 19.** The major characteristics of this smooth strain are well known. Inoculation produces a transient infection, and postvaccinal antibodies persist for varying times. The protection from vaccination is not complete in all cattle. Controlled experiments have usually shown that about 70% of the vaccinates will resist a challenge that will infect approximately 90% of the controls. Under field conditions, the protection is higher when prevaccination and postvaccination prevalences are compared. For example, in 153 dairy herds in Florida and Puerto Rico, there was a 95% reduction in cattle considered infected 1 year following vaccination [27].

The immunogenic response to strain 19 is largely cell mediated, and the degree of postvaccinal serologic response is poorly equated with protection. Several workers have concluded that there is no benefit from repeated doses of strain 19 and that immunity lasts several years. However, field experience has shown that revaccination is sometimes necessary. A summary of experiments on differing doses and ages of vaccinates concluded that a booster dose in young animals was required and better protection was obtained in mature cattle than in calves [28]. A reduced dose did not diminish the level of immunity.

The rapidity of seroconversion from positive to negative following strain 19 vaccination is inversely related to age and dose. Fewer reactions are found using supplemental tests such as rivanol, 2-mercaptoethanol, or complement fixation than with the tube or plate agglutination tests. Field studies in Florida used strain 19 in large infected dairy herds by differing doses and methods of administration [29]. There was a large reduction in brucellosis regardless of dose and method of vaccination. A subcutaneous dose of approximately 1/20th the standard dose was equally effective and greatly reduced the postvaccinal serologic reactions [30]. The author concluded the problem of postvaccinal titers can be largely eliminated by using reduced doses of strain 19 and supplemental serologic tests. The effectiveness of reduced subcutaneous dose of strain 19 was shown in controlled studies [31]. There was a linear relationship of postvaccinal serologic responses to dose of vaccine, and no differences in immunity were seen among groups of cattle.

Further evidence from field studies in other countries of the effectiveness of a reduced dose of strain 19 has been published. In Mexico, over 18,000 dairy cattle were vaccinated with subsequent large reductions in brucellosis.

The authors concluded that revaccination at 18–24 months was necessary and successful [32]. In another study in Spain, vaccination prevented the spread of brucellosis, even though infected cattle were retained in the herd [33].

In recent years, numerous methods of administering strain 19 have been considered with the primary goal of reducing postvaccinal serologic reactions. Strain 19 was given by the conjunctival route, first in guinea pigs and then in cattle [34]. There was little serologic response, and a dose of approximately $5 \times 10^9$ was effective against challenge. A subsequent field trial in an infected herd confirmed these findings [35].

Strain 19 persists in a small percentage of vaccinated cattle. A summary of several publications and data from herds vaccinated by differing methods has been published [36]. Using positive serologic tests as a criterion for possible infection, shedding from the udder was found to be highest among cows that received the vaccine by the conjunctival route and varied among cows according to dose in those vaccinated subcutaneously. The persistence of infection beyond 1 year was low. The author concluded that persistence of strain 19 is very low and is related to age at time of vaccination and method of administration. There is no evidence of transmission of strain 19 from cow to cow or of public health risk from ingestion of contaminated dairy products.

Vaccination of bulls with strain 19 is generally not recommended as serum agglutinins appear to persist longer than in female cattle. Localization of strain 19 in the genital tract may also occur. A syndrome of arthritis caused by strain 19 in heifers has been described by several workers [37–40]. The femorotibial joint is most often affected with macroscopic erosions of the femoral trochlea and/or tibial condyle with antibodies in the joint fluids.

**2. Strain 45/20.** Strain 45 was first isolated in England and is a rough strain developed by 20 passages in guinea pigs. It was used for some years as a live strain, but was shown to revert to virulent smooth mutants. Because killed rough cells alone produce poor immunity, many types of adjuvants have been used. These have been shown to be very important in effectiveness of 45/20 and in local reactions following inoculation.

It is generally recommended that strain 45/20 vaccines be administered in two doses. A recent study on a limited number of cows found a single vaccination gave good protection [28]. The authors concluded that data on the necessity of two or more doses of strain 45/20 are scarce. They further suggested the possibility of developing a superior vaccine using an improved adjuvant with incorporation of an additional immunostimulant.

The immunity following strain 45/20 vaccination has been studied by many workers and results have varied. Some authors have reported immunity similar to that from strain 19 vaccination, whereas others have reported poor immunity. A recent study found that strain 19 protected 75% of heifers challenged with *B. abortus* strain 544 while a similar challenge infected 70% of strain 45/20 vaccinates [41].

Strain 45/20 can safely be used on cattle of any age and regardless of pregnancy. They should be at least 6 months of age. Local reactions often develop at the site of inoculation, especially following repeated doses. The serologic response is usually small, except to the complement fixation (CF) or antiglobulin (Coombs) tests, especially in cattle previously given strain 19.

Strain 45/20 has been used as a diagnostic reagent to hopefully detect incubative infections. The concept is based upon differences between primary and secondary responses. If the primary dose causes a secondary response, it is assumed the animal has previously experienced an exposure to brucella antigens. In a previously unvaccinated animal, a response is considered to be from natural exposure. The serum of cattle is tested approximately 6 weeks following the first dose of vaccine by the CF or Coombs tests and, if negative, are then given the second dose. The procedure should not be used in cattle previously vaccinated with strain 19.

**3. Other vaccines.** *B. melitensis* Rev. 1 has been considered for use in cattle, especially where cattle are at risk to *B. melitensis* infections. Because Rev. 1 is a potent immunogen, perhaps lower doses of this strain could be more effective than strain 19. Two studies [42,43] have shown Rev. 1 to be superior to strain 19 in cattle. However, use of Rev. 1 in cattle has thus far been primarily limited to experimental studies.

*B. suis* strain 2 has been widely used in China [1]. A dose of 25–50 $\times 10^9$ live cells in cattle gave a protection rate of over 70% when all controls became infected. Excellent results have been reported on dairy farms, and the vaccine did not induce persisting antibodies.

Following use of extracts of strain 45/20 with and without adjuvants in guinea pigs, a bacterin with trehalose dimycolate adjuvant was tested in heifers [44]. Abortions occurred in 16 of 19 heifers vaccinated with the experimental product compared to 6 of 12 and 17 of 18 strain 19 vaccinates and controls, respectively. In another study, cattle were vaccinated twice with either a salt-extractable protein preparation (CSP) or chemically modified CSP (dCSP) from strain 19 in Freund's complete adjuvant [45]. Both preparations stimulated marked antibody responses that persisted for many months. Following challenge with pathogenic strain 2308, 60 to 83% were infected. The authors concluded that neither preparation was an effective immunogen and that preparation of a nonliving vaccine that has the desirable property of immunogenesis without adverse effects will be difficult.

Experimental vaccines composed of native cell envelopes or outer-membrane proteins of *B. abortus* strain 2308 were compared with a vaccine (PG) composed of the insoluble residue of cell envelopes that had been extracted with hot sodium dodecyl sulfate [46]. Heifers were vaccinated with 5 mg of antigens in adjuvant; PG was administered without adjuvant. Humoral and cell-mediated immune responses were tested at monthly intervals. PG without adjuvant induced negligible immune responses. Native cell-envelope antigens induced significantly higher titers of whole-cell agglutinins over a 3 month period than did PG. Lymphocyte blastogenetic responses and delayed hypersensitivity reactions to porin and group 3 proteins were stimulated by both native and PG vaccines. Although the heifers were not challenged, the authors concluded the findings provide a basis for further evaluations of nonliving *B. abortus* vaccines in cattle.

### C. Sheep and Goats

Brucellosis has only been eradicated from small ruminants on a national scale by slaughtering the herd. Few countries can afford this control method. The vast majority of affected countries will need to rely on vaccination, at least for the foreseeable future [47]. Of the vaccines now used for immunizing sheep and goats, *B. melitensis* Rev. 1 is generally preferred and H38 adjuvant vaccine is now little used.

**1. Rev. 1.** A review of the use of Rev. 1 *B. melitensis* vaccine in experimental and field studies has recently been published [48] and further summarized [47]. Rev. 1 is highly immunogenic and one dose induces protection for several years [49]. There are few local reactions, but pregnant goats may sometimes abort if a full dose of vaccine is used. Rev. 1 may occasionally be excreted in the milk for a brief time following inoculation, but no human cases have been reported from ingestion of contaminated dairy products.

Vaccination with Rev. 1 induces humoral antibodies. The titer depends on the dose, age of animal at time of vaccination, and type of serologic test used. The complement fixation test is negative prior to the tube agglutination test, which may be positive for several years.

The original recommendation for use of Rev. 1 was to vaccinate only young animals with a dose of $1–2 \times 10^9$. This was to lessen the persistence of postvaccinal antibodies and not cause abortions. Subsequent studies have shown that similar protection can be obtained with much-reduced doses. A dose of $5 \times 10^4$ did not cause abortion, organisms were not excreted in the milk, and it did not interfere with serologic tests, but significant resistance was provided [50]. Another trial examined the use of $5 \times 10^8$ Rev. 1 in a large infected flock with good success [51].

Following the successful trials with strain 19 conjunctival vaccination in cattle [34], Rev. 1 vaccine was given by the same route in sheep. A conjunctival inoculation of either $1.4 \times 10^8$ or $1.4 \times 10^9$ was compared with $1.6 \times 10^9$ given subcutaneously. The ewes were challenged 7 months postvaccination and when approximately 85 days pregnant [52]. There was similar protection among the groups, but much less serologic response following conjunctival inoculation. A subsequent study in goats found good protection against a challenge that caused all controls to abort [53]. The authors concluded that conjunctival vaccination is an alternative to subcutaneous vaccination for control of brucellosis in small ruminants.

**2. H38.** The killed vaccine H38 is prepared from a virulent strain of *B. melitensis* emulsified in an oily adjuvant. The vaccine contains $10^{11}$ bacteria per dose. Two doses are given 6 to 12 weeks apart. The immunity is similar to that induced by vaccination using Rev. 1 [47]. The disadvantages of H38 vaccine are that it causes a prolonged serologic response and a local reaction at the site of inoculation. H38 has not received wide acceptance in field use.

**3. *B. suis* strain 2.** A dose of $10^{10}$ live cells of *B. suis* strain 2 produces good immunity in sheep and goats and is effective for up to 4–5 years. Serologic responses are reported to be low and transient [1].

**4. *B. ovis* infection.** After ram epididymitis was recognized to be a major problem in sheep, efforts were made to control the disease by vaccination. Different preparations have been tried experimentally and in field use in several countries. These include *B. ovis* water-in-oil, alum-adsorbed bacterin, *B. abortus* strain 19 or 45/20 with *B. ovis* bacterin, and *B. melitensis* Rev. 1. Many of these results were recently summarized [54]. Certain preparations of *B. ovis* water-in-oil adjuvant vaccines have provided good protection while others have failed. Protection usually ranges from 40 to 65% under experimental and field conditions [54]. A study that used a *B. ovis*-vitamin E adjuvant vaccine found 78% protection as compared to 56% for one containing Freund's incomplete adjuvant and 56% for a commercially available alum-precipitated bacterin [55].

Rev. 1 vaccine has been shown to be effective against *B. ovis* infection. Twenty-five rams were vaccinated with $5 \times 10^8$ Rev. 1 and 25 were unvaccinated controls [56]. All but two controls became infected, while only a few bacterial colonies could be recovered from a vaccinated ram challenged with a high dose of a virulent strain. In another study, 15 rams were vaccinated with $8.5 \times 10^8$ Rev. 1 at 4 months [57]. These and 15 controls were challenged 9 months later and then necropsied after 3 months. Only a few organs and lymph nodes of three vaccinated rams were infected, whereas all controls were heavily infected. An experiment to examine the necessity of revaccination found that *B. ovis* was isolated from 27% of rams vaccinated once, 21% from those revaccinated, and from all controls following challenge [58]. The authors concluded that vaccination of young rams, followed or not by revaccination, is a satisfactory method for the prophylaxis of *B. ovis* infections and that vaccination is the only viable means of control in countries with high levels of infection.

## D. Swine

It is generally concluded that trials with attenuated *B. suis* or *B. abortus* strain 19 have been unsuccessful in producing a persistent immunity. Bacterins or extracts of killed *B. suis* have generally been ineffective in stimulating immunity, or there has been no conclusive evidence that the persistence of immunity is adequate [59].

*B. suis* strain 2 vaccine is recommended for pigs in two oral doses of $2 \times 10^{10}$ live cells given 2 to 3 months apart [1]. In a trial cited by the author, after culling pigs that had aborted, the vaccine was given once a year for 2 years. No subsequent abortions were observed and by the third year the serologic reactors had been reduced on two farms from 62.8 and 75% to 2.4 and 0%, respectively. A protection rate of 75% was reported in challenge studies on 16 vaccinated pigs and 7 controls.

## E. Other Animals

**1. Buffaloes.** Apparently few studies have been made in the use of vaccines in buffaloes *(Bubalus bubalis)*. Strains 19 and 45/20 vaccines were compared in 612 and 125 buffaloes, respectively [60]. There were similar results between the two groups in effectiveness when abortions and retention of placenta were the criteria. The authors suggested that revaccination may have increased the effectiveness. Tests on 120 buffaloes showed that strain 19 used in the same or half the cattle dose was effective in immunizing those of various ages and that the immunity lasted for 18 to 20 months [61].

The live attenuated *B. abortus* strain 82 was reported to be as effective in young buffaloes as in cattle [62].

**2. Yaks.** In Mongolia, yaks of all age groups were first vaccinated in 1967 with strain 19 with subsequent inoculations limited to young animals [63]. The infection rate in one aimak (province) was reduced by 81.4% in 2 years; in another, the reduction was 95%. The author concluded that vaccination reduces brucellosis so that complete elimination is possible. In China, strain 19 was less effective and experiments with *B. suis* strain 2 were conducted [64]. The vaccine was administered by aerosol to 30 yaks, subcutaneously to 7, to 7 by oral route, and 26 were unvaccinated. Following mating and 3 to 4 months later, all were challenged with a virulent strain. Various tissues and fluids were cultured following abortion or parturition. There was an infection index (number of tissues culture positive) in vaccinated yaks of 6.7% as compared to 37.6% of controls. The authors concluded that strain 2 engenders a good immunity in yaks.

**3. Camels.** There are apparently no published reports of vaccines in any species of camels.

**4. Bison *(Bison bison)*.** Studies are underway in the United States to evaluate the pathogenicity and effectiveness of strain 19 in American bison. Preliminary data suggest that strain 19 is far more pathogenic for bison than for cattle.

## V. IMMUNIZATION IN HUMANS

### A. General

Brucellosis is a true zoonosis, and prevention of the disease in humans is directly affected by the prevalence in animal hosts. Therefore, primary efforts at control should be directed toward livestock. The incidence of human brucellosis is a good indicator of that in animals, especially where *B. melitensis* exists.

Few countries have used vaccination of humans even among subpopulations at high risk. Live vaccines have caused adverse effects and it has been difficult to develop and evaluate cell fraction products.

Some of the problems in evaluating vaccines in humans include the effects of varying exposure. If those at risk are separated into vaccinees and controls, it is difficult to determine the amount and frequency of exposure and the impact on each group. The populations may be small and diverse. Frequently there are concurrent efforts to reduce the incidence of brucellosis among livestock, so the impact of each phase of the studies is difficult to assess. Humans in certain occupations are exposed to low and repeated contacts with brucella organisms. They may develop hypersensitivity to brucella antigens, subclinical infections, or clinical disease with recovery and immunity. These conditions may not be distinct, and vaccine product assessment or use may be severely affected.

A vaccine should be innocuous and not produce sensitization and preferably not cause antibodies that will interfere with subsequent diagnostic tests.

### B. Types of Vaccines and Results

**1. Live vaccines.** The live attenuated vaccine strains used in livestock have received the most attention as immunizing products in humans. Because cross-immunity is developed to *Brucella* spp. and because *B. abortus* is less pathogenic for humans than *B. melitensis,* the attenuated *B. abortus* 19BA in the USSR and *B. abortus* 104M in China were selected for study [65]. Studies on Rev. 1 with human volunteers and in monkeys showed that the margin between an innocuous dose and one leading to febrile and symptomatic reactions is too narrow to recommend its use in humans [66].

In the USSR, 19BA (2–3 $\times$ $10^6$ cells) is injected subcutaneously or scarified (1 $\times$ $10^9$) into the skin. These doses were immunogenic and protection was considered to last 1 year [65]. The vaccine results in a positive skin test in a high percentage of humans. To avoid adverse reactions caused by previous sensitization, intradermal and serologic tests are performed prior to vaccination.

A live vaccine, the attenuated 104M, containing 7–10 $\times$ $10^9$ cells, has been widely used in China since 1965 [65]. It is placed on the skin of the upper arm and the area scarified. In 24–48 h in sensitized humans, the vaccination site may show erythema and infiltration. A generalized reaction such as headache and weakness may be observed in a low percentage of vaccinated persons after 1 to 3 days [65]. Severe adverse effects were found if the vaccine was given subcutaneously.

**2. Nonliving vaccines.** In the USSR, a protein-polysaccharide complex derived from the cell wall of brucella S-forms by acetic acid hydrolysis has been developed. It was prepared as an alternative vaccine to 19BA, which sometimes causes severe reactions, especially following revaccination [67]. The antigenic potency of different doses of the new product was evaluated and was higher in humans who had previously received live vaccine [68]. The optimal dose of 1 mg of the vaccine was recommended. The optimal interval between primary and booster immunizations is 11 to 12 months [69]. The authors reported the vaccine is safe and of low reactogenicity. Local or systemic reactions do not depend on the dose of the preparation or previous immunization. The sensitizing activity was several times lower than the allergenic action of live (19BA) vaccine [67].

In France, an immunogenic phenol-insoluble fraction obtained by phenol extraction (PI) of delipidated *B. abortus* and *B. melitensis* has been developed [70]. The fraction is nontoxic. Vaccination is by two injections each of 1 mg 15 days apart. It is recommended for those with occupational risk. The efficacy and duration of immunity have not been established.

The effect of the PI fraction on antibody formation was studied in 63 humans [71]. Serum IgE anti-brucella antibodies developed in 43% and a significant association was found between these and the presence of mild reactions of primary vaccination. The PI fraction is not a good inducer of IgG antibodies. In 51 of the 63 subjects, primary vaccination induced mild local and/or general reaction during the first 24 h. Other workers examined dermal and serologic reactions to the PI fraction in 57 humans exposed to brucellosis [72]. Four serologic and an intradermal injection of PI fraction were studied and used as criteria for immunity. It was concluded that immunity wanes and a booster injection 1 year after primary vaccination is recommended.

## VI. THE FUTURE

While many countries have eradicated brucellosis and others have made great progress toward that goal, this success has largely been with *B. abortus* infections in cattle. Vaccines have been enormously important in these efforts. In many countries, especially those with *B. melitensis* infections, the incidence has increased. Therefore, it is certain that vaccines will play a major role in brucellosis control for many years.

Decades of research have failed to produce superior vaccines to live attenuated smooth strains. These vaccines have multiple and well-known advantages. They are effective, relatively inexpensive to produce and use, and one inoculation induces an immunity that persists for years. The major disadvantages of their use, the development of postvaccinal antibodies, can largely be overcome by reducing the dose, using alternate methods of inoculation, and using serologic procedures that discriminate antibody classes. Research should continue in all these areas.

It has been well established that killed vaccines can induce protective immunity. However, these have many disadvantages, including the need for adjuvants and revaccination.

Much current research has concentrated on isolation of fractions of brucella or even separate antigens in these fractions that are responsible for immunogenesis. Active immunogens have been extracted. S-lipopolysaccharides are known to be immunogenic in mice, but cause several host reactions, including development of antibodies. Proteins linked to the cell wall or outer-membrane proteins are potential candidates for vaccines that are devoid of S-LPS [18]. If the determinants (epitopes) of protective antigens are characterized, these could be sequenced and peptides obtained by enzymatic cleavage. The ultimate goal would be to produce effective genetically engineered vaccines by cloning the genes of protective antigens and incorporating them within a nonpathogenic bacterial or viral carrier.

Recent studies have identified a 31 kDa protein with potential as a subunit vaccine. The gene for this protein has been cloned and expressed in *Escherichia coli* [73]. The availability of this 31 kDa protein now allows rigorous study of its immunogenic properties.

A successful vaccine must receive extensive experimental and field studies and hopefully protect against most, or preferably all, naturally occurring strains of a pathogen in natural hosts. Immune mechanisms involved in vaccine production may differ according to the route of challenge and from one species to another. As long as these mechanisms are not clearly understood, great care should be taken before extrapolating results obtained in one species to the natural host [18].

## VII. SUMMARY

Vaccination has played an enormous role in reducing brucellosis in many countries. It is certain to continue to be the preeminent factor in control of the disease in others. The search for an ideal vaccine continues.

Live vaccines have proved to be superior to inactivated products. They are effective, inexpensive, and immunity is more persistent. The disadvantages of postvaccinal antibodies can be minimized by reduction of previously recommended doses and through use of supplemental diagnostic tests. These procedures now make entire population vaccination of great practical significance with many advantages over limited use of the strains 19 and Rev. 1. Adult animal vaccination should be much more extensive in many countries.

A live *B. suis* strain 2 vaccine developed in China deserves much additional evaluation, including use in swine, for which no satisfactory vaccine exists.

It is generally agreed that cell-mediated responses are the dominant aspect of immunogenesis. However, the correlates that have frequently been used—dermal hypersensitivity and lymphocyte stimulation in vitro—appear to be poor indices of cell-mediated immunity in brucellosis. Many studies have shown that postvaccinal antibodies do not predict subsequent immunity. There is a great need for simple in vivo or in vitro methods to measure CMI.

While vaccination of humans may be useful in control of brucellosis in some high-risk occupations, the ultimate success is dependent upon reduction of this very important zoonosis in natural hosts. This is most effectively accomplished by widespread use of vaccination.

## REFERENCES

1. Xie Xin: Orally administrable brucellosis vaccine: *Brucella suis* strain 2 vaccine. Vaccine 4:212, 1986.
2. Shumilov KV, Kas'yanov AN, Romakhov VA, et al.: Properties of vaccine strain 104M of *Brucella abortus*. Trudy Vses Inst Eksper Vet 61:10, 1986. Abstract 5815 in Vet Bull 1984.
3. Canning PC, Deyoe BL, Roth JA: Opsonin-dependent stimulation of bovine neutrophil oxidative metabolism by *Brucella abortus*. Am J Vet Res 49:160, 1988.
4. Chukwu CC: Comparison of cell-mediated immune responses to *Brucella abortus* strain 19 vaccine and *Brucella abortus* killed 45/20 adjuvant vaccine in cattle. Microbios Lett 34:93, 1987.
5. Kaneene JMB, Angus RD, Johnson DW, et al.: Temporal cell-mediated immune responses of cattle following experimental and natural exposure to living *Brucella abortus*. Can J Comp Med 43:132, 1979.
6. Kaneene JMB, Anderson RK, Johnson DW, et al.: *Brucella* antigen preparations for *in vitro* lymphocyte immunostimulation assays in bovine brucellosis. Infect Immun 22:486, 1978.
7. de Ropp RS: Comparison of the immunising value in guinea-pigs of living avirulent *Brucella abortus* vaccines, strains 45(20) and S.19. J Comp Pathol 55:70, 1945,
8. Thorton DH, Muskett JC: The use of laboratory animals in the potency test of *Brucella abortus* S.19 vaccine. Response of guinea-pigs to graduated doses of vaccine and challenge. J Comp Pathol 82:201, 1972.
9. Cameron CM: Comparative efficacy of six *Brucella* vaccines in guinea-pigs. Onderstepoort J Vet Res 46:229, 1979.

10. Alton GG: A trial in guinea-pigs of four methods of vaccination against *Brucella melitensis* infection. Res Vet Sci 10:329, 1969.
11. Garcia-Carrillo C: Comparacion de la immunidad en cabayos vacunados con cepa 19 y Rev. 1 desafiados *B. abortus* un mes y un ano despues. J Vet Med B 33:231, 1986.
12. Chen TH, Elberg SS: Immunization against *Brucella* infections: Immune responses of mice, guinea pigs, and *Cynomolgus philipinensis* to live and killed *Brucella melitensis* Rev. 1 administered by various methods. J Infect Dis 122:489, 1970.
13. Rasooly G, Olitzki AL, Sulitzeanu D: Immunization against brucella with killed vaccines. II. Immunizing activity of killed brucella vaccines in guinea pigs. Isr J Med Sci 3:814, 1967.
14. Keppie J, Kitt K, Smith H: The immunization of guinea-pigs and mice with a whole-culture extract of a smooth and a rough strain of *Brucella abortus*. Br J Exp Pathol 53:518, 1972.
15. Bosseray N, Plommet M, Dubray G: Immunogenic activity of a cell wall fraction extracted from *Brucella abortus* in guinea-pigs. Ann Microbiol 129B:571, 1978.
16. Woodard LF, Toone NM, Jasman RL: *Brucella abortus* vaccines. Comparison of protection by immunopotentiated 45/20 bacterins and live strain 19 vaccine in guinea pigs. Am J Vet Res 42:1959, 1981.
17. Panangala VS, Haynes TB, Schultz RD, et al.: Immunomodulation with killed *Propionibacterium acnes* in guinea pigs simultaneously vaccinated with *Brucella abortus* strain 19. Vet Immunol Immunopathol 13:71, 1986.
18. Plommet M, Serre A, Fensterbank R: Vaccines, vaccination in brucellosis. Ann Inst Pasteur/Microbiol 138:117, 1987.
19. Cheers C: Pathogenesis and cellular immunity in experimental murine brucellosis. Dev Biol Stand 56:237, 1983.
20. Bascoul S, Cannat A, Huguet M, et al.: Studies on the immune protection to murine experimental brucellosis conferred by *Brucella* fractions. I. Positive role of immune serum. Immunology 35:213, 1978.
21. Dubray G: Protective antigens in brucellosis. Ann Inst Pasteur/Microbiol 138:84, 1987.
22. Dubray G, Bezard G: Isolation of three cell-wall antigens protective in murine experimental brucellosis. Ann Rech Vet 11:367, 1980.
23. Montarez JA, Winter AJ: Comparison of living and nonliving vaccines for *Brucella abortus* in BALB/c mice. Infect Immun 53:245, 1986.
24. Vanderwagen LC, Sharp J, Meyer ME: A retrospective study on the relationships of vaccination status of reactor animals, management practices at calving and herd size to eradicating brucellosis in 79 dairy herds. Proceedings of the 81st Annual meeting, US Animal Health Association, 83, 1977.
25. Koh, SH. Morley FHW: The effect of calfhood vaccination with strain 19 on the serological diagnosis and eradication of bovine brucellosis. Aust Vet J 57:551, 1981.
26. Enright FM, Hugh-Jones ME: Effects of reactor retention on the spread of brucellosis in strain 19 adult vaccinated herds. Prev Vet Med 2:505, 1984.
27. Nicoletti P: The effects of adult cattle vaccination with strain 19 on the incidence of brucellosis in dairy herds in Florida and Puerto Rico. Proceedings of the 83rd Annual Meeting, US Animal Health Association, 83, 1979.
28. Alton GG, Corner LA, Plackett P: Vaccination against brucellosis. Dev Biol Stand 56, 643, 1984.
29. Nicoletti P: A preliminary report on the efficacy of adult cattle vaccination using strain 19 in selected dairy herds in Florida. Proceedings of the 80th Annual Meeting, US Animal Health Association, 91, 1976.
30. Nicoletti P, Jones LM, Berman DT: Adult vaccination with standard and reduced doses of

*Brucella abortus* strain 19 vaccine in a dairy herd infected with brucellosis. J Am Vet Med Assoc 173:1445, 1978.

31. Deyoe BL, Dorsey TA, Meredith KB, et al.: Effect of reduced dosages of *Brucella abortus* strain 19 in cattle vaccinated as yearlings. Proceedings of the 83rd Annual Meeting, US Animal Health Association, 92, 1979.
32. Flores-Castro R, Fernandez de Cordoba L, Trejo-Solomon J, et al.: Adult cattle vaccination and revaccination with strain 19 reduced doses for the control of brucellosis: A field experience in Mexico. Int J Zoonoses 12:299, 1985.
33. Marin CM, de Bagues M, Blasco JM: Vacunacion de vacas adultas con dosis reducidas de vacuna B-19. Med Vet 3:301, 1986.
34. Plommet M, Fensterbank R: Vaccination against brucellosis with a low dose of strain 19 administered by the conjunctival route. III. Serological response and immunity in the pregnant cow. Ann Rech Vet 7:9, 1976.
35. Nicoletti P, Jones LM, Berman DT: Comparison of the subcutaneous and conjunctival route of vaccination with *Brucella abortus* strain 19 vaccine in adult cattle. J Am Vet Med Assoc 173:1450, 1978.
36. Nicoletti P: Prevalence and persistence of *Brucella abortus* strain 19 infections and prevalence of other biotypes in vaccinated adult dairy cattle. Am J Vet Res 178:143, 1981.
37. Bracewell CD, Corbel MJ: An association between arthritis and persistent serological reactions to *Brucella abortus* in cattle from apparently brucellosis-free herds. Vet Rec 106:99, 1980.
38. Wyn-Jones G, Baker JR, Johnson PM: A clinical and immunological study of *Brucella abortus* strain 19-induced arthritis in cattle. Vet Rec 107:5, 1980.
39. Nicoletti P, Crowley AM, Richardson JA, et al.: Suspected *Brucella abortus* strain 19-induced arthritis in a dairy cow. Agri-Practice 7:5, 1986.
40. Rogerson BA, Morgan IR: Investigation of aberrant positive reactions to serological tests for bovine brucellosis. Aust Vet J 63:227, 1986.
41. Worthington RW, Howell FD, Mulders MSG, et al.: An investigation of the efficacy of three *Brucella* vaccines in cattle. J S Afr Vet Med Assoc 45:87, 1976.
42. Horwell FD, Van Drimmelen GC: *Brucella melitensis* strain Rev. 1 as a vaccine for cattle. J S Afr Vet Med Assoc 42:233, 1971.
43. Garcia-Carrillo C: Comparison of *B. melitensis* Rev. 1 and *B. abortus* strain 19 as a vaccine against brucellosis in cattle. Zentralbl Veterinarmed [B] 27:131, 1980.
44. Woodard LF, Jasmin RL: Comparative efficacy of an experimental S45/20 bacterin and a reduced dose of strain 19 vaccine against bovine brucellosis. Am J Vet Res 44:907, 1983.
45. Confer, AW, Tabatabai LB, Deyoe BL, et al.: Vaccination of cattle with chemically modified and unmodified salt-extractable proteins from *Brucella abortus*. Vet Microbiol 15:325, 1987.
46. Winter AJ, Rowe GE: Comparative immune responses to native cell envelope antigens and the hot sodium dodecyl sulfate insoluble fraction (PG) of *Brucella abortus* in cattle and mice. Vet Immunol Immunopathol 18:149, 1988.
47. Alton GG: Control of *Brucella melitensis* infection in sheep and goats—A review. Trop Anim Health Prod 19:65, 1987.
48. Elberg SS: Rev. 1 *Brucella melitensis* vaccine. Part II. 1968–1980. Vet Bull 51:67, 1981.
49. Alton GG: Further studies on the duration of the immunity produced in goats by the Rev. 1 *Brucella melitensis* vaccine. J Comp Pathol 78:173, 1968.
50. Alton GG: Vaccination of goats with reduced doses of Rev. 1 *Brucella melitensis* vaccine. Res Vet Sci 11:54, 1970.
51. Blasco JM, Estrada A, Mercadal M: A note on adult sheep vaccination with reduced dose of *Brucella melitensis* Rev. 1. Ann Rech Vet 15:553, 1984.

52. Fensterbank R, Pardon P, Marly J: Vaccination of ewes by a single conjunctival administration of *Brucella melitensis* Rev. 1 vaccine. Ann Rech Vet 16:351, 1985.
53. Fensterbank R, Verger JM, Grayon M: Conjunctival vaccination of young goats with *Brucella melitensis* Rev. 1. Ann Rech Vet 18:397, 1987.
54. Afzal M, Kimberling CV: How to control *Brucella ovis*-induced epididymitis in rams. Vet Med 81:364, 1986.
55. Afzal M, Tengerdy RP, Ellis RP, et al.: Protection of rams against epididymitis by a *Brucella ovis*-vitamin E adjuvant vaccine. Vet Immunol Immunoapthol 7:293, 1984.
56. Garcia-Carrillo C: Protection of rams against *Brucella ovis* infection by *Brucella melitensis* Rev. 1 vaccine. Zentralbl Veterinarmed [B] 28:425, 1981.
57. Fensterbank R, Pardon P, Marley J: Efficacy of *Brucella melitensis* Rev. 1 vaccine against *Brucella ovis* infection in rams. Ann Rech Vet 13:185, 1982.
58. Blasco JM, Marin CM, Barberan M, et al.: Immunization with *Brucella melitensis* Rev. 1 against *Brucella ovis* infection in rams. Vet Microbiol 14:381, 1987.
59. Deyoe BL: Brucellosis. In Leman AD, Straw B, Glock RD, et al. (eds): "Disease of Swine." Ames: Iowa State University Press, 1986, p 599.
60. Bonaduce A, Quesada A, Martone F, et al.: Osservazioni e considerazioni sulla profilassi antibrucellare effettuata con un vaccino altenuato (Buck 19) e con un vaccino inattivato (45/20) in un allevamento di bufali dell'italia meridionale. Acta Med Ved 31:3, 1985.
61. Dashdamirov DM: Vaccination of buffaloes against brucellosis. Veterinariya (Moscow) 10:58, 1974. Abstract 985 in Vet Bull 45, 1975.
62. Aliev EA, Sultanova FG: Use in buffaloes of the live *Brucella abortus* strain 82. Veterinariya (Moscow) 3:30, 1980. Abstract 7097 in Vet Bull 50, 1980.
63. Tcherendash T: Brucellosis in yaks. WHO Background Paper 33, 1970.
64. Qi Han-jun, Qi Wen-guan: Immunization of yaks with *Brucella suis* strain 2 vaccine against Brucella infection (personal communication).
65. Anonymous: In Elberg SS (ed): "A Guide to the Diagnosis, Treatment and Prevention of Human Brucellosis." Geneva: WHO, VPH/81.31, p 31.
66. Pappagianis D, Elberg SS, Crouch D: Immunization against brucella infection. Effects of graded doses of viable attenuated *Brucella melitensis* in humans. Am J Epidemiol 84:21, 1966.
67. Sumarokov AA, Karinskaya GA, Dranovskaya EA, et al.: Comparative study of the safety, reactogenicity and antigenic potency of chemical and live brucellosis vaccines under the conditions of a controlled epidemiological trial. Zh Mikrobiol Epidemiol Immunobiol 2:58, 1984.
68. Dranovskaya EA, Vershilova PA, El'shina EA, et al.: Comparative study of the antigenic activity and allergic transformation of the body after revaccination of humans with different doses of chemical brucellosis vaccine. Zh Mikrobiol Epidemiol Immunobiol 11:88, 1985. Abstract 111762 in Biol Abstr, 1986.
69. Vershilova PA, El'shina EA, Dranovskaya EA, et al.: Comparative study of the safety and reactinogenicity of different doses of chemical brucellosis vaccine in human revaccination. Zh Mikrobiol Epidemiol Immunobiol 11:56, 1985. Abstract 11764 in Biol Abstr, 1986.
70. Roux J: La vaccination humaine contre les brucellosis. Bull Acad Natl Med (Paris) 170: 289, 1986.
71. Escande A, Serre A: IgE anti-brucella antibodies in the course of human brucellosis and after specific vaccination. Int Arch Allergy Appl Immunol 68:172, 1982.
72. Arnaud-Bosq C, Brun M, Roux J: Profil immunologique d'une population exposee a la brucellose, non vaccinee ou vaccinee par la fraction PI depuis 1 an ou moins. Med Mal Infect 1:31, 1986.

73. Mayfield JE, Bricker BJ, Godfrey H, et al.: The cloning, expression, and nucleotide sequence of a gene coding for an immunogenic *Brucella abortus* protein. Gene 63:1, 1988.

Bacterial Vaccines, pages 169–199


# Pertussis Vaccines: Present Status

**Myron Christodoulides**

Department of Pathology, Royal Victoria Infirmary, University of Newcastle-upon-Tyne, Newcastle-upon-Tyne, Tyne and Wear, NE1 4LP, England



## I. INTRODUCTION

Pertussis (whooping cough) remains one of the least understood of all infectious diseases. *Bordetella pertussis* is the etiological agent of human pertussis, a highly contagious disease of the respiratory tract that is most severe for very young infants. Control of the disease has invariably relied on vaccine prophylaxis, and the logical design of a vaccine against pertussis requires an understanding of the components of *B. pertussis* essential to the pathogenesis of the disease. Pertussis vaccines have been used for more than 50 years, and their effectiveness has been demonstrated by the remarkable decline in both morbidity and mortality of clinical whooping cough in those countries where immunization programs have been established. Despite the fact that "present vaccines are compromises that accept a certain level of toxicity in order to achieve an acceptable level of potency" [1], the balance of risks and benefits strongly favors vaccine prophylaxis. In this chapter, an assessment of current whole-cell and acellular pertussis vaccines will be presented. Also, the properties of potential vaccine antigens of *B. pertussis* and their roles in the development of future acellular vaccines will be explored. But to emphasize the serious nature of this childhood disease, a brief description of clinical pertussis is first presented.

Pertussis is a worldwide disease that has been reported from all continents and most countries, except for some remote or island populations. It is currently estimated that 606,000 children less than 5 years of age die from the disease annually, with many more suffering permanent sequelae [2]. The clinical disease can be divided into four stages: 1) an incubation stage, normally 6–20 days; 2) a catarrhal stage, normally 10–15 days; 3) a paroxysmal stage, which usually lasts 1–4 weeks; and 4) a convalescent stage. Pertussis starts with the aerial transmission of *B. pertussis* from the respiratory tract of an infected individual to that of a new host. The first stage in the pathogenesis of pertussis begins during incubation, with adherence of the organism to the ciliated epithelial cells of the repiratory tract and its subsequent multiplication and colonization of the ciliated mucosa. Onset of clinical disease during the incubation and catarrhal stages is usually mild with nonspecific symptoms and not recognized as pertussis. Symptoms include inflammation of mucosal membranes, nasal discharge, tearing, sneezing, a mild cough, but little fever. These continue to worsen, with local disease manifested by ciliostasis and other pathological disturbances in the respiratory tract [3]. As the disease progresses, the cough becomes more spasmodic and violent, and the number of coughing episodes increases. At no point does *B. pertussis* invade tissue, including blood, and rapid disappearance of the organism is usually associated with the appearance of coughing.

The disease now enters the paroxysmal stage, which is characterized by numerous episodes of coughing, each attack consisting of 10–30 forceful coughs per spasm. Every effort from the infant is concentrated on clearing the mucus-blocked air passages such that with each cough the effect is more violent. After considerable exertion, the mucus is expelled and inspiration follows through a narrowed glottis, resulting in the characteristic "whoop." Vomiting usually follows the coughing spasms, and other consequences of these violent attacks include hemorrhagic events, pulmonary complications, and complications secondary to increased intrathoracic and intra-abdominal pressure. In addition, secondary bacterial and viral infections, seizures, encephalopathy, and death are observed. The major systemic effects of pertussis include leuko-lymphocytosis, metabolic alterations, vascular permeability changes, and neurological complications, but the underlying pathogenic mechanisms are still far from clear, particularly the pathophysiological changes that give rise to the distressing "paroxysmal coughing" syndrome. Permanent sequelae of severe pertussis include damage to the lungs and central nervous system, the latter including blindness, deafness, and developmental retardation, which are probably due to cerebral anoxia and local hemorrhaging.

Convalescence is characterized by coughing episodes of diminished severity and frequency, and the time until spontaneous paroxysms disappear may be up to 6 months. The patient is usually afebrile and susceptible to secondary infections, which may suggest an alteration of the host's normal inflammatory responses and immune systems by infection with *B. pertussis*. For further background and information on clinical pertussis and the possible pathogenic mechanisms involved, several reviews have been published [1, 4–9].

## II. WHOLE-CELL PERTUSSIS VACCINES

Current regulations governing the production of whole-cell pertussis vaccine (WCPV) have been published by the Expert Committee on Biological Standardization of the World Health Organization [10]. Essentially, WCPV is defined as a "saline suspension of killed *B. pertussis* usually made from a mixture of several strains so as to include agglutinogens 1, 2 and 3" [11]. The organisms are killed and detoxified with different time-temperature incubations, long-term storage at 2–8°C, chemical reagents [12–14], or a combination of these methods. Inactivation also serves to remove heat-labile toxin (HLT) activity. At the bulk stage, WCPV is usually mixed with diphtheria and tetanus toxoids that have already been adsorbed onto aluminum hydroxide or phosphate, or calcium phosphate, as adjuvant, this final batch material is tested for purity, sterility, shelf life, potency, and safety [10].

Not surprisingly, WCPV is a complex biological preparation and vaccines from different manufacturers are demonstrably not identical [15,16]. Also, little is known of the relative content of biologically active components present in current WCPVs [17].

### A. Efficacy of Whole-Cell Pertussis Vaccines

The efficacy of pertussis vaccine is usually defined as the percentage reduction in risk, attributable to vaccination, in vaccinated individuals as compared to equally exposed, unvaccinated individuals. Numerous clinical trials with WCPV have been made and have varied greatly in their design and conduct; reference to these trials may be found in many comprehensive reviews [3,18–21]. These early studies showed some clinical efficacy for vaccine prophylaxis, and in the late 1940s and through the 1950s, the Medical Research Council (MRC) of the United Kingdom [22–24] conducted randomized, double-blind field trials to assess the protective activity of several pertussis vaccines in children and correlate this activity with laboratory procedures for the determination of vaccine potency. These trials showed that the vaccines varied greatly in their ability to confer protection against whooping cough, the best vaccines reducing the home-exposure rate to approximately 4% while the least effective allowed an attack rate of approximately 87%. A further valuable result was the significant correlation shown between the protective efficacy of vaccines in children and in the intracerebral mouse-protection test (ICMPT) and the production of agglutinins in both mice and children.

Information from the controlled trials in the United States and the United Kingdom during the 1940s and 1950s indicated that WCPV protected children from whooping cough, and vaccination became a common health policy. Assessments of vaccine efficacy since 1957 have had to be made on statistically much-less-satisfactory data [11], for no further double-blind field trials have been conducted since those of the MRC. Extensive surveillance of pertussis case rate and severity in relation to vaccination status has been made in Japan [25], the United States [26], Sweden [27], and especially in the United Kingdom [21,28–36]. Estimates of vaccine effectiveness have usually been in the range of 70–90%. Although social and economic factors may influence assessments of vaccine effectiveness [20,37], all these studies supported the view that WCPV in current use provided a worthwhile, though not fully satisfactory, level of protection against the disease. Indeed, convincing evidence for vaccine effectiveness comes from data that show an increase in the incidence of the disease following a period of reduced vaccine acceptance, as in the United Kingdom [32,35], or the complete abandonment of routine immunization, as in Japan [25,38] and Sweden [27].

### B. Toxicity of Whole-Cell Pertussis Vaccines

Despite the belief that the benefits outweigh the risks, an extensive literature exists on adverse reactions to vaccination and the safety of present WCPVs [39]. Reactions to WCPV in the child may be considered under two headings: those that are "mild or moderate" and those that are in some degree "severe" [11]:

1. Mild or moderate reactions: the minor, transient reactions include fever

and irritability, redness, swelling, and tenderness at the injection site [40–42], and do not usually cause excessive concern.

2. Severe reactions: shock, collapse, extreme pallor, and persistent screaming have been occasionally seen as rare responses occurring between 2 and 12 h postimmunization with various pertussis vaccines [16,43,44].

With spasms, febrile and nonfebrile convulsions, and seizures, the implication of WCPV is less clear because of the relatively high background incidence of such conditions occurring spontaneously (or for other reasons) in the infant population [16,39]. In fact, many complications of infancy, including "sudden infant death syndrome" (SIDS) and encephalopathic illnesses, occur most commonly during the first 6 months of life, a period concomitant with many vaccinations [45].

Of all reactions associated with infantile administration of WCPV, the most serious and alarming are those that result in permanent neurological sequelae or death. In Britain, from mid-1976 to mid-1979, a nationwide inquiry into all cases of serious neurological disease in early childhood was carried out [46,47]. This inquiry, by the National Childhood Encephalopathy Study (NCES), was to assess the risk of severe neurological reactions to vaccination. There has been much variation in the reported incidence of neurological reactions associated with pertussis vaccination, and the NCES 1) concluded that over 95% of the cases of severe neurological disorders reported were not associated with recent (within 28 days) vaccination with pertussis-containing vaccines and 2) estimated that the attributable risk of a serious neurological disorder within 7 days after DTP vaccination, in previously "normal" children, was 1 in 110,000 immunizations (95% confidence limits (CL) 1 in 44,000, 1 in 360,000). For permanent (1 year later) brain damage, the figure was 1 in 310,000 immunizations (95% CL 1 in 54,000, 1 in 5,310,000). However, recent reports [39,48] have stressed that the frequency of more serious episodes of high-pitched screaming, pallor, convulsions, neurological disorders, and SIDS cannot emphatically be attributed to the pertussis component of vaccines. Also, the pathogenesis of common and severe reactions after pertussis vaccination is obscure [11,45], and evidence presented at the recent test case (Loveday and Renton) at the High Court of London further negated the association of pertussis vaccine and vaccination with permanent brain damage.

## III. POTENTIAL VACCINE ANTIGENS OF *BORDETELLA PERTUSSIS*

*B. pertussis* produces an array of biologically active components that individually or collectively may have important roles in the pathogenesis of pertussis. Much effort has been directed toward the characterization of those components, and they have been reviewed extensively elsewhere [1,6,7,9,17,49]. For the purpose of this review, the biological properties of these components, and their importance for inclusion in acellular vaccines, will be briefly discussed.

### A. Toxins of *B. pertussis*

**1. Pertussis toxin.** The component of *B. pertussis* that has been most extensively studied and that appears unique to this bacterial species is pertussis toxin (PT). The toxin has been isolated and purified from the cells and/or culture fluid of *B. pertussis* by many procedures [50,51], and its structure and mode of action have been elucidated. Tamura et al [52,53] demonstrated that the molecular structure and mode of catalytic activity of PT was in accord with the A-B model of toxins [54]. The toxin is a hexamer of five dissimilar subunits—S1, S2, S3, S4, and S5—with molecular weights of approximately 28, 23, 22, 11.7, and 9.3 kDa and present in the molar ratios of 1:1:1:2:1, respectively. The S2, S3, S4, and S5 subunits combine to form a pentamer—termed the binding or B-oligomer—that interacts with the S1 subunit—termed the active or A-protomer—to form the intact holotoxin. The A-protomer has both NAD-glycohydrolase and ADP-ribosyltransferase activities, catalyzing the transfer of intracellular NAD to a membrane-bound protein. This protein is the GTP-binding protein (Gi or Ni), which acts as a transducer to couple receptors to the adenylate cyclase catalytic protein, in an inhibitory fashion, in mammalian cell membranes [55]. Consequently, membrane signal transduction is interrupted and cellular function(s) altered. The B-oligomer is considered to bind the holotoxin to the cell surface and allow entry of the A-protomer, as well as being a potent mitogen for T cells [55,56].

PT manifests a remarkably wide range of biological activities in experimental animals and has consequently been known by the following names: histamine-sensitizing factor, HSF (enhances sensitivity to histamine and other agents); leukocytosis-promoting factor, LPF (produces a marked leukocytosis with a predominant lymphocytosis); islet-activating protein, IAP (increases insulin secretion); heat-labile adjuvant, HlAd (has adjuvant activity and enhances the induction of autoimmune diseases and immediate and delayed-type hypersensitivity); and pertussigen (''pertussis-inducing substance'') [57]. Many in vitro effects of PT on cells have also been described and reviewed [58].

PT is seen as the factor responsible for the cardinal symptoms of pertussis, although the mechanism of this exotoxinosis remains unclear [5,7]. PT is believed to play an ancillary role in the adherence of *B. pertussis* to ciliated epithelial cells in the respiratory tract [59,60], and it may also be responsible for the initial alteration of host function. The toxin affects macrophage [61], neutrophil [62], and lymphocyte [63,64] function and migration. Consequently, the toxin may alter essential cell functions of the host's immune system required to remove the infecting organism. Some of the systemic effects of PT in experimental animals are similar to symptoms of clinical pertussis [6,7,50,65].

Many studies have confirmed PT as a major protective antigen. Toxoided PT protected mice against intracerebral [66–69] and respiratory infections [68,70]. Also, anti-PT polyclonal or monoclonal antibodies protected mice against intracerebral infection [66,71], and anti-PT polyclonal [68–70,72,73] or monoclonal antibodies [68,71] protected mice against respiratory infection with *B. pertussis*. Because PT is probably the major toxin involved in the pathogenesis of pertussis, and because antibodies to PT may be important for protection against systemic disease in man, a suitably nontoxic, antigenic, and immunogenic preparation of PT is likely to be an essential component of an effective acellular vaccine.

**2. Adenylate cyclase toxin.** *B. pertussis* produces an adenylate cyclase toxin (ACT) that is primarily extracytoplasmic [74] and activated by the eukaryotic regulatory protein calmodulin [75]. The toxin can enter a variety of mammalian cells to catalyze the production of cAMP [76] and thus interfere with normal cellular function.

The isolation, purification, and characterization of ACT has proved difficult and several data indicate that it is a complex molecule that may appear in multiple forms depending upon the method of purification [77]. Little is known of the toxin structure, the target-cell receptor, and the mechanism of entry into the target cell [77]. The lack of purified material has hampered the consideration of ACT as a vaccine antigen. Circumstantial data indicate a role in the pathogenesis of pertussis: the toxin has been implicated in the lethal infection of suckling mice with *B. pertussis* [78] and was lost from avirulent or modulated strains of the bacterium [79,80]. Also, the most extensively studied category of cell types intoxicated by ACT has been immune effector cells [77], indicating that the toxin may contribute to pathogenesis by interfering with the host's defense mechanisms [81]. Recently, a crude preparation containing ACT was shown to be immunogenic in mice [82]. Nevertheless, further studies are required to establish the importance of this toxin as a vaccine antigen.

**3. Heat-labile toxin.** Heat-labile toxin (HLT), also known as dermonecrotizing toxin (DNT), mouse-lethal toxin, or lienotoxin, was first described by Bordet and Gengou [83]. HLT is a heat-sensitive (56°C for 10 min) toxin that is located intracellularly [84], although a part of the molecule may be surface exposed [85]. Because of the instability, poor recovery, and insolubility of HLT, few reliable methods for its purification to homogeneity have been reported [86–92]. Consequently, most early studies have been done with crude cell extracts or partially purified preparations.

The pure toxin or crude cell lysate was lethal and dermonecrotizing for suckling mice [84,85,93] and also induced spleen atrophy [89,94]. Recent findings suggested that ischemia or edema with subsequent hemorrhagic necrosis were characteristic reactions to HLT [95]. HLT also had no reproducible effects on isolated cells or tissues, despite its severe toxicity in vivo [95]. The mechanism of HLT action remains largely unknown. Highly purified toxin has been reported to cause vasoconstriction, with a specific effect on vascular smooth muscle [96]. Although a definitive role in pathogenesis is obscure [6], HLT-induced vasoconstriction could result in respiratory and peripheral circulatory failure, as well as affecting highly vascularized organs and tissues such as spleen and lungs [94,95].

For many years after its discovery, HLT was considered to be nonantigenic. However, in the 1940s, several investigators showed that crude, detoxified HLT stimulated, both in children and experimental animals, antibodies that neutralized the lethal and dermonecrotizing activities [97]. Native HLT [97] and a formalin-toxoided preparation [87] failed as protective antigens in experimental intracerebral infections of mice. However, toxoided HLT did produce some antitoxin in rabbits and guinea pigs, but not in mice [85]. The question of HLT being able to function as a protective antigen in the mouse or child should be regarded as unanswered [6], and further studies are required to determine the specific role of HLT in pathogenesis and the development of immunity in pertussis.

**4. Tracheal cytotoxin.** Tracheal cytotoxin (TCT) was isolated from culture supernates of *B. pertussis* as a small glycopeptide containing diaminopimelic acid and muramic acid [98], and the presence of these components indicated that TCT was derived from the peptidoglycan of the cell envelope. TCT has been shown to inhibit DNA synthesis in monolayer cultures of hamster tracheal epithelial cells and to cause ciliostasis and specific ciliated-cell damage [98]. Consequently, TCT may play a role in pathogenesis by effecting the primary lesion in pertussis, i.e., cytopathology in the respiratory tract. Ciliostasis may also contribute to a reduction in the effectiveness of mucociliary clearance mechanisms and thus promote secondary infection of the respiratory tract [99].

TCT appears to be a virulence determinant common to all species of the genus *Bordetella* [100] and is similar to muramyl peptides produced by other bacteria, e.g. *Neisseria gonorrhoeae*, which can promote ciliated cell-specific damage. In addition, TCT is absolutely identical in structure to FSu, a potent sleep-promoting factor of human origin. However, no immune response toward TCT in animal or human infections has been reported, and the relevance of this toxin to pathogenicity and inclusion in vaccine formulations requires further study.

**5. Endotoxin.** The lipopolysaccharide (LPS) endotoxin of *B. pertussis* possesses many properties in common with LPS from other gram-negative bacteria, notably toxicity, pyrogenicity, Shwartzman reactivity, antigenicity, and adjuvanticity [6,101]. Chemical analyses have shown that it contains two different polysaccharides—PS-I, which has nonphosphorylated KDO (3-deoxy-2-octulosonic acid) and PS-II, which has phosphorylated KDO—and two different lipids—lipid A and lipid X [102].

In the pathogenesis of pertussis, LPS probably contributes to some inflammation and local tissue damage and could exert some effects on the immunological response. The LPS is antigenic, and anti-LPS antibodies have been stimulated in animals [103] and in humans after vaccination with WCPV or infection [104–106]. However, it does not appear to be an important antigen for protection, and the consensus of opinion is that LPS is an undesirable element in pertussis vaccines that contributes to toxicity and reactogenicity [17]. However, traces of suitably detoxified LPS may enhance, by its adjuvant and nonspecific activities, the antigenicity and immunogenicity of other antigens in acellular vaccines [101].

### B. Adhesions and Other Virulence Factors

**1. Filamentous hemagglutinin.** *B. pertussis* possesses two chemically and biologically distinct hemagglutinins, PT and filamentous hemagglutinin (FHa). FHa has been purified from both static liquid culture supernates [107–110] and cell extracts [111,112] and was found to be heterogeneous when examined by SDS-PAGE. Major polypeptides of molecular weight 126 [107], 127 and 95 [111], and 160, 115, and 95 kDa [109] were revealed, and these various fragments probably arose from a polypeptide of high molecular weight, 220 kDa [112]. Although the original fimbrial origin of FHa [113–115] has been questioned [17,116,117], it is undeniably involved in bacterial adhesion [59,60]. FHa is one of the most likely candidates for inclusion in any acellular vaccine because of its apparent lack of toxicity and its protective activity in certain animal models of infection with *B. pertussis*.

FHa has been found to protect mice against intracerebral challenge with live *B. pertussis*, but only when contaminated with trace amounts of active PT [67,69]. When rigorously purified, FHa was nonprotective against intracerebral challenge [66,68], but did protect against aerosol or intranasal infection [68–70,72]. In passive protection tests for mice, anti-FHa antibodies were nonprotective against intracerebral infection [66], but protected against respiratory infection [69,70]. FHa also stimulated a protective immune response in rabbits [103]. However, antibody to FHa has not been uniformly detected in human convalescents from pertussis, indicating that serum antibody to FHA may not be essential for recovery and long-term protection against infection [106].

**2. Agglutinogens.** In *B. pertussis,* agglutinogens (AGGs) can be operationally defined as surface antigens that stimulate the production of antibodies that cause bacterial cell agglutination [17]. Based on agglutination and agglutinin adsorption, the genus *Bordetella* was differentiated by AGGs unique to each species [118,119]. Agglutinogens 1 and 7 are common to all *B. pertussis* strains, with AGG-7 common to all species of the genus. Agglutinogens 2–6 and 13 vary in individual *B. pertussis* isolates, and AGG-8, 9, and 10 are found in *B. parapertussis*, with number 14 being specific for the latter species. All strains of *B. bronchiseptica* have AGG-12 with individual isolates containing 8, 9, 10, and/or 11. It is now believed that there are three major AGGs in *B. pertussis*—1, 2, and 3—while 4, 5, and 6 are probably minor antigens [120]. Agglutinogens 2, 3, and 6 have been identified as fimbrial antigens [117,121–124], whereas AGG-1 has not [117].

The protective role of AGGs has been discussed for many years and still remains to be clarified [11]. For example, although AGGs in WCPV were unimportant for protection of mice against intracerebral infection [125,126], recent studies have shown that isolated fimbriae were protective in the presence of a trace of native PT [67]. In addition, isolated AGGs, devoid of PT, protected mice against respiratory infection [70,127,128].

Fimbriae have been suggested to play a role in adhesion of *B. pertussis* to ciliated epithelial cells of the respiratory tract. However, direct evidence that pertussis fimbriae are adhesins is lacking. Monoclonal antibodies to AGGs have inhibited bacterial adherence to nonciliated Vero cells [129] and HeLa cells [114,130]. But, conversely, nonfimbriated *B. pertussis* strains adhered to nonciliated WiDr and human ciliated cells, and some fimbriated cells failed to adhere [131,132]. Nevertheless, there is still sufficient epidemiological evidence for the inclusion of AGGs in acellular pertussis vaccines. Agglutinogens 1, 2, and 3 are considered important in the epidemiology of pertussis [4], and in the United Kingdom MRC trials of the 1950s, serum agglutinin titers of human vaccinees correlated with clinical protection from pertussis [24]. Their importance is further reflected in the WHO recommendation that (representative strains of *B. pertussis* in) pertussis vaccines contain serotype antigens 1, 2, and 3 [10].

**3. Outer-membrane proteins.** The outer-membrane proteins (OMPs) of several gram-negative bacteria have been recognized as playing important roles in pathogenesis, and several antigens are known to be located on the surface of *B. pertussis* [4,133]. In fact, six polypeptides associated with the outer membrane were found to be specific to virulent, phase 1, X-mode *B. pertussis* [134]. Detergent-extracted OMPs protected mice against intracerebral and intranasal infection with *B. pertussis*, but this depended on the presence of small amounts of native PT [67,70,134]. Also, OMPs were found to protect rabbits against intranasal infection [103]. No immune response toward specific OMPs has been reported, and more details are required of the biological activities of purified OMPs to determine their role in pathogenesis and their inclusion as vaccine antigens.

**4. Hemolysin.** Genetic studies with *B. pertussis* mutants have confirmed the importance of hemolysin as a virulence factor. *B. pertussis* mutants deficient in hemolysin and ACT had reduced virulence in the infant mouse respiratory model [78], and recent studies have shown that, although hemolysin was important for causing disease in mice, it was not essential for colonization of the respiratory tract [135]. It has recently been postulated that hemolysin activity may be associated with an unprocessed and nonsecreted precursor of ACT [136]. Nevertheless, hemolysin remains a wholly uncharacterized component of *B. pertussis* and is exempt from inclusion in acellular vaccines.

## IV. DEVELOPMENT OF ACELLULAR PERTUSSIS VACCINES

Whole-cell pertussis vaccines have been effective in preventing disease, but have also been the subject of considerable debate with regard to safety and efficacy. What are the reasons for the development of acellular pertussis vaccines (APVs)? The fundamental criteria for assessing new acellular vaccines should be related to the disadvantages and imperfections of current WCPVs, consequently, acellular vaccines should 1) be uniform and reproducible in composition, 2) be nontoxic, thereby ensuring the absence of local and systemic reactions, severe neurological reactions, or permanent sequelae in immunized children, 3) stimulate a more predictable and long-lasting immune response, and 4) give a higher incidence of clinical protection against both infection and disease.

A fuller, though far from complete, understanding of the roles of various *B. pertussis* components in the complex pathogenesis of the disease has greatly improved the logical design of APV. However, even as WCPVs were being introduced for mass vaccination, acellular vaccines of crude formulation were being developed. The earliest APV was the stromata protective antigen (SPA) of Pillemer et al. [137], which was prepared from extracts of sonically disrupted *B. pertussis* cells adsorbed onto human erythrocytes. The vaccine was protective in the ICMPT and showed clinical efficacy in the MRC field trials [24], but it was also the most reactogenic of all vaccines tested and consequently not licensed for clinical use. Felton and Verwey [138] conducted a limited clinical trial of a vaccine prepared by sonication [139] and demonstrated that the acellular vaccine was less toxic and induced higher levels of agglutinins than a cellular preparation. A trisodium extract of

*B. pertussis* was sold from 1962 to 1977 as Tri-Solgen by the Eli Lilly Company as an acellular component of their DTP vaccine. In the initial investigation [140], the extracted vaccine was antigenic and produced a markedly lower incidence of local and systemic reactions in children when compared to a whole-cell preparation. Also, in the early 1960s, a soluble pertussis extract was developed that was potent in the ICMPT but not tested for clinical efficacy [141]. More recently, Schwick et al. [142] reported that a vaccine developed at Behringwerke was potent in the ICMPT and had clinical use in children. This vaccine was reported to induce fewer local reactions in children of 3–4 months of age.

Most of the recent studies to develop experimental APVs have concentrated on the isolation and (partial) purification of *B. pertussis* antigens, especially PT and FHa, followed by chemical detoxification. Examples of these partially characterized vaccines include those of Garcia-Sanz et al. [143] and Christodoulides et al. [144]. But without doubt, the major contribution to the development and clinical use of APV has come from Japan. The present status of these licensed vaccines, as well as other acellular preparations currently in clinical studies, will be discussed below.

### A. Present Status of Acellular Pertussis Vaccines

There are currently two rationales for the design of APV and convincing arguments have been presented for both. The argument for a monocomponent vaccine is based on the belief that pertussis is essentially an exotoxinosis and that PT is the fundamental element in the understanding of *B. pertussis*, of the pathogenesis and prolonged immunity of pertussis, and of the reactivity of pertussis vaccine(s) [7]. Conversely, a multicomponent vaccine has been proposed, since pertussis may not be simply an exotoxinosis, but rather a disease involving a complex interaction between the host, bacterium, and a plethora of toxins and virulence factors. Consequently, an optimal vaccine would include factors protective against infection and also toxoids protective against the local and systemic effects of liberated toxin(s). This controversial issue of vaccine composition is currently being addressed by the clinical testing of vaccines prepared from the various antigens present in different ratios.

The Japanese acellular vaccines are prepared essentially according to the method of Sato et al. [38]. Briefly, static culture fluid of *B. pertussis* is concentrated by ammonium sulfate precipitation and fractionated by sucrose density gradient centrifugation. Fractions from the sucrose gradient containing hemagglutinating activity, but with reduced endotoxin, are isolated, pooled, and detoxified with formaldehyde. The final vaccines contain primarily PT and FHa and are adsorbed onto aluminum hydroxide as adjuvant. The ratio of PT to FHa is variable and dependent on the condition of the culture and culture medium and the (co-)purification procedure used. There are basically two types of vaccines:

1. The "Takeda-type" vaccine, which consists of PT, excess FHa, and agglutinogen-2 present in the ratios of 10:90:1 w/w [128]. It is currently manufactured by the Takeda, Kitasato, Chiba, Denka, and Kaketsu companies.
2. The "Biken-type" vaccine, which consists of equal proportions of PT and FHa and no detectable agglutinogens [145].

These acellular vaccines had reduced LPS content, histamine-sensitizing and leukocytosis-promoting activities, and mouse toxicity, compared to WCPV [38,146]. However, an acellular vaccine did manifest a late histamine-sensitizing activity [147], and this may have been due to incomplete detoxification of PT or in vivo reversion of the toxoid. (The use of chemical reagents that predominantly modify lysine amino groups, e.g., formaldehyde and glutaraldehyde, both currently used for vaccine preparation, may be unsuitable for the detoxification of PT since the S1 subunit does not contain any lysine residues [148].) The individual components were antigenic, and the vaccines immunogenic, in laboratory animals [38].

In a departure from the production method of Sato et al. [38], a recent acellular vaccine from Kaketsu contained PT and FHa purified with affinity and ion-exchange column chromatography. Both PT and FHa were treated with formalin in the presence of an amino acid and then mixed in a ratio of 1:4. In limited clinical field trials, this vaccine caused fewer local reactions than the Kaketsu "Takeda-type" vaccine and had comparable antigenicity [149].

Since 1981, these vaccines have been used for mass immunization in Japan, although initially only 2-year-old children were vaccinated. More than 30 million doses have been administered, largely of the "Takeda-type" vaccines. Information concerning the safety and efficacy of these vaccines has become available in the last few years. In general, transient local and systemic reactions caused by APV were significantly less frequent and milder when compared to the Japanese WCPV [45]. A small number of children in the United States received a Japanese "Takeda-type" vaccine and similar mild reactions were observed [150–155]. In the Japanese clinical studies, local reactions appeared at a much later time after the first dose than following subsequent doses, suggesting a delayed hypersensitivity [156]. The rate of "severe reactions" did not significantly differ between the acellular and whole-cell vaccines when used at 24 months of age [45]. However, Japanese data indicated that delaying the initial vaccination until a child was 24 months old, regardless of the type of vaccine, reduced most of the temporally associated severe adverse reactions. Cases of presumed vaccine-associated reactions decreased significantly in Japan with the introduction of immunization at 24 months instead of 3 months with WCPV, as was done in 1975, and with the use of acellular vaccines beginning in 1981 [45,156]. Consequently, assessment of the "encephalopathic potential" of acellular vaccines in children less than 2 years of age cannot be made from the Japanese data.

The vaccines stimulated high antibody responses to both PT and FHa and the "Takeda-type" vaccine produced an agglutinin response. Several studies have shown that after vaccination with APV, antibody levels to individual components were similar to those found in children immunized with WCPV [150,151]. In some cases a higher response to FHa was observed after vaccination with APV than with WCPV [154,155]. Also, antibody levels after vaccination with APV were equivalent to or greater than those in children recovering from pertussis [157,158].

There are no published large formal surveillance studies of field efficacy of APV in Japan. Efficacy has been inferred from epidemiological evidence. Immunization acceptance increased rapidly in Japan beginning in 1978, initially with WCPV-DTP, which was replaced with APV-DTP in 1981 [159]. Disease incidence declined rapidly [160,161], including in unimmunized children from birth to 2 years of age, indicating some "herd immunity." However, pertussis incidence has remained high in children less than 2 years of age compared with incidence levels attained with earlier immunization [45]. In household contact studies carried out in the period from 1978 to 1985, the effectiveness of acellular vaccines was between 78 and 95% [156,160–163]. In a study from 1981 to 1987, 442 households of patients with pertussis were surveyed [164], and the secondary attack rates in children of 2 to 8 years of age were assessed as follows: 61.3% in unimmunized children, 7.5% and 14.3% in children immunized with the "Takeda-type" and "Biken-type" vaccines respectively, and 13.5% in children immunized with WCPV. Statistically, there was no difference in the effectiveness of the Takeda (87.7%) and Biken (76.7%) vaccines, nor between these acellular vaccines and WCPV. However, although more data are being reported on the use of acellular vaccines in Japan (most recently at the Fifth International Symposium on Pertussis in Copenhagen, Denmark, in 1988), no effectiveness data are available for children younger than 1 year of age.

Information has also become available from clinical studies of Japanese APVs in Sweden. Two acellular vaccines were evaluated: JNIH-6, which contained equal amounts of toxoided PT and FHa, and JNIH-7, which was a monocomponent pertussis toxoid. Neither vaccine contained agglutinogens. In a phase I clinical trial in 6–10-month-old infants [165], both vaccines stimulated an antitoxin seroconversion with a rate of 100%, and adverse reactions were few and mild. However, local reactions occurred more often after administration of the second dose and among recipients of the two-component vaccine. In a phase II trial, the antigenicity and rate of adverse reactions of JNIH-6 were compared with WCPV and placebo for primary immunization of 6-month-old infants [166]. The APV produced fewer and milder local and systemic reactions, and antitoxin reponses were obtained in 97–100% of infants receiving either two of three doses of APV compared to 59% after three doses of WCPV. This formed the basis for using a two-dose immunization schedule in the phase III efficacy trial.

The carefully planned, randomized, double-blind, placebo-controlled phase III efficacy trial was conducted from February 1986 to October 1987 [45,167]. In this study, the first dose of vaccine or placebo was given at 5–11 months of age, followed by the second dose some 8–12 weeks later. Of the 3,801 children enrolled, 954 received the placebo (vaccine solvent), 1,419 received the two-component vaccine (JNIH-6), and 1,428 received the monocomponent toxoid (JNIH-7). The immediate side effects of vaccination were mild, although small local reactions occurred more often in vaccinated infants than in those receiving placebo, especially after the second vaccine dose. During 15 months of follow-up from 30 days after the second dose, culture-confirmed whooping cough occurred in 40 infants receiving placebo and in 17 and 18 infants receiving JNIH-7 and JNIH-6 respectively. Thus, vaccine efficacy was 54% (95% CL 26, 72) for the monocomponent toxoid JNIH-7 and 69% (95% CL 44, 82) for the two-component vaccine JNIH-6. Protection against culture-confirmed whooping cough of over 30 days' duration was 80% (95% CL 59, 91) for JNIH-7 and 79% (95% CL 57, 90) for JNIH-6. The two-component vaccine gave a significantly lower antitoxin response than the monocomponent toxoid, but offered the same degree of protection. However, a notable feature was the lack of correlation between postvaccination serum concentrations of anti-PT and/or anti-FHa antibodies and subsequent protection against whooping cough.

Despite the lower vaccine effectiveness in the Swedish trials, the course of illness was modified and less severe in vaccine failures. Nevertheless, four children died of severe invasive bacterial infection 2 weeks to 5 months after administration of either JNIH-6 or JNIH-7 [167]. Epidemiological data from Japan, however, suggest that a correlation between the administration of acellular vaccines and death is unlikely [168].

Several other APVs have recently been developed by various manufacturers and are currently in clinical studies [146,159]. These are as follows:

1. A vaccine developed by Connaught, and containing FHa and PT in the ratio of 6:1, was detoxified with glutaraldehyde. In clinical studies, the

vaccine produced minor local and systemic reactions in children but was not highly antigenic.

2. Two vaccines have been produced by Merieux: a monocomponent vaccine of glutaraldehyde-detoxified PT and a bivalent vaccine containing equal amounts of detoxified PT and FHa. The monocomponent vaccine was safe and antigenic in 3–6-month-old infants [169], and the two-component vaccine, when compared with WCPV in children of 4–6 years and 18–24 months of age, induced fewer local reactions and had comparable antigenicity [170].

3. Another monocomponent vaccine, of hydrogen peroxide–detoxified PT, was safe and antigenic in adults [171] and 18-month-old children [172].

4. An acellular vaccine from the Centre for Applied Microbiology and Research, England, consists of equal amounts of separately purified and formalin-toxoided PT, FHa, and AGGs-2 and -3; this multicomponent vaccine was antigenic in adults and had low reactogenicity [173].

## B. Design of Future Acellular Pertussis Vaccines

The APVs described in the preceding section can be considered as first- and second-generation vaccines [1]. From the essentially crude manufacturing process described by Sato et al. [38], the production of these vaccines has become more efficient and their composition more uniform and reproducible. Nevertheless, they still contain individually purified antigens of bacterial origin, which have been subjected to the uncertain principles of chemical detoxification. The third-generation vaccines would be a natural progression from the preceding generations, but may be prepared by recombinant DNA and hybridoma technologies and chemical synthesis. In this section, the design of future acellular pertussis vaccines is considered.

**1. Genetically engineered vaccines.** Modern techniques in mutagenesis have provided the means for identifying the important virulence-associated factors of *B. pertussis* [135]. Recombinant DNA technology allows for the genetic determinants of virulence-associated factors to be cloned, sequenced, and expressed. Such techniques have been applied to *B. pertussis* and much progress has been made toward the genetic engineering of vaccine antigens.

Recently, several investigators [148,174] have successfully cloned DNA fragments containing PT structural genes, and the entire nucleotide sequence of the PT operon has been determined. The amino acid compositions, molecular weights, and isoelectric points of each of the toxin subunits were virtually indistinguishable from the values determined by Tamura et al. [52,53]. Recent studies have shown that the *ptx* operon was conserved in *B. pertussis, B. parapertussis* and *B. bronchiseptica,* but only transcribed in *B. pertussis* [175,176]. Gross and Rappuoli [177] have constructed a pLARF2 vector for *Bordetella* species that contained the PT promoter fused to the coding region of the chloramphenicol acetyltransferase gene. It was used to identify the regulatory elements involved in the transcription of the PT operon. Efficient transcription required at least two features—a 170-base-pair DNA sequence upstream from the start of transcription and a trans-acting factor encoded by the *vir* locus. *B. parapertussis* and *B. bronchiseptica* also have trans-activating systems, but did not produce PT because of selective mutations within their PT promoter regions. The conservation of such "cryptic genes" may have profound implications for the future eradication of pertussis.

Initial attempts to detect expression of PT (or other *B. pertussis* virulence factors) in *E. coli* were unsuccessful unless an expression vector was used. Reiser and Germanier [178] constructed a library of *B. pertussis* strain 165 DNA in the bacteriophage vector λgt11, which enabled DNA insertion into the *E. coli lac*Z gene. Consequently, structural genes, including those for PT, were expressed as hybrid proteins. The individual structural genes for the five PT subunits have also been cloned into expression vectors pEX31 and pEX34 [179]. Because these vectors encoded for the DNA polymerase of phage MS2, cloned structural genes were expressed as fusion proteins containing the N-terminal portion of the polymerase. However, fusion proteins of subunits S1 and S5 accounted for only 5 and 20% respectively of the total cell protein. The S1 fusion protein had similar ADP-ribosyltransferase activity to native PT. The S1 gene has also been clonsed into expression vector pUC18, and the subunit expressed as a β-galactosidase fusion protein [180]. Similarly, the S1 genes from *B. parapertussis* and *B. bronchiseptica* have been cloned in *E. coli,* with the expressed fusion proteins having the same ADP-ribosyltransferase activity as the S1 subunit from *B. pertussis* [175]. Randomly generated fragments of the *ptx* operon have also been cloned into an *E. coli tac* promotor expression vector pRI133 [181], and expression of the S4 subunit achieved.

Recently, Locht et al. [182] cloned *Sau*3A fragments from the *Pst*1 fragment A of the *ptx* operon [183] into pUC18: the cloned structural genes were transcribed by way of the inducible *lac* promoter and translated as *lac*Z fusion proteins. From studying several derived plasmids and the interaction of their expressed fusion proteins with anti-S1 monoclonal antibodies, the sequence of amino acids 109–187 of the S1 subunit was described as the immunodominant region. An even more discrete region of the S1 protein, delineated by tyrosine 8 and proline 14, was defined by recombinant analysis and mapping as possessing at least a portion of the enzyme active site and the protective antigenic epitope [184].

Recently, portions of the *ptx* operon have been cloned into bacterial expression vectors and high levels of directly expressed, i.e., nonfusion, protein subunits have been obtained in recombinant *E. coli* [185]. Also, a DNA fragment for S1 has been inserted into a *Bacillus* α-amylase secretion vector pKTH114, and the recombinant plasmid pKTH212 introduced into *Bacillus subtilis* [186]. This resulted in high expression of the subunit, with approximately 100 mg/liter of intact and truncated S1 proteins secreted into the culture supernatant. The major advantage of this expression system is that of *Bacillus subtilis,* unlike *B. pertussis* and *E. coli,* is entirely free of endotoxin. Nevertheless, pertussis vaccine antigens produced from this system (although free of other *B. pertussis* or *E. coli* components) would still require rigorous purification to remove contaminating bacillar proteins.

In addition to Pt, several other virulence factors have been cloned and expressed:

1. The gene for immunologically reactive FHa was cloned and expressed in *E. coli* as a β-galactosidase fusion protein using the expression vector λgt11 [187,188]. A similar expression was obtained in *E. coli* after cloning into the $Km^r$ gene of the broad-host-range vector pCP13 [189].
2. The gene for AGG-2 from *B. pertussis* strain Wellcome 28 was cloned in pBR328, and the complete amino acid sequence determined [190].
3. Brownlie et al. [191] constructed a gene library of *B. pertussis* strain Tab1 in *E. coli* using a broad-host-range cosmid vector pLAFR1. One recombinant plasmid, pLAFR1138, restored hemolysin and adenylate cyclase activities when transferred by conjugation to a mutant *B. pertussis* strain (BP348). Recently, the AC toxin has been cloned and expressed in an *E. coli cya* strain, in the absence of any *B. pertussis* trans-activating factor [136].
4. The *vir* locus of *B. pertussis* encodes a trans-acting positive regulator that is required for the expression of genes for several virulence factors, notably PT, FHa, ACT, and hemolysin. The *vir* gene has been cloned on vector pLAFR1 [135] and a recombinant plasmid, pLAFR1279, restored all virulence factors to a mutant *B. pertussis* strain 347 ($Vir^-$). Cloning of the *vir* and *fha* structural genes has recently been reported [192].
5. Shareck and Cameron [193] reported the cloning of *B. pertussis* OMP determinants in pHC79 and the expression of two polypeptides of 30 and 33 kDa in *E. coli.* Also, the gene coding for P.69, a *vir*-controlled OMP and protective antigen in the ICMPT, has been cloned, its DNA sequence determined, and protein fusion products expressed in *E. coli* [194].

The main problem with engineering pertussis vaccine antigens is to obtain adequate prokaryotic and/or eukaryotic expression. Nevertheless, recombinant DNA technology allows for the determination of immunodominant epitopes of pertussis antigens that may form the basis for safer APVs. One approach is to produce "in vivo" toxoids by reintroducing cloned genes, modified so that the products lack biological activity, back into the host of origin or a suitable recipient. For example, the ADP-ribosyltransferase activity of the PT S1 subunit may be modified by site-directed mutagenesis to produce a nontoxic derivative that still retains the tertiary structure important for stimulating protective immunity. Furthermore, recent studies may permit the construction of recombinant site-specific S1 mutants that retain the important protective epitope(s), yet eliminate enzyme-related toxicity. (However, recent data suggest that ADP-ribosyltransferase activity is critical for optimum immunoprotection, and consequently should be an important consideration in the formulation of APVs [195]. Also, PT subunits expressed in *E. coli* or *Bacillus subtilis* as fusion proteins failed as protective antigens in the ICMPT, possibly indicating that protective epitopes were mostly in the assembled toxin molecule [196].) Despite the obvious attractions, pertussis vaccines derived through genetic manipulation remain at present a distant choice for prophylaxis and require further work before licensing for clinical use. Nevertheless, it is possible to use *B. pertussis* mutants to prepare vaccine antigens for inclusion in current APVs, e.g., FHa purified from a *B. pertussis* mutant $PT^-$ $FHa^+$ would not need to be detoxified, which invariably affects antigenicity, to remove trace amounts of PT activity.

**2. Synthetic peptide vaccines.** The attraction of synthetic peptide vaccines is clear. Essentially, they would eliminate the need to compose vaccines from inactivated whole-cell or acellular preparations. Also, the problems of contamination associated with the purification of antigens from *B. pertussis* culture fluid and cells, as well as from biotechnological processes that use genetically manipulated cells, would be avoided. Specific peptide vaccines would use small, chemically synthesized peptides representing limited but antigenically important epitopes of the relevant protective antigens. The identification of immunodominant epitopes would come from using recombinant DNA and hybridoma technology. Recently, the nucleotide sequence of the PT operon was used for the synthesis of peptides [15–22 amino acids) of the S1 subunit [197]. Five sequences were identified as carrying epitopes that reacted with hyperimmune sera from mice immunized with WCPV and with human serum samples from infants vaccinated with either WCPV or a monocomponent pertussis toxoid.

However, it is unlikely that any peptide vaccine will be useful in the prevention of human disease unless given in conjunction with an immunostimulatory adjuvant. The antigenicity and immunogenicity of such peptide sequences (haptens) would be enhanced by efficient coupling to carrier proteins, which may include tetanus and diphtheria toxoids or nontoxic pertussis antigens such as FHa, fimbriae, and OMPs. Or, the use of immunopotentiating conjugates, with a backbone of synthetic carrier immunopotentiator and peptide antigen(s) can be envisaged [198].

**3. Pertussis anti-idiotypic antibodies as vaccines.** Jerne [199] first suggested that anti-idiotypic antibodies could express an "internal image" of the original antigen. Since anti-idiotypes can be used to stimulate or enhance an immune response, this concept can be used for vaccine preparation. For simplicity, the anti-idiotype is depicted as sharing an epitope with the original antigen. Also, it is not necessary for the anti-idiotype to bear any marked resemblance to the antigen, e.g., any of the pertussis toxins, beyond the sharing of an epitope. Consequently, immunization with anti-idiotypic antibodies carries none of the problems of toxicity and reactogenicity associated with the original pertussis antigens.

The stimulation of antibodies to pathogens by anti-idiotypic antibodies has been achieved in many systems [200]. At present no experimental data have been published on anti-idiotypic antibodies mimicking pertussis antigens. However, monoclonal antibodies have been prepared against *B. pertussis* and a wide range of its purified antigens [201]. Therefore, the potential exists for the preparation of anti-idiotypic antibodies using monoclonal antibodies that neutralize active toxins of *B. pertussis* and/or passively protect experimental animals against infection. Such an approach is currently under study, using an anti-PT $IgG_1$ monoclonal antibody as the idiotypic antigen (personal communication, B. Brodeur, Hybridoma Lab., Bureau of Microbiology, Health and Welfare, Canada). The monoclonal antibody, specific for the S1 subunit, neutralized active toxin and passively protected mice against *B. pertussis* infection.

Pertussis anti-idiotypic vaccines are unlikely to replace immunogenic peptides derived by recombinant DNA technology, chemical synthesis, or directly from the pathogen. However, they could be useful as "priming antigens," especially in very young infants for whom there is considerable concern when vaccinating with incompletely detoxified or unstable acellular vaccines. They may also be replacements for adjuvants and may be capable of influencing the quality of the immune response. Therefore, anti-idiotypic antibodies may well be useful adjuncts to pertussis vaccine prophylaxis.

## V. SUMMARY

The eradication of pertussis as a worldwide disease is unforeseeable for the present and immediate future. Mortality and morbidity from clinical pertussis are still commonly reported, especially in underdeveloped countries where mass immunization programs are virtually nonexistent. To achieve control of the disease, a high level of immunization coverage is necessary, and pertussis WCPV still remains one of the most effective bacterial vaccines. But, primarily because of its reactogenicity for infants, much effort has been directed toward the characterization of bacterial components important in pathogenesis and for the development of acellular vaccines.

Progress in the last decade has resulted in the production and use of such vaccines for routine vaccination, and their use in Japan, as well as the recent clinical trials in Sweden and several phase I/II studies in other countries, has shown that these preparations are safer than conventional WCPV, and equally effective, in preventing disease.

The development of future acellular pertussis vaccines by gene manipulation may finally inspire public confidence for vaccine prophylaxis, eventually leading to eradication of the disease. However, the production and use of such sophisticated vaccines is dependent on many factors, and consequently conventional WCPV may still be used in many countries for several years to come.

## ACKNOWLEDGMENTS

I thank Dr. Roger Parton of the Department of Microbiology, University of Glasgow, for his many helpful suggestions and for critically reviewing this manuscript.

## REFERENCES

1. Manclark CR, Cowell JL: Pertussis. In Germanier R (ed): ''Bacterial Vaccines,'' London: Academic Press, 1984, p 69.
2. Mortimer E: Pertussis in the developing world. Abstract from the Fifth International Symposium on Pertussis, Copenhagen, Denmark, Sept. 22–23, 1988; unpublished.
3. Pittman M: *Bordetella pertussis*—Bacterial and host functions in the pathogenesis and prevention of whooping cough. In Mudd S (ed): ''Infectious Agents and Host Reactions.'' Philadelphia: Saunders, 1970, p 239.
4. Olson LC: Pertussis. Medicine (Baltimore) 54:427, 1975.
5. Pittman M: Pertussis toxin: The cause of the harmful effects and prolonged immunity of whooping cough. Rev Infect Dis 1:401, 1979.
6. Wardlaw AC, Parton R: *Bordetella pertussis* toxins. Pharmacol Ther 19:1, 1983.
7. Pittman M: The concept of pertussis as a toxin-mediated disease. Pediatr Infect Dis 3:467, 1984.
8. Walker E: Clinical aspects of pertussis. In Wardlaw AC, Parton R (eds): ''Pathogenesis and Immunity in Pertussis.'' Chichester: John Wiley and Sons, 1988, p 273.
9. Wardlaw AC, Parton R: The host-parasite relationship in pertussis. In Wardlaw AC, Parton R (eds): ''Pathogenesis and Immunity in Pertussis.'' Chichester: John Wiley and Sons, 1988, p 327.
10. World Health Organization: Requirements for diphtheria toxoid, pertussis vaccine, tetanus toxoid, and combined vaccines. WHO Tech Rep Ser 638:37, 1979.
11. Wardlaw AC, Parton R: Pertussis vaccine. In Easmon CSF, Jeljaszewicz J (eds): ''Medical Microbiology, Vol. II: Immunization Against Bacterial Disease.'' London: Academic Press, 1983, p 207.
12. Cameron J: Pertussis vaccine: Formalin as preferred toxoiding agent. Lancet 1:880, 1983.

13. Gupta RK, Sharma SB, Ahiya S, Saxena SN: The effects of different inactivating agents on the potency, toxicity and stability of pertussis vaccine. J Biol Stand 15:87, 1987.
14. Iida T, Horiucji Y: The detoxification of *Bordetella pertussis* with glutaraldehyde. J Biol Stand 15:17, 1987.
15. Cameron J, Stainer DW: Pertussis vaccine—Fact and fiction. Acta Microbiol Acad Sci Hung 24:195, 1977.
16. Griffith AH: Reactions after pertussis vaccine: A manufacturer's experiences and difficulties since 1964. Br Med J 1:809, 1978.
17. Robinson A, Irons LI, Ashworth LAE: Pertussis vaccine: Present status and future prospects. Vaccine 3:11, 1985.
18. Lapin JH: "Whooping Cough." Springfield, IL: C.C. Thomas, 1943.
19. Joint Committee on Vaccination and Immunization: In: "DHSS. Whooping Cough Vaccination." London: HMSO, 1977, p 1.
20. Miller DL, Alderslade R, Ross EM: Whooping cough and whooping cough vaccine: The risks and benefits debate. Epidemiol Rev 4:1, 1982.
21. Griffiths E: Efficacy of whole-cell pertussis vaccine. In Wardlaw AC, Parton R (eds): "Pathogenesis and Immunity in Pertussis." Chichester: John Wiley and Sons, 1988, p 353.
22. Medical Research Council: Report: The prevention of whooping cough by vaccination. Br Med J 1:1463, 1951.
23. Medical Research Council: Report: Vaccination against whooping cough: Relation between protection in children and results of laboratory tests. Br Med J 2:454, 1956.
24. Medical Research Council: Report: Vaccination against whooping cough: The final report to the Whooping Cough Immunization Committee of the Medical Research Council. Br Med J 1:944, 1959.
25. Kanai K: Review: Japan's experience in pertussis epidemiology and vaccination in the past thirty years. Jpn J Med Sci Biol 33:107, 1980.
26. Broome CV, Fraser DW: From the Centers for Disease Control. Pertussis in the United States, 1979: A look at vaccine efficacy. J Infect Dis 144:187, 1981.
27. Romanus V, Jonsell R, Bergquist SO: Pertussis in Sweden after the cessation of general immunization in 1979. Pediatr Infect Dis 6:364, 1987.
28. Miller CL, Fletcher WB: Severity of notified whooping cough. Br Med J 1:117, 1976.
29. Noah ND: Attack rates of notified whooping cough in immunized and unimmunized children. Br Med J 1:128, 1976.
30. Pollard R: Relation between vaccination and notification rates for whooping cough in England and Wales. Lancet 1:1180, 1980.
31. Grob PR, Crowder MJ, Robbins JF: Effect of vaccination on severity and dissemination of whooping cough. Br Med J 282:1926, 1981.
32. Joint Committee on Vaccination and Immunization: In: "DHSS. Whooping Cough." London: HMSO, 1981, p 170.
33. Report from the Swansea Research Unit of the Royal College of General Practitioners: Effect of a low pertussis vaccination uptake on a large community. Br Med J 282:23, 1981.
34. PHLS: Efficacy of pertussis vaccination in England: Report from the PHLS Epidemiological Research Laboratory and 21 area health authorities. Br Med J 285:357, 1982.
35. PHLS: Communicable disease quarterly. PHLS Microbiol Digest 4:9, 1987.
36. Fine PEM, Clarkson JA: Reflections on the efficacy of pertussis vaccines. Rev Infect Dis 9:866, 1987.
37. Stewart GT: Whooping cough in relation to other childhood infections in 1977–1979 in the United Kingdom. J Epidemiol Community Health 35:139, 1981.

38. Sato Y, Kimura A, Fukumi H: Development of a pertussis component vaccine in Japan. Lancet 1:122, 1984.
39. Ross EM: Reactions to whole-cell pertussis vaccine. In Wardlaw AC, Parton R (eds): "Pathogenesis and Immunity in Pertussis." Chichester: John Wiley and Sons, 1988, p 373.
40. Barkin RM, Pichichero ME: Diphtheria-pertussis-tetanus vaccine: Reactogenicity of commercial products. Pediatrics 63:256, 1979.
41. Cody CL, Baraff LJ, Cherry JD, Marcy SM, Manclark CR: Nature and rates of adverse reactions associated with DTP and DT immunizations in infants and children. Pediatrics 68:650, 1981.
42. Feery BJ, Finger WK, Kortus Z, Jones GA: The incidence and type of reactions to plain and adsorbed DTP vaccines. Aust Paediatr J 21:91, 1985.
43. Haire M, Dane DS, Dick G: Reactions to combined vaccines containing killed *Bordetella pertussis*. Med Officer 117:55, 1967.
44. Hannik CA: Major reactions after DTP-polio vaccination in the Netherlands. Symp Ser Immunobiol Stand 13:161, 1970.
45. Cherry JD, Brunell PA, Golden GS, Karzon DT: Report of the task force on pertussis and pertussis immunization—1988. Pediatrics 81(Suppl):939, 1988.
46. Alderslade R, Bellman MH, Rawson NSB, Ross EM, Miller DL: The National Childhood Encephalopathy Study. In: "DHSS. Whooping Cough." London: HMSO, 1981, p 79.
47. Miller DL, Ross EM, Alderslade R, Bellman MH, Rawson NSB: Pertussis immunization and serious acute neurological illness in children. Br Med J 282:1595, 1981.
48. Pollock TM, Miller E, Mortimer JY, Smith G: Post-vaccination symptoms following DTP and DT-vaccination. Dev Biol Stand 61:407, 1985.
49. Weiss AA, Hewlett EL: Virulence factors of *Bordetella pertussis*. Annu Rev Microbiol 40:661, 1986.
50. Munoz JJ, Bergman RK: "*Bordetella pertussis*. Immunology and Other Biological Properties." New York: Marcel Dekker, 1977.
51. Irons LI, Gorringe AR: Pertussis toxin: Production, purification, molecular structure and assay. In Wardlaw AC, Parton R (eds): "Pathogenesis and Immunity in Pertussis." Chichester: John Wiley and Sons, 1988, p 95.
52. Tamura M, Nogimori K, Murai S, Yajima M, Ito K, Katada T, Ui M, Ishii S: Subunit structure of islet-activating protein, pertussis toxin, in conformity with the A-B model. Biochemistry 21:5516, 1982.
53. Tamura M, Nogimori K, Yajima M, Ase K, Ui M: A role of the B-oligomer moeity of islet-activating protein, pertussis toxin, in the development of the biological effects on intact cells. J Biol Chem 258:6756, 1983.
54. Gill DM: Seven toxin peptides that cross cell membranes. In Jeljaszewicz J, Wadstrom T (eds): "Bacterial Toxins and Cell Membranes." New York: Academic Press, 1978, p 291.
55. Ui M: The multiple biological activities of pertussis toxin. In Wardlaw AC, Parton R (eds): "Pathogenesis and Immunity in Pertussis." Chichester: John Wiley and Sons, 1988, p 121.
56. Ui M, Nogimori K, Tamura M: Islet-activating protein, pertussis toxin: Subunit structure and mechanism for its multiple biological activities. In Sekura RD, Moss J, Vaughan M (eds): "Pertussis Toxin." New York: Academic Press, 1985, p 19.
57. Munoz JJ: Action of pertussigen (pertussis toxin) on the host immune system. In Wardlaw AC, Parton R (eds): "Pathogenesis and Immunity in Pertussis." Chichester: John Wiley and Sons, 1988, p 173.

58. Munoz JJ: Biological activities of pertussigen (pertussis toxin). In Sekura RD, Moss J, Vaughan M (eds): ''Pertussis Toxin.'' New York: Academic Press, 1985, p 1.
59. Tuomanen E, Weiss A, Rich R, Zak F, Zak O: Filamentous haemagglutinin and pertussis toxin promote adherence of *Bordetella pertussis* to cilia. Dev Biol Stand 61:197, 1985.
60. Tuomanen E: *Bordetella pertussis* adhesins. In Wardlaw AC, Parton R (eds): ''Pathogenesis and Immunity in Pertussis.'' Chichester: John Wiley and Sons, 1988, p 75.
61. Meade BD, Kind PD, Manclark CR: Altered mononuclear phagocyte function in mice treated with the lymphocytosis-promoting factor of *Bordetella pertussis*. Dev Biol Stand 61:63, 1985.
62. Becker EK, Kermode JC, Naccache PH, Yassin NR, Munoz JJ, Marsh ML, Huang CK, Sha'afi RI: Pertussis toxin as a probe of neutrophil activation. Fed Proc 45:2151, 1986.
63. Sugimoto M, Nakanishi Y, Otokawa M, Uchida N, Yasuda T, Sato H, Sato Y: Effect of *Bordetella pertussis* leucocytosis (lymphocytosis)-promoting factor (LPF) on the physical lympo-epithelial cell association studied with the use of an *in vitro* model of mouse thymus. J Immunol 130:2767, 1983.
64. Spangrude GJ, Braaten BA, Daynes RA: Molecular mechanisms of lymphocyte extravasation. 1. Studies of two selective inhibitors of lymphocyte recirculation. J Immunol 132:354, 1984.
65. Furman BL, Sidey FM, Smith M: Metabolic disturbances produced by pertussis toxin. In Wardlaw AC, Parton R (eds): ''Pathogenesis and Immunity in Pertussis.'' Chichester: John Wiley and Sons, 1988, p 147.
66. Munoz JJ, Arai H, Cole RL: Mouse protective and histamine-sensitizing activities of pertussigen and fimbrial haemagglutinin from *Bordetella pertussis*. Infect Immun 32: 243, 1981.
67. Robinson A, Irons LI: Synergistic effect of *Bordetella pertussis* lymphocytosis-promoting factor on protective activities of isolated *Bordetella* antigens in mice. Infect Immun 40:523, 1983.
68. Oda M, Cowell JL, Burstyn DG, Manclark CR: Protective activities of the filamentous haemagglutinin and lymphocytosis-promoting factor of *Bordetella pertussis* in mice. J Infect Dis 150:823, 1984.
69. Sato H, Sato Y: *Bordetella pertussis* infection of mice: Correlation of specific antibodies against two antigens, pertussis toxin and filamentous haemagglutinin with mouse protectivity in an intracerebral or aerosol challenge system. Infect Immun 46:415, 1984.
70. Robinson A, Ashworth LAE, Baskerville A, Irons LI: Protection against intranasal infection of mice with *Bordetella pertussis*. Dev Biol Stand 61:165, 1985.
71. Sato H, Ito A, Chiba J, Sato Y: Monoclonal antibody against pertussis toxin: Effect on toxin activity and pertussis infection. Infect Immun 46:422, 1984.
72. Sato Y, Izumiya K, Sato H, Cowell JL, Manclark CR: Role of antibody to leucocytosis-promoting factor haemagglutinin and to filamentous haemagglutinin in immunity to pertussis. Infect Immun 31:1223, 1981.
73. Oda M, Cowell JL, Burstyn DG, Thaib S, Manclark CR: Antibodies to *Bordetella pertussis* in human colostrum and their protective activity against aerosol infection of mice. Infect Immun 47:441, 1985.
74. Hewlett EL, Urban MA, Manclark CR, Wolff J: Extracytoplasmic adenylate cyclase of *Bordetella pertussis*. Proc Natl Acad Sci USA 73:1926, 1976.
75. Wolff J, Cook GH, Goldhammer AR, Berkowitz SA: Calmodulin activates prokaryotic adenylate cyclase. Proc Natl Acad Sci USA 77:3841, 1980.
76. Confer DL, Eaton JW: Phagocyte impotence caused by an invasive bacterial adenylate cyclase. Science 217:948, 1982.
77. Hewlett EL, Gordon VM: Adenylate cyclase toxin of *Bordetella pertussis*. In Wardlaw

AC, Parton R (eds): "Pathogenesis and Immunity in Pertussis." Chichester: John Wiley and Sons, 1988, p 193.
78. Weiss AA, Hewlett EL, Myers GA, Falkow S: Pertussis toxin and extracytoplasmic adenylate cyclase as virulence factors of *Bordetella pertussis*. J Infect Dis 150:219, 1984.
79. Brownlie RM, Coote JG, Parton R: Adenylate cyclase activity during phenotypic variation of *Bordetella pertussis*. J Gen Micro 131:27, 1985.
80. Brownlie, RM, Parton R, Coote JG: The effect of growth conditions on adenylate cyclase activity and virulence-related properties of *Bordetella pertussis*. J Gen Microbiol 131:7, 1985.
81. Hewlett EL: Adenylate cyclases of *Bordetella pertussis* and *Bacillus anthracis*: Mediators of bacterial toxin action in disease. Ann Intern Med 101:653, 1984.
82. Novotny P, Chubb AP, Cownley K, Monterez JA: Adenylate cyclase activity of a 68,000 molecular-weight protein isolated from the outer membrane of *Bordetella bronchiseptica*. Infect Immun 50:199, 1985.
83. Bordet J, Gengou O: L'endotoxine coquelucheuse. Ann Inst Pasteur 23:415, 1909.
84. Cowell JL, Hewlett EL, Manclark CR: Intracellular localization of the dermonecrotic toxin of *Bordetella pertussis*. Infect Immun 25:896, 1979.
85. Livey I, Wardlaw AC: Production and properties of *Bordetella pertussis* heat-labile toxin. J Med Microbiol 17:91, 1984.
86. Onoue K, Kitagawa M, Yamamura Y: Chemical studies on cellular components of *Bordetella pertussis*. III. Isolation of highly potent toxin from *Bordetella pertussis*. J Bacteriol 86:648, 1963.
87. Nakase T, Takatsu K, Tateishi M, Sekiya K, Kasuga T: Heat-labile toxin of *Bordetella pertussis* purified by preparative acrylamide gel electrophoresis. Jpn J Microbiol 13:359, 1969.
88. Iida T, Okonogi T: Lienotoxicity of *Bordetella pertussis* in mice. J Med Microbiol 4:51, 1971.
89. Sekiya K, Munoz JJ, Nakase Y: Effect of dermonecrotic toxin of *Bordetella pertussis* on the spleen of CFW and C57 BL/10 ScN mice. Microbiol Immunol 26:971, 1982.
90. Zhang YL, Sekura RD: Isolation and characterization of heat labile toxin (HLT) from *Bordetella pertussis*. Fed Proc 44:674, 1985.
91. Kume K, Nakai T, Samejima Y, Sugimoto C: Properties of dermonecrotic toxin prepared from sonic extracts of *Bordetella bronchiseptica*. Infect Immun 52:370, 1986.
92. Endoh M, Amitani M, Nakase Y: Purification and characterization of heat-labile toxin from *Bordetella bronchiseptica*. Microbiol Immunol 30:659, 1986.
93. Kurokawa M, Ishida S, Asakawa S: Attempts at analysis of toxicity of pertussis vaccine. II. Quantitative determination of the heat-labile toxin by skin reaction. Jpn J Med Sci Biol 22:293, 1969.
94. Nakase Y, Endoh M: *Bordetella* heat-labile toxin: Further purification, characterization and mode of action. Dev Biol Stand 61:93, 1985.
95. Nakase Y, Endoh M: Heat-labile toxin of *Bordetella pertussis*. In Wardlaw AC, Parton R (eds): "Pathogenesis and Immunity in Pertussis." Chichester: John Wiley and Sons, 1988, p 211.
96. Endoh M, Amitani M, Nakase Y: Effect of purified heat-labile toxin of *Bordetella bronchiseptica* on the peripheral blood vessels in guinea pig or suckling mice. Microbiol Immunol 30:1327, 1986.
97. Munoz JJ: Protein toxins from *Bordetella pertussis*. In Kadis S, Montie TC, Ajl SJ (eds): "Microbial Toxins, Vol IIa. Bacterial Protein Toxins." New York: Academic Press, 1971, p 271.

98. Goldman WE, Klapper DG, Baseman JB: Detection, isolation and analysis of a released *Bordetella pertussis* product toxic to cultured tracheal cells. Infect Immun 36:782, 1982.
99. Goldman WE: Tracheal cytotoxin of *Bordetella pertussis*. In Wardlaw AC, Parton R (eds): "Pathogenesis and Immunity in Pertussis." Chichester: John Wiley and Sons, 1988, p 231.
100. Cookson BT, Goldman WE: Tracheal cytotoxin: A conserved virulence determinant of all *Bordetella* species. J Cell Biochem IIb(Suppl):124, 1987.
101. Chaby R, Caroff M: Lipopolysaccharides of *Bordetella pertussis* endotoxin. In Wardlaw AC, Parton R (eds): "Pathogenesis and Immunity in Pertussis." Chichester: John Wiley and Sons, 1988, p 247.
102. Le Dur A, Chaby R, Szabo L: Isolation of two protein-free and chemically different lipopolysaccharides from *Bordetella pertussis* phenol-extracted endotoxin. J Bacteriol 143:78, 1980.
103. Ashworth LAE, Fitzgeorge RB, Irons LI, Morgan CP, Robinson A: Rabbit nasopharyngeal colonization by *Bordetella pertussis*: The effects of immunization on clearance and on serum and nasal antibody levels. J Hyg (Camb) 88:475, 1982.
104. Ashworth LAE, Robinson A, Irons LI, Morgan CP, Isaccs D: Antigens in whooping cough vaccine and antibody levels induced by vaccination of children. Lancet 2:878, 1983.
105. Nagel J, DeGraaf S, Schijf-Evers D: Improved serodiagnosis of whooping cough caused by *Bordetella pertussis* by determination of IgG and anti-LPF antibody levels. Dev Biol Stand 61:325, 1985.
106. Winsnes R: Serological responses to pertussis. In Wardlaw AC, Parton R (eds): "Pathogenesis and Immunity in Pertussis." Chichester: John Wiley and Sons, 1988, p 283.
107. Arai H, Sato Y: Separation and characterization of two distinct haemagglutinins contained in purified leucocytosis-promoting factor from *Bordetella pertussis*. Biochim Biophys Acta 444:765, 1976.
108. Arai H, Munoz JJ: Fimbrial haemagglutinin in stationary and static cultures of *Bordetella pertussis*. Infect Immun 25:704, 1979.
109. Cowell JL, Sato Y, Sato H, an der Lan B, Manclark CR: Separation, purification and properties of the filamentous haemagglutinin and the leucocytosis-promoting factor-haemagglutinin from *Bordetella pertussis*. In Robbins JB, Hill JC, Sadoff JC (eds): "Seminars in Infectious Disease, Vol. 4. Bacterial Vaccines; International Symposium 1980." New York: Thieme-Stratton, 1982, p 371.
110. Imaizumi A, Suzuki Y, Ginnaga A, Sakuma S, Sato Y: A new culturing method for the production of filamentous hemagglutinin of *Bordetella pertussis*. J Microbiol Methods 2:339, 1984.
111. Irons LI, MacLennan AP: Substrate specificity and the purification by affinity combination methods of two *Bordetella pertussis* haemagglutinins. In Manclark CR, Hill JC (eds): "International Symposium on Pertussis." Washington, DC: US DHEW Pub. No. (NIH)79-1830, 1979, p 338.
112. Irons LI, Ashworth LAE, Wilton-Smith P: Heterogeneity of the filamentous haemagglutinin of *Bordetella pertussis* studied with monoclonal antibodies. J Gen Microbiol 129: 2769, 1983.
113. Morse SI, Morse JH: Isolation and properties of the leucocytosis- and lymphocytosis-promoting factor of *Bordetella pertussis*. J Exp Med 143:1483, 1976.
114. Sato Y, Izumiya K, Oda MA, Sato H: Biological significance of *Bordetella pertussis* fimbriae or hemagglutinin: A possible role of the fimbriae or hemagglutinin for pathogenesis and antibacterial immunity. In Manclark CR, Hill JC (eds): "International Sym-

posium on Pertussis.'' Washington, DC: US DHEW Pub. No. (NIH)79-1830, 1979, p 51.

115. Blom J, Hansen GA, Poulsen FM: Morphology of cells and haemagglutinogens of *Bordetella* species: Resolution of substructural units in fimbriae of *Bordetella pertussis*. Infect Immun 42:308, 1983.
116. Ashworth LAE, Irons LI, Dowsett AB: Antigenic relationship between serotype-specific agglutinogen and fimbriae of *Bordetella pertussis*. Infect Immun 37:1278, 1982.
117. Ashworth LAE, Dowsett AB, Irons LI, Robinson A: The location of surface antigens of *Bordetella pertussis* by immuno-electron microscopy. Dev Biol Stand 61:143, 1985.
118. Anderson EK: Serological studies on *H. pertussis, H. parapertussis* and *H. bronchiseptica*. Acta Pathol Microbiol Scand 33:202, 1953.
119. Eldering G, Hornbeck C, Baker J: Serological study of *Bordetella pertussis* and related species. J Bacteriol 74:133, 1957.
120. Preston NW, Surapatana N, Carter EJ: A reappraisal of serotype factors 4, 5 and 6 of *Bordetella pertussis*. J Hyg (Camb) 88:39, 1982.
121. Carter EJ, Preston NW: Association between *Bordetella pertussis* agglutinogen 2 and fimbriae. J Med Microbiol 18:87, 1984.
122. Irons LI, Ashworth LAE, Robinson A: Release and purification of fimbriae from *Bordetella pertussis*. Dev Biol Stand 61:153, 1985.
123. Zhang JM, Cowell JL, Steven AC, Carter PH, McGrath PP, Manclark CR: Purification and characterization of fimbriae isolated from *Bordetella pertussis*. Infect Immun 48:422, 1985.
124. Cowell JL, Zhang JM, Urisu A, Suzuki A, Steven AC, Liu T, Liu TY, Manclark CR: Purification and characterization of serotype 6 fimbriae from *Bordetella pertussis* and comparison of their properties with serotype 2 fimbriae. Infect Immun 55:916, 1987.
125. Eldering G, Holwerda J, Baker J: *Bordetella pertussis* culture having only species factor 1. J Bacteriol 91:1759, 1966.
126. Eldering G, Holwerda J, Baker J: Mouse-protective properties of *Bordetella pertussis* serotypes in passive tests. J Bacteriol 93:1758, 1967.
127. Zhang JM, Cowell JL, Steven AC, Manclark CR: Purification of serotype 2 fimbriae of *Bordetella pertussis* and their identification as a mouse protective antigen. Dev Biol Stand 61:173, 1985.
128. Shukuda Y: Manufacturing methods for the Japanese acellular vaccines. In The Interagency Group to Monitor Vaccine Development, Production and Usage: ''Workshop on Acellular Pertussis Vaccines.'' Washington, DC: US DHHS & PHS, 1986, p 23.
129. Gorringe A, Ashworth LAE, Irons LI, Robinson A: Effect of monoclonal antibodies on the adherence of *Bordetella pertussis* to Vero cells. FEMS Microbiol Lett 26:5, 1985.
130. Redhead K: An assay of *Bordetella pertussis* adhesion to tissue-culture cells. J Med Microbiol 19:99, 1985.
131. Tuomanen E, Weiss A: Characterization of two adhesins of *Bordetella pertussis* for human ciliated respiratory epithelial cells. J Infect Dis 152:118, 1985.
132. Urisu A, Cowell JL, Manclark CR: Filamentous haemagglutinin has a major role in mediating adherence of *Bordetella pertussis* to human WiDr cells. Infect Immun 52:695, 1986.
133. Parker CD, Branes LV, Armstrong SK, Frank DW, Cole A: Cell surface antigens of *Bordetella pertussis*. Dev Biol Stand 61:123, 1985.
134. Robinson A, Hawkins DC: Structure and biological properties of solubilized envelope proteins of *Bordetella pertussis*. Infect Immun 39:590, 1983.
135. Coote JG, Brownlie RM: Genetics of virulence of *Bordetella pertussis*. In Wardlaw AC,

Parton R (eds): ''Pathogenesis and Immunity in Pertussis.'' Chichester: John Wiley and Sons, 1988, p 39.
136. Glaser P, Ladant D, Sezer O, Pichot F, Ullmann A, Danchin A: The calmodulin-sensitive adenylate cyclase of *Bordetella pertussis*: Cloning and expression in *Escherichia coli*. Mol Microbiol 2:19, 1988.
137. Pillemer L, Blum L, Lepow IH: Protective antigen of *Haemophilus pertussis*. Lancet 1:1257, 1954.
138. Felton HM, Verwey WF: The epidemiological evaluation of a non-cellular pertussis antigen. Pediatrics 16:637, 1955.
139. Pennell RB, Thiele EH: Studies on the fractionation of *Haemophilus pertussis* extracts. J Immunol 66:627, 1951.
140. Weihl C, Riley HD, Lapin JH: Extracted pertussis antigen. Am J Dis Child 106:210, 1963.
141. Millman I, Schuchardt LF, Gray A: Purification and concentration of pertussis protective antigen. Round Table Conf Pertussis Immun, 1962, p 43 (unpublished).
142. Schwick HG, Helting TB, Zwister O: Bacterial split vaccines. In Mizrahi A, Hertman I, Klinberg MA, Kohn A (eds): ''New Developments With Human and Veterinary Vaccines.'' New York: Alan R. Liss, 1980, p 143.
143. Garcia-Sanz JA, Ruiz-Puente J, Jimenez-Paredes J, Gonzalez-Pacheco M, Villalva-Dosada H: Comparative biological activities of whole-cell pertussis vaccine and a new acellular preparation. Vaccine 3:23, 1985.
144. Christodoulides M, Sidey FM, Parton R, Stewart-Tull DES: Acellular pertussis vaccine prepared by a simple extraction and toxoiding procedure. Vaccine 5:199, 1987.
145. Chazano M, Yoshida I, Konobe T, Fukai K: The purification and characterization of an acellular pertussis vaccine. J Biol Stand 16:83, 1988.
146. Robinson A, Ashworth LAE: Acellular and defined-component vaccines against pertussis. In Wardlaw AC, Parton R (eds): ''Pathogenesis and Immunity in Pertussis.'' Chichester: John Wiley and Sons, 1988, p 399.
147. Isawa S, Ishida S, Asakawa S, Akama K: A curious histamine-sensitizing activity shown by the newly-developed Japanese acellular pertussis vaccine. Dev Biol Stand 61:453, 1985.
148. Nicosia A, Perugini M, Franzini C, Casagli MC, Borri MG, Antoni G, Almoni M, Neri P, Ratti G, Rappuoli R: Cloning and sequencing of the pertussis toxin genes: Operon structure and gene duplication. Proc Natl Acad Sci USA 83:4631, 1986.
149. Ginnaga A, Morokuma K, Aihara K, Sakou M, Suzuki Y, Imaizumi A, Sato Y, Sato H, Ueda K, Kuno-Sakai H, Kimura M: Characterization and a clinical study on a Kaketsuken's acellular pertussis vaccine produced by combination of column purified pertussis toxin and filamentous haemagglutinin. Abstract from the Fifth International Symposium on Pertussis, Copenhagen, Denmark, Sept. 22–23, 1988; unpublished.
150. Edwards KM, Lawrence E, Wright PF: Diphtheria, tetanus and pertussis vaccine: A comparison of the immune response and adverse reactions to conventional and acellular pertussis components. Am J Dis Child 140:867, 1986.
151. Lewis K, Cherry JD, Holroyd HJ, Baker LR, Dudenhoeffer FE, Robinson RG: A double-blind study comparing an acellular pertussis-component diphtheria tetanus pertussis vaccine with a whole-cell pertussis-component diphtheria tetanus pertussis vaccine in 18-month-old children. Am J Dis Child 140:872, 1986.
152. Anderson EL, Belshe RB, Bartram J, Gurwith M, Hung P, Levner M, Vernon SK: Clinical and serological responses to acellular pertussis vaccine in infants and young children. Am J Dis Child 141:949, 1987.
153. Pichichero ME, Badgett JT, Rodgers GC, McLinn S, Trevino-Scatterday B, Nelson JD:

Acellular pertussis vaccine: Immunogenicity and safety of an acellular pertussis vs. a whole-cell pertussis vaccine combined with diphtheria and tetanus toxoids as a booster in 18- to 24-month-old children. Pediatr Infect Dis 6:352, 1987.

154. Anderson EL, Belshe RB, Bartram J: Differences in reactogenicity and antigenicity of acellular and standard pertussis vaccines combined with diphtheria and tetanus in infants. J Infect Dis 157:731, 1988.
155. Dekker CL, Reisinger KS, Congeni BL, Cherry JD, Stout MG, Study Group: A randomized double-blind study of the safety and immunogenicity of an acellular pertussis-DT vaccine compared to whole cell pertussis DT in 17–24-month-old children. Pediatr Res 23:367A, 1988.
156. Noble GR, Bernier RH, Esber EC, Hardegree MC, Hinman AR, Klein D, Saah AJ: Acellular and whole-cell pertussis vaccines in Japan; report of a visit by US scientists. JAMA 257:1351, 1987.
157. Sato H, Sato Y: Anti-pertussis toxin IgG and anti-filamentous haemagglutinin IgG production in children immunized with pertussis acellular vaccine and comparison of those titres with the sera of pertussis convalescent children. Dev Biol Stand 61:367, 1985.
158. Aoyama T, Hagiwara S, Murase Y, Kato T, Iwata T: Adverse reactions and antibody responses to acellular pertussis vaccine. J Pediatr 109:925, 1986.
159. Interagency Group to Monitor Vaccine Development, Production and Usage: ''Workshop on Acellular Pertussis Vaccines.'' Washington, DC: US DHHS & PHS, 1986.
160. Kimura M, Hikino N: Results with a new DTP vaccine in Japan. Dev Biol Stand 61:545, 1985.
161. Isomura S, Suzuki S, Sato Y: Clinical efficacy of the Japanese acellular pertussis vaccine after intrafamiliar exposure to pertussis patients. Dev Biol Stand 61:531, 1985.
162. Aoyama T, Murase Y, Kato T, Iwata T: Efficacy of an acellular pertussis vaccine in Japan. J Pediatr 107:180, 1985.
163. Isomura S: Efficacy and safety of acellular pertussis vaccine in Aichi Prefecture, Japan. Pediatr Infect Dis 7:258, 1988.
164. Aoyama T, Murase Y, Gonda T, Iwata T: Type-specific efficacy of acellular pertussis vaccine. Am J Dis Child 142:40, 1988.
165. Hedenskog S, Granstrom M, Olin P, Tiru M, Sato Y: A clinical trial of a monocomponent pertussis toxoid vaccine. Am J Dis Child 141:844, 1987.
166. Blennow M, Granstrom M, Jaatmaa E, Olin P: Primary immunization of infants with an acellular pertussis vaccine in a double-blind randomized clinical trial. Pediatrics 82:293, 1988.
167. Ad Hoc Group for the Study of Pertussis Vaccines: Placebo-controlled trial of two acellular pertussis vaccines in Sweden—Protective efficacy and adverse reactions. Lancet 1:955, 1988.
168. Kimura M, Kuno-Sakai H: Acellular pertussis vaccines and fatal infections. Lancet 1:881, 1988.
169. Reinert P, Amzallag M, Danve B, Arminjon F, Quentin-Millet MJ, Cadoz M: Reactogenicity and immunogenicity of an acellular pertussis vaccine candidate APV composed of glutaraldehyde-detoxified pertussis toxin. Comparison of 3 dosages in 3–6 month old infants. Pediatr Res 23:379A, 1988.
170. Edwards KM, Bradley R, Van Savage J, Palmer P, Wright PF: Evaluation of a new acellular pertussis vaccine. Pediatr Res 23:367A, 1988.
171. Sekura RD, Schneerson R, Robbins JB: Safety and antigenicity of a pertussis toxoid vaccine evaluated in adult volunteers. In Interagency Group to Monitor Vaccine Development, Production and Usage: ''Workshop on Acellular Pertussis Vaccines.'' Washington, DC: US DHHS & PHS, 1986, p 135.

172. Thakrar N, Siber GR, Yancey BA, Todd CW, Cohen ND, Herzog LW, Bithoney WG, Roskey AM, Lowe CU: Safety and immunogenicity of pertussis toxoid vaccine in 18-month-old children. Pediatr Res 23:367A, 1988.
173. Rutter DA, Ashworth LAE, Day A, Funnell S, Lovell F, Robinson A: Trial of a new acellular pertussis vaccine in healthy adult volunteers. Vaccine 6:30, 1988.
174. Locht C, Barstad PA, Coligan JE, Mayer L, Munoz JJ, Smith SG, Keith JM: Molecular cloning of pertussis toxin genes. Nucleic Acids Res 14:3251, 1986.
175. Arico B, Rappouli R: *Bordetella parapertussis* and *Bordetella bronchiseptica* contain transcriptionally silent pertussis toxin genes. J Bacteriol 169:2847, 1987.
176. Marchitto KS, Smith SG, Locht C, Keith JM: Nucleotide sequence homology to pertussis toxin gene in *B. bronchiseptica* and *B. parapertussis*. Infect Immun 55:497, 1987.
177. Gross R, Rappuoli R: Positive regulation of pertussis toxin expression. Proc Natl Acad Sci USA 85:3913, 1988.
178. Reiser J, Germanier R: Expression in *Escherichia coli* of a *Bordetella pertussis* toxin-coding region. Cold Spring Harbor Symp Quant Biol 51:235, 1986.
179. Nicosia A, Bartoloni A, Perugini M, Rappuoli R: Expression and immunological properties of the five subunits of pertussis toxin. Infect Immun 55:963, 1987.
180. Barbieri JT, Rappuoli R, Collier RJ: Expression of the S1 catalytic subunit of pertussis toxin in *Escherichia coli*. Infect Immun 55:1321, 1987.
181. Black WJ, Falkow S: Construction and characterization of *Bordetella pertussis* toxin mutants. Infect Immun 55:2465, 1987.
182. Locht C, Cieplak W, Marchitto KS, Sato H, Keith JM: Activities of complete and truncated forms of pertussis toxin subunits S1 and S2 synthesized by *Escherichia coli*. Infect Immun 55:2546, 1987.
183. Locht C, Keith JM: Pertussis toxin gene: Nucleotide sequence and genetic organization. Science 232:1258, 1986.
184. Burnette WN, Mar VL, Cieplak W, Morris CF, Kaljot K, Sato H, Keith JM: Toward development of a recombinant pertussis vaccine. J Cell Biochem (Suppl)12B:4, 1988.
185. Burnette WN, Mar VL, Cieplak W, Morris CF, Kaljot KT, Marchitto KS, Sachdev RK, Locht C, Keith JM: Direct expression of *Bordetella pertussis* toxin subunits to high levels in *Escherichia coli*. Biotechnology 6:699, 1988.
186. Runeberg-Nyman K, Engstrom O, Lofdahl S, Ylostalo S, Sarvas M: Expression and secretion of pertussis toxin subunit S1 in *Bacillus subtilis*. Microbiol Pathol 3:461, 1987.
187. Reiser J, Friedman RL, Germanier R: *Bordetella pertussis* filamentous haemagglutinin gene: Molecular cloning of a potential coding sequence. Dev Biol Stand 61:265, 1985.
188. Mattei D, Pichot F, Bellalou J, Mercereau-Puijalon O, Ullman A: Molecular cloning of a coding sequence of *Bordetella pertussis* filamentous haemagglutinin gene. FEMS Microbiol Lett 37:73, 1986.
189. Brown DR, Parker CD: Cloning of the filamentous hemagglutinin of *Bordetella pertussis* and expression in *Escherichia coli*. Infect Immun 55:154, 1987.
190. Livey I, Duggleby CJ, Robinson A: Cloning and nucleotide sequence analysis of the serotype 2 fimbrial subunit gene of *Bordetella pertussis*. Mol Microbiol 1:203, 1987.
191. Brownlie RM, Coote JG, Parton R: Complementation of mutations in *Escherichia coli* and *Bordetella pertussis* by *B. pertussis* DNA clones in a broad-host-range cosmid vector. J Gen Microbiol 132:3221, 1986.
192. Stibitz S, Weiss AA, Falkow S: Genetic analysis of a region of the *Bordetella pertussis* chromosome encoding filamentous hemagglutinin and the pleiotropic regulatory locus *vir*. J Bacteriol 170:2904, 1988.
193. Shareck F, Cameron J: Cloning of *Bordetella pertussis* outer membrane proteins in *Escherichia coli*. Dev Biol Stand 61:273, 1985.

194. Charles IG, Dougan G, Pickard D, Chatfield S, Smith M, Novotny P, Morrissey P, Fairweather N: Molecular cloning and analysis of P.69 a *vir*-controlled protein from *Bordetella pertussis*. Abstract from the Fifth International Symposium on Pertussis, Copenhagen, Denmark, Sept. 22–23, 1988; unpublished.
195. Black WJ, Munoz JJ, Peacock MG, Schad PA, Cowell JL, Burchall JJ, Lim M, Kent A, Steinman L, Falkow S: ADP-ribosyltransferase activity of pertussis toxin and immunomodulation by *Bordetella pertussis*. Science 240:656, 1988.
196. Bartolini A, Pizza MG, Rappuoli R: Recombinant molecules and the development of new vaccines against whooping cough. Abstract from the Fifth International Symposium on Pertussis, Copenhagen, Denmark, Sept. 22–23, 1988; unpublished.
197. Askelop P, Rodmalm K, Abens J, Unden A, Bartfai T: Use of synthetic peptides to map antigenic sites of *Bordetella pertussis* toxin subunit S1. J Infect Dis 157:738, 1988.
198. Stewart-Tull DES: Immunopotentiating conjugates. Vaccine 3:40, 1985.
199. Jerne NK: Towards a network theory of the immune system. Ann Immunol 125C:373, 1974.
200. Finberg RW, Ertl H: The use of antiidiotypic antibodies as vaccines against infectious agents. CRC Crit Rev Immunol 7:269, 1987.
201. Parker CD: Monoclonal antibodies to *Bordetella pertussis*. In Parker CD (ed): ''Monoclonal Antibodies Against Bacteria.'' London: Academic Press, 1986, Vol. III, p 165.

Bacterial Vaccines, pages 201–222


# Vaccines Against Leprosy

**Duncan E.S. Stewart-Tull**

Department of Microbiology, University of Glasgow, Glasgow, G12 8QQ
Scotland



## I. INTRODUCTION

In a recent comprehensive text on vaccines it was significant that leprosy was not included [1], even though the latest technical report from the WHO Expert Committee on Leprosy [2] states that there are currently an estimated 10–12 million cases in the world. Seventy-five percent of the world's population live in countries where the disease is endemic, so it is surprising that there is an overall estimated prevalence of 1–5 cases per 1,000 individuals. The number of registered cases has increased in the past 20 years, due in part to the efficiency and vigilance of the agencies involved in the control of leprosy: in 1966, 2,850,000 cases were recorded; in 1976, 3,600,000; and in 1985, 5,400,000.

In 1984, I concluded that it would be worthwhile for pharmaceutical companies to develop a combined leprosy-tuberculosis vaccine [3]. The past

**TABLE I. Effectiveness of *M. bovis* BCG in Leprosy Field Trials**

| Country | Incidence of lepromatous leprosy | Mean value of BCG vaccine effectiveness (%) | Reference |
|---|---|---|---|
| Uganda | 8.0 | 80.0 | 7–9 |
| Burma | 70.0 | 17.0 | 10 |
| New Guinea | 10.0 | 40.0 | 11 |
| Malawi | ? | 57.0 | 12 |
| South India | ? | 23.0 | 13 |

5 years have seen considerable advances in the study of both the immunological mechanisms involved in clinical leprosy and the antigens of the *Mycobacterium leprae* organism and in manipulation of the genes into other organisms. These advances will assuredly lead to the development of new vaccines, urgently needed in the fight against leprosy, in the not-too-distant future. This is a common goal of mycobacteriologists and funding agencies worldwide, and it is to their credit that the state-of-the-art technology has been made freely available to all.

## II. ZERO-GENERATION VACCINES

For many years general vaccine technology has relied on the use of killed or live whole cells in which the protective epitopes were ill defined. These vaccines contained organisms that were easily cultivable in large quantities and were thus comparatively inexpensive to produce. With regard to leprosy, a whole-cell vaccine was not initially envisaged because there is no indisputable evidence that cultivation of *M. leprae* on a laboratory medium has been achieved [4–6]. There is no doubt that an early success in this venture would have advanced the control of leprosy. Other lines of research may inevitably lead to a different type of vaccine, as discussed later in this chapter, but growth of the organism still remains as an important challenge to be accepted by bacteriologists.

The use of *M. bovis* bacillus Calmette-Guérin (BCG) as a vaccine in leprosy endemic areas has had varied success (Table I). This may be due to 1) previous exposure to local strains of *M. leprae* and 2) various geographical and environmental factors, for example, exposure to sunlight and skin pigmentation, host genetics, and prevalence of "atypical" mycobacteria in the environment prior to the administration of BCG [12,14].

It was suggested by others that the strain of *M. bovis* BCG used in the preparation of different commercial vaccines might be variable in regard to the array of protective antigens presented to the immune system. The Copenhagen vaccine of the Statens Serum Institut was used in the South India trial and the Glaxo vaccine was used in Uganda and Burma. Consequently, the variations in the protective efficacy found in the different areas might have been due to the differences in the BCG vaccine strain. Stanford [15] stated that there was no real evidence to support this contention, and most workers no longer implicate the variable responses of the vaccines for the different levels of protection.

### A. Uganda—Teso District

The first large-scale trial of the use of BCG vaccination was conducted in the Teso District of Uganda by the Leprosy Committee of the Medical Research Council in London [7]. Prior to the trial, the prevalence of leprosy in Uganda was assessed: only 8% of the total cases were shown to have lepromatous leprosy. After consideration of the possible BCG adverse effects, all children were tuberculin tested by the Heaf method. Children with negative or weakly positive tuberculin reactions and who showed no leprosy skin lesions were assigned at random to the test or unvaccinated control groups.

Each child in the test group (8,149 children) was vaccinated intradermally (i.d.). with 0.1 ml of the Glaxo BCG reconstituted, freeze-dried vaccine. There were 8,152 children in the unvaccinated group. In all, there were 14,024 under the age of 10 years and 1,361 under the age of 1 year. The eitelas (rural population units) were visited periodically for the long-term monitoring of the children for signs of leprosy. Some 2 years later, the incidence of leprosy was 11.0/1,000 in the unvaccinated group and 2.2/1,000 in the vaccinated group—an 80% reduction, attributable to BCG, in an area of low incidence of lepromatous leprosy. A less-vigilant surveillance was continued between 1970 and 1975; during this period, one initially tuberculin-positive child, one BCG-vaccinated child, and six children in the control group developed leprosy [9].

### B. Northern Malawi—Karonga District

In the Karonga District of northern Malawi, mass BCG vaccination with the Glaxo freeze-dried vaccine was given intradermally to children under the age of 15 years, commencing in 1974. The effectiveness of the scheme in protecting against leprosy was assessed by case control methods. Fine et al. [12] concluded that the effectiveness of the Glaxo freeze-dried BCG against leprosy was about 57% in northern Malawi, with a rate of 62% in males and 52% in females; ages ranged from 1 to 34 years. The total number of cases in the BCG-vaccinated group of 45,619 was 23; 47 were reported in the control group of 34,182. These results and those from the Uganda trial

TABLE II. Comparison of BCG Vaccines Used in Burma Trial

| | Protective efficacy (%) | | |
|---|---|---|---|
| BCG vaccine | Males | Females | Total |
| F20J | 16.0 | 5.0 | 11.0 |
| F53H | 35.0 | 23.0 | 30.0 |

suggest that BCG was an effective prophylactic against leprosy in East and Central Africa.

### C. Burma—Mandalay/Sagaing

A controlled trial was conducted in an area of Burma where the incidence of lepromatous leprosy was 70%, namely, Singu Township, Mandalay District, and Shwebo Township, Sagaing Division; both towns are adjacent to the Irrawaddy River. The initial intake in 1964 involved 13,066 children, aged 0 to 14 years, who were given 0.1 ml of the reconstituted Glaxo BCG vaccine i.d.; another 13,176 children formed the control group. After 9 years, the BCG-vaccinated children, 0–4 years of age at intake, showed a lower incidence of leprosy than did children in the control group. However, the BCG did not effectively protect household contacts nor the children in the 5–14-year age group. It did appear however that BCG-vaccinated children had a less serious form of leprosy than the nonvaccinated and tuberculin-negative responders [10].

Annual checks continued until 1978–1979. At this time, 663 cases of leprosy had been recorded in the BCG-vaccinated group and 831 cases in the control group; the rates were 4.4 and 5.5 per 1,000 per year respectively. The observed protection level was only 20.4%; when analyzed by sex it was calculated to be 26% for males and 14% for females [16].

Two batches of BCG vaccine were used and there was an obvious difference between them (Table II). The F2OJ vaccine was estimated to have 3.8 $\times 10^6$ viable cells ml$^{-1}$ and the F53H vaccine, 4.5 $\times$ $10^6$ viable cells ml$^{-1}$; both were reported to be suitable for routine tuberculosis vaccination. Nevertheless, the F53H batch had a protective efficacy against leprosy—40% among infants and 30% among children under 15 years of age.

The authors cautioned that 1) some of the children may have been infected at intake and 2) the BCG may have accelerated the disease in the early years. As Kinnear Brown and Stone [7] pointed out, the development of clinical leprosy must reach a stage when vaccination would no longer arrest the process. The BCG was not deemed to produce a significant level of protection, and Lwin et al. [16] stressed the need to develop a more effective antileprosy vaccine.

### D. South India—Chingleput

The Copenhagen BCG was tested for protective activity against tuberculosis and leprosy in a trial carried out in South India. At the end of 10 years, the protection afforded by the BCG vaccination was low, about 23%. Mitchison [17] noted that strains of *M. tuberculosis* from this region possessed low virulence levels against guinea pigs, which could be a possible reason for the poor activity of BCG, as individuals may have been exposed to the less-virulent *M. tuberculosis*. However, it is possible that environmental mycobacteria were responsible for the vaccine failure. In South India, repeated contact with *M. avium,* an environmental organism, is a more likely cause, since this will act to stimulate a cell-mediated immune response, as found in primary tuberculosis. Further exposure to BCG will enhance the Koch-type response, and as Stanford [15] suggests, the individual would still be susceptible to adult tuberculosis and to leprosy. A similar explanation was proposed for the failure of BCG in Burma, but the environmental organism here was *M. scrofulaceum* [15].

## III. FIRST-GENERATION VACCINES

As *M. leprae* could not be grown on a laboratory medium, the only real advances in leprosy research came from organisms grown in animals. Initially Shepard [18] collected secretions from patients with lepromatous leprosy and inoculated these into the footpads of CFW strain white mice. After periods ranging from 3 to 12 months, the *M. leprae* multiplied and were detected in granulomas in the feet. The technique was advanced by the use of thymectomy with total-body irradiation before the mice were inoculated with *M. leprae* [19]; this process yielded $9.8 \times 10^9$ *M. leprae* after 19 months. The idea was extended by the use of congenitally athymic *nunu* mice [20], in which $1 \times 10^{5.3}$ *M. leprae* were found in the liver and spleen after 10 months. The bacterial yields were still too small to produce an *M. leprae* whole-cell vaccine, and it was not until 1971 that the multiplication of the organism in the nine-banded armadillo, *Dasypus novemcinctus,* was demonstrated [21–23]. *M. leprae* organisms were inoculated into the armadillo, and after 15 months a disseminated infection was obvious with lesions in the liver, spleen, lung, heart, stomach, spinal cord, and brain [24]. An infected animal yields approximately $1 \times 10^{10}$ organisms $g^{-1}$ (wet weight) of tissue [25].

Subsequently, it has been shown that other species are susceptible to *M. leprae* inoculation: the eight-banded armadillo *(D. sabanicola)* [26], the seven-banded armadillo *(D. hybridus)* [27], the hairy armadillo *(Chaetophractus villosus)* [23], and the hedgehog *(Erinaceus europaeus)* [28]. The transmission of leprosy has now been reported in the chimpanzee *(Pan trog-*

lodytes) [29], mangabey monkey [30], African green monkey *(Cercopithecus aethiops)* [31], and the rhesus monkey *(Macaca mulatta)* [31].

### A. Armadillo-Derived *M. leprae*

In an initial study [32], healthy individuals were inoculated with a total of $2 \times 10^8$ killed armadillo-derived *M. leprae*. Six of seven showed a positive delayed-hypersensitivity reaction, mean value 10.4 mm in diameter, against the *M. leprae* ultrasonicated antigen (MLSA). Subsequently, the armadillo-derived *M. leprae* organisms were used in a heat-killed, whole-cell vaccine in a trial in healthy human volunteers living in Norway, a leprosy nonendemic country. Although the 31 vaccinees (12 males and 19 females) were volunteers, ethical clearance was obtained from the National Norwegian Drug Agency, the Radium Hospital, Oslo, and the WHO Secretariat Committee on Research Involving Human Subjects in this carefully controlled trial [33].

The vaccinees, aged between 23 and 28 years, were pretested with purified protein derivative (PPD) and seven to nine volunteers were placed in groups A–D with varying degrees of reaction to PPD in each group. Twenty-eight days later, group A volunteers were skin-retested with coded preparations of PPD and MLSA. After reading the reactions, the volunteers were injected with 0.1 ml vaccine i.d. three times at the intersection of an equilateral triangle of 3 cm; a total of $1.5 \times 10^7$ *M. leprae* whole organisms were injected. The injection sites were examined 1 month later; acceptable reactions were observed, so the B group volunteers were injected with $5 \times 10^7$ *M. leprae*. This procedure of successively inoculating groups C and D, with $1.5 \times 10^8$ and $5 \times 10^8$, respectively, continued at monthly intervals if the reactions to the inoculum were acceptable in the previously inoculated group [33]. Three months postvaccination, the volunteers were skin-tested once again with the coded preparations of PPD and MLSA.

The pre- and postvaccination delayed-type hypersensitivity responses did not differ significantly with PPD in all groups (A–D) ($P > 0.5$) nor with MLSA in group A ($P > 0.5$), but they increased significantly in groups B–D ($P < 0.001$). The mean reaction to MLSA was approximately 22.5 mm (range 19.5–25.5 mm) in group D.

This study avoided any possible interference from a natural exposure to *M. leprae*, although in Norway the population is vaccinated with BCG. Nevertheless, the *M. leprae* induced a strong delayed-type hypersensitivity to MLSA that was as good as the reaction to PPD. It is known that BCG and *M. leprae* share some antigens [34], and the authors realized that the latter could be enhancing an existing response. In this instance such a boosting effect was thought to be minimal because the MLSA responses were equivalent in prevaccination negative and positive responders.

## B. BCG and Armadillo-Derived *M. leprae*

In 1979, Convit et al. [35] injected leprosy patients intradermally with 0.1 ml of a mixture containing armadillo-derived, autoclaved *M. leprae* ($6.4 \times 10^8$ acid-fast bacilli $ml^{-1}$) plus *M. bovis* BCG (1.0 mg dry weight $ml^{-1}$). Each patient was given six to eight injections over a period of 18 to 30 months. This treatment had a therapeutic effect in patients with borderline or lepromatous leprosy, as they had a lepromin-positive skin reaction at 48 h and after 30 days and clearance of the acid-fast bacilli from their skin lesions. The in vitro lymphocyte transformation test to *M. leprae* from both human and armadillo lesions was positive. The combined use of *M. leprae* and BCG was recommended for highly susceptible individuals or those already infected, after very careful selection.

The dosage mentioned above was not proposed as the optimum dose to achieve a maximal response with minimal side effects. This is an extremely important point because the innate adjuvant effect of these mycobacterial organisms should be considered. Although the peptidoglycolipid from *M. bovis* BCG was adjuvant-inactive at the 200 μg dose [36], the cell walls were highly adjuvant-active at the 200μg dose [37]. In other studies [38] the adjuvant activity of *M. leprae* whole organisms (200–300 μg) derived from the liver and spleen of a lepromatous leprosy patient was demonstrated; this organism does contain mycolic acid [39–41], so it is likely that the organism contains the adjuvant-active peptidoglycolipid. There is now sufficient documentation to show that excessive doses of adjuvant in an inoculum will depress rather than stimulate the adjuvant effect; repeated doses of an inoculum containing adjuvant may have a similar effect [42–44].

An evaluation of 531 Venezuelan patients given this form of immunotherapy since 1979 showed that, after 7 years, lepromatous and borderline leprosy patients, 164/227 and 54/77 respectively, no longer had any clinically detectable lesions. Thus it seemed that a cell-mediated immune response was induced in a group of unresponsive patients and this was associated with protection [45].

The study was extended in 2,659 contacts with an immunoprophylactic, single injection of the combined vaccine; after 8 months, 85% gave positive skin reactions to an MLSA preparation. Subsequently, under the auspices of WHO and the Venezuelan Council of Scientific and Technological Research, a vaccine trial was initiated in 1984 with 65,000 contacts. In this trial, the vaccine was injected in two sites, but the dose of BCG was varied according to the tuberculin sensitivity of the contact. The protective activity of the vaccine in this double-blind trial awaits evaluation.

There is no real evidence to suggest that one type of BCG provides greater protection against leprosy than another [46]. The Glaxo BCG (batch D698)

**TABLE III. Skin Reaction to *M. leprae* Sonicate Antigens After BCG Vaccination**

| *M. leprae* sonicate antigen | BCG preparation | Skin test conversion (%): negative on day 0 to positive at | |
|---|---|---|---|
| | | 90 days | 9 months |
| CD 19 | Glaxo | 78 | 36 |
| (Rees-type) | Copenhagen | 61 | 15 |
| WEL-1 | Glaxo | 25 | 15 |
| (Convit-type) | Copenhagen | 20 | 14 |

Data from Pönnighaus and Fine [47].

was compared to the Copenhagen BCG (batch 616) prior to the trial that began in 1986 in the Karonga District of northern Malawi. The vaccine (0.1 ml) was injected i.d. into the right deltoid region; the Glaxo BCG was noted to produce slightly larger ulcers than the Copenhagen BCG in all vaccinees [47].

It seemed that the Copenhagen BCG would have an advantage over the Glaxo BCG, since large ulcers were unacceptable to the patients. The data in Table III indicate a greater conversion of skin reactivity to MLSA after 3 months with the Glaxo preparation; however, this may be related to the type of antigen used. After 9 months the conversion rate was generally low, and Pönnighaus and Fine questioned the relationship between this reaction and protective immunity. Vaccination with Glaxo BCG has resulted in 50% protective immunity against leprosy, and they queried the role of ulcer formation and protective immunity in the Malawi population. Consequently, in the trial that began in 1986, the Glaxo BCG was used to test whether BCG plus killed armadillo-derived *M. leprae* ($5 \times 10^7$ or $6 \times 10^8$ organisms) would give greater protection against leprosy [48]. The vaccine containing the low and high dose of *M. leprae* + BCG was given in a single injection dose to 36,000 participants with no evidence of a previous BCG scar. There was some argument about the inclusion of a placebo group in those individuals without a BCG scar; for ethical reasons the control group was given BCG as this alone would be expected to give approximately 50% protection against leprosy [48]. The trial is also intended to show whether further protection is afforded by the use of a repeat BCG injection.

In both of these trials it is unlikely that the code for the different vaccination groups will be broken in the next 3 to 5 years. The Venezuelan trial is in an advanced state, and in Malawi the intake of participants is expected to be complete by 1989.

A large-scale trial of the armadillo-derived vaccine was planned for some 300,000 healthy contacts of leprosy patients in the Chingleput District of

Tamil Nadu in South India. Some unfortunate disputes and misunderstandings have delayed authorization of the trial by the Indian Council for Medical Research. In view of the great support provided by the Immunology of Leprosy (IMMLEP) Committee of the UNDP/World Bank/WHO special program for Research and Training in Tropical Diseases, the British Leprosy Relief Association (LEPRA), and Wellcome Laboratories, U.K., for the production of the vaccine, it was disappointing to read that "India entered the race to develop an anti-leprosy vaccine" [49] and "India has insisted on technology transfer as a condition for testing a WHO product" [50]. The latter argument lacks substance as the vaccine production method is not patented and is available from the scientific community [51]. Another misconception was the idea that patients might be harmed by armadillo proteins and DNA together with unknown retroviruses that might be present in the WHO vaccine [49]. Rees [52] completely discounted this in view of the rigid protocol for preparing the *M. leprae,* including treatment with DNase, 2.5 M rad gamma radiation, and autoclaving.

### C. Cultivable Mycobacteria

**1. ICRC vaccine.** A vaccine was developed at the Indian Cancer Research Centre (ICRC) by Bapat and Deo and their colleagues during the 1970s [53,54]. Leprosy-derived slow-growing mycobacteria were isolated and cultivated; these were termed ICRC bacilli and these belong to the *M. avium intracellulare* group. A vaccine was prepared from gamma-irradiated mycobacteria, and after phase 1 and 2 clinical trials this was cleared for use in 1984. A trial began in the state of Maharashtra, and by 1991 it is anticipated that some 50,000 patients with leprosy or healthy contacts will have been vaccinated.

**2. NII vaccine.** A vaccine was developed at the National Institute of Immunology (NII) by Talwar and Fotedar [55] and contained mycobacterium "W", another strain of *M. intracellulare.* This is being tested as an immunotherapeutic vaccine with a small group of 200 leprosy patients in New Delhi [56].

Although both the ICRC and NII vaccines produce skin-test conversion responses to lepromin in vaccinees, the protective activity is uncertain and remains to be established. Agreement has now been reached to test the WHO armadillo-derived vaccine and both the ICRC and NII vaccines in a comparative vaccine trial in Chingleput. It will take at least 10 years to evaluate the results (R.J.W. Rees, personal communication).

**3. *Mycobacterium vaccae.*** *M. vaccae* can be isolated from habitats supporting domestic cattle and is placed in the *M. parafortuitum* complex [57]. The organism is similar to *M. leprae* in the lack of group ii and group iii mycobacterial antigens, but it contains the common group i antigens [58]. A single i.d. immunotherapeutic injection of $1 \times 10^9$ *M. vaccae* gamma-irradiated organisms caused positive skin-test conversion to leprosin in 50% of lepromatous patients for 2 years. Stanford [59] suggests that *M. leprae, M. intracellulare,* and *M. vaccae* group i antigens stimulate cell-mediated mechanisms that may kill leprosy bacilli; as such, they are potentially useful immunotherapeutic agents.

There are some 4 million leprosy cases in India, so it is important to monitor without delay all of the vaccines for their immunotherapeutic or immunoprophylactic activity. This is especially important when one considers the primary drug resistance of *M. leprae* to dapsone and the period in years before a critical assessment of vaccine efficacy can be achieved. A leading leprologist informed me that "it is important to keep the various agencies in a state of readiness to test any new vaccine if and when it arises. At present we still do not know whether you can successfully vaccinate against leprosy but contingency plans are available to escalate existing commitments in the event that positive results are obtained from the trials."

## IV. SECOND-GENERATION VACCINES

There is a population of some 10 million armadillos in the southern United States; each of these could produce 4,000 vaccine doses of *M. leprae*. But because there are some 2,000 million people at risk, this would deplete the natural armadillo population. In addition, it is doubtful whether the continued use of armadillos or primates would be ethically acceptable. Consequently, attention has been turned to the use of the techniques of genetic engineering. A number of interesting approaches have been described, and some of these are given below. Clark-Curtiss et al. [60] reported that the DNAs of *M. leprae* and *M. vaccae* were unrelated (G + C ratios of 56% and 65%, genomic sizes $2.2 \times 10^9$ and $3.1 \times 10^9$ daltons, respectively), although there was evidence of some conserved sequences between the DNAs. A genomic library of *M. leprae* was prepared with the cosmid vector pHC79. The DNA was partially digested with *Pst*I and ligated to pHC79 DNA, also digested with the same enzyme; the ratio of chromosomal DNA to vector DNA was 2:1 (mol:mol) and phage T4 ligase was added to the reaction mixture, which was left at 16°C for 16 h. The recombinant cosmids, with mycobacterial inserts of 40–45 kilobases, were packaged into phage that were used to infect *Escherichia coli* K12. The expression of cloned *M. leprae* DNA was enhanced by digesting DNA from the total pHC79::*M. leprae* library clones with *Pst*I. The fragments were subcloned into the *Pst*I site of an expression vector pYA626, a derivative of pBR322 that contains the powerful promoter of the *asd* (aspartate semialdehyde dehydrogenase) gene and T4 ligase. The recombinant pYA626::*M. leprae* molecules were transformed into *E. coli* K12 cells that synthesized polypeptides encoded by the inserted *M. leprae* DNA; this was confirmed by hybridization studies with *M. leprae* DNA probes. On the other hand, there was no reaction with sera from leprosy patients or with monoclonal antibodies against specific *M. leprae* proteins, so it was apparent that no antigen-expressing clones were isolated. In another study, several clones that showed complementation of a mutation in the citrate synthase gene of *E. coli* were detected; the recombinant showed the production of a 46.0 kDa polypeptide coded for by the *M. leprae* DNA [61]. Jacobs et al. [61] used the same expression vector pYA626 containing the promoter region of the *S. mutans asd* gene to isolate clones that complement a mutation in the citrate synthase gene of *E. coli.* Expression of the *M. leprae* DNA was presumably from the vector promoter because the recombinant pYA626 with the *M. leprae* in the reserve orientation did not complement the mutation in the citrate synthase gene. When the *M. leprae* DNA fragment was cloned in the *Pst*I site of pHC79, to allow expression from the β-lactamase *(bla)* promoter, or into pYA626 to allow expression from the *asd* promoter, the resulting recombinant did not complement the citrate synthase gene but still produced the 46.0 kDa polypeptide, although in smaller amounts. It was apparent that this *M. leprae* polypeptide was not a fusion protein and the authors concluded that *M. leprae* translational sequences could be recognized in *E. coli.*

Young and his colleagues in various centers [62,63] collaborated to construct an *M. leprae* DNA library using bacteriophage λgt11 as a vector that will drive the expression of foreign-insert DNA with *E. coli* transcription and translation signals. The *M. leprae* DNA was mechanically sheared to produce 1–7 kilobase fragments that were ligated to *Eco*RI linkers. The fragments were inserted at the *Eco*RI site of λgt11, within the coding region of β-galactosidase (*lac*Z) gene, and packaged into phage, which were used to infect *E coli.* When the *E. coli* λgt11 was treated with an inducer of the β-galactosidase enzyme, the *M. leprae* coding sequences were expressed as a fusion protein. The strategy of this scheme, to produce random end points in the *M. leprae* DNA and recombinant phage in sufficient numbers, ensured that at least some of the coding sequences were in the correct transcriptional orientation and translational reading frame for expression. Murine monoclonal antibodies, prepared against crude extracts of armadillo-derived *M. leprae,* were used to probe plaques. Recombinant DNA clones were detected with monoclonal antibodies directed against 65.0, 36.0, 28.0, 18.0, and 12.0 kDa antigens.

It is interesting to note that the mycobacterial antigen A (a group i antigen) was expressed in *E. coli* transduced with phage λ vector EMBL 3 containing fragments of *M. bovis* BCG DNA, 9–22 kilobases. Rocket immunoelectro-

**TABLE IV. Expression of *M. leprae* Antigen on *E. coli* Surface**

| Serum | *E. coli* (*M. lep-pho*E[nru]) | *E. coli* *pho*E |
|---|---|---|
| Lepromatous leprosy patient | + + + | ∓ |
| Tuberculoid leprosy patient | ± | · |

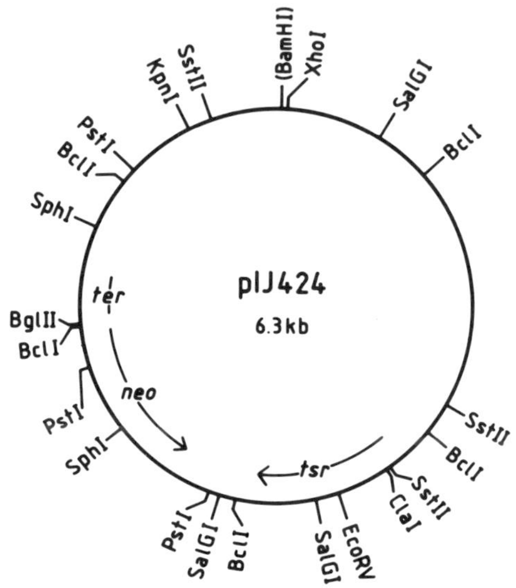


**Fig. 1.** Restriction map of the *Streptomyces* promoter-probe plasmid pIJ424. Reproduced from Kieser et al. [67] with permission of the American Society for Microbiology.

phoresis indicated a cross-reaction between this antigen A and a soluble protein extract of armadillo-derived *M. leprae* [64].

A neat method of expressing an antigen of *M. leprae* on the surface of *E. coli* has been described by Marja Agterberg (personal communication). The polypeptide of *pho*E from the pore-forming protein of *E. coli* K12 traverses the membrane 16 times in a β-sheet exposing eight regions at the cell surface. The sequence KLKLTGDEA of the 65 kDa *M. leprae* antigen [62] was inserted in the *Nru*I site in the *pho*E gene (*M. lep-pho*E[nru]). The foreign epitope was exposed at the surface of *E. coli* and was antigenic in an ELISA assay (Table IV). This procedure may also allow a convenient approach to the search for diagnostic aids in leprosy.

*Streptomyces* species are more closely related to mycobacteria but have a higher G + C content of about 73%. The cloning systems in *Streptomyces* are well characterized [65,66]. Plasmid-free *S. lividans* was the host for

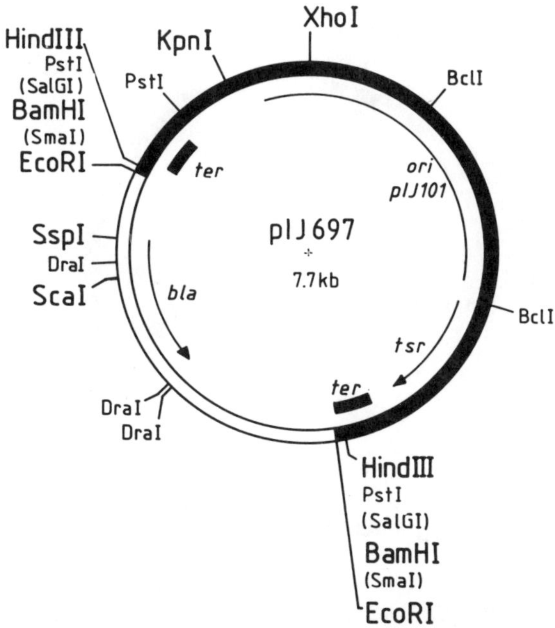


**Fig. 2.** Map of plasmid pIJ697, which can be used for cloning with *Eco*RI, *Bam*HI, and *Hin*d III. Blackened segment, vector part; open segment, "spacer" fragment. Unique sites or sites designed for cloning are drawn in larger letters. The two black boxes marked *ter* form two arms of a palindrome (inverted repeat) and carry the fd terminator; *bla,* ampicillin resistance gene; *tsr,* thiostrepton resistance gene from *Streptomyces azureus; ori,* origin of DNA replication. Reproduced from Kieser and Melton [69] with permission of Elsevier Science Publishers BV.

transformation with the 6.3 kilobase promoter probe plasmid pIJ424 (Fig. 1), which has a promoterless kanamycin resistance *(neo)* determinant. If a piece of DNA with promoter activity is inserted in the *Bgl*II site, the $Km^r$ gene is activated. The strength of the promoter activity is reflected in the level of kanamycin resistance. The results indicated that *S. lividans* used the translational signals of *M. bovis* BCG [67] and *M. leprae* [68] and as a nonpathogenic species could be used to express mycobacterial antigens.

The λgt11 library of Young and his colleagues was used to select the *M. leprae* of 65.0 kDa antigen clone (Y3178) and 28 kDa antigen clone (Y3164). The mycobacterial insert was excised with *Eco*RI and ligated with the 7.7 kilobase positive selection plasmid vector pIJ697. This consists of a *Streptomyces* plasmid and *E. coli* plasmid joined by two polylinker restriction sites and two terminator sequences. The plasmid pIJ697 is only stable when the terminator sequences are separated by inserted DNA (Fig. 2). Both

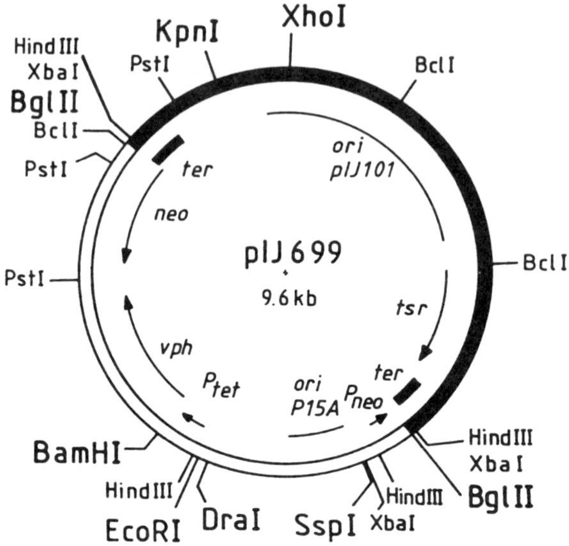


**Fig. 3.** Map of pIJ699, a multicopy positive-selection vector for *Streptomyces*. Symbols are the same as in Figure 2. $P_{neo}$, promoter region of the *neo* gene; $P_{tet}$, promoter region of the *tet* gene; *neo,* kanamycin resistance gene; *vph,* viomycin resistance gene from *Streptomyces vinaceus*. Reproduced from Kieser and Melton [69] with permission of Elsevier Science Publishers BV.

the 65.0 kDa and 28.0 kDa antigenic proteins were subcloned into *Streptomyces* using this vector [68]. A 9.6 kilobase plasmid vector, pIJ699 (Fig. 3), which provides positive selection for cloned DNA, has been constructed for the *M. leprae* studies (T. Kieser, personal communication) using the replication functions of the *Streptomyces* multicopy plasmid pIJ101 [69]. pIJ699

**Fig. 4.** Diagram of the construction of the shuttle phasmids. TM4 phage DNA was ligated at a concentration of 250 μg $ml^{-1}$ and then partially digested with *Sau*3A to obtain fragments that averaged 30–50 kilobases. These fragments were ligated at a 1:2 molar ratio of TM4 fragments to pHC79 that had been cleaved with *Bam*HI and alkaline phosphatased. The packaging of an aliquot of this ligation with in vitro packaging mix (Gigapack Plus, Stratagene) and subsequent transduction into ER1381 (*hsd*R *mcr*A + *mcr*B +, yielded $10^6$ ampicillin colonies per microgram of TM4 DNA insert when plated on L-agar containing ampicillin at 50 μg $ml^{-1}$. A pool of 40,000 ampicillin-resistant clones was prepared by homogenizing colonies in L-broth from the selection plates with a glass spreader. Plasmid was isolated from the pool of clones by alkaline-SDS treatment followed by phenol-chloroform extraction and concentration with ethanol. Covalently closed circular plasmid DNA was transfected into *M. smegmatis*. Reproduced from Jacobs et al. [70] with permission of the publishers of *Nature*.

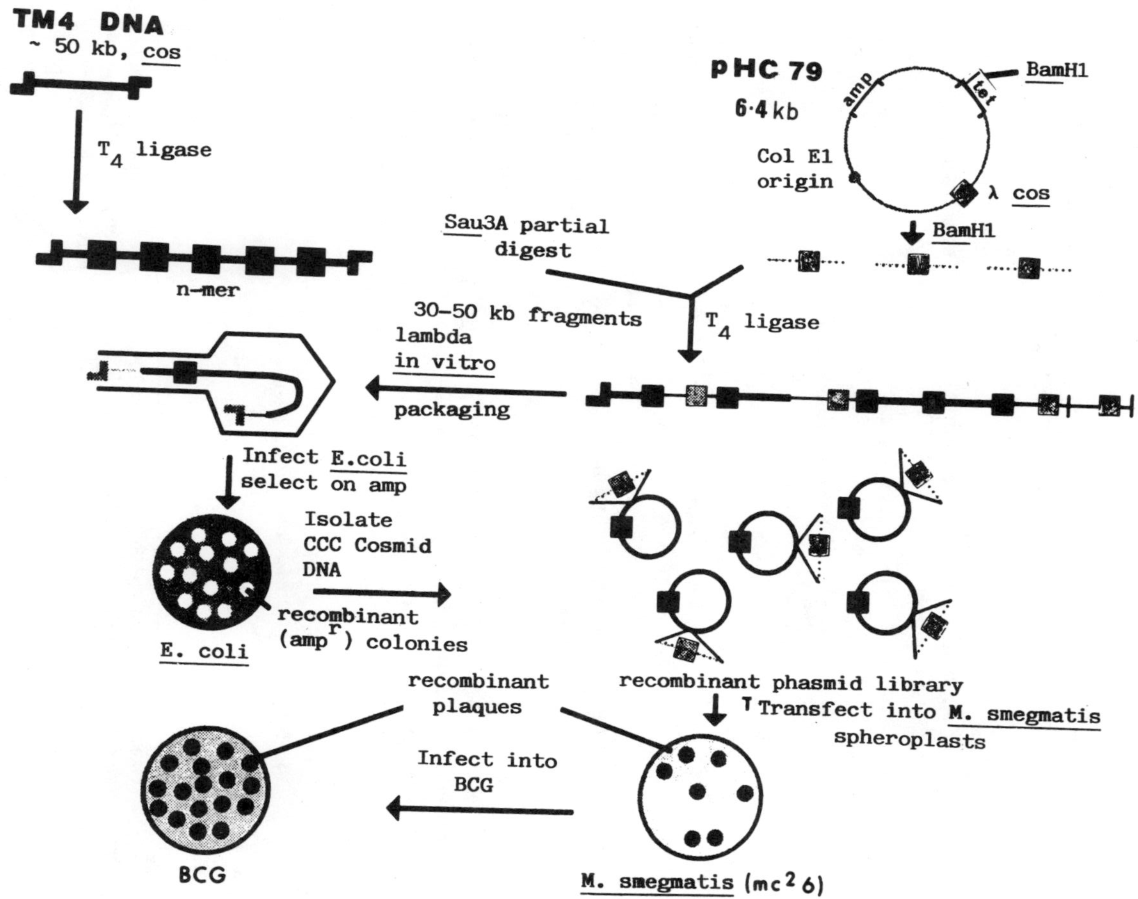
TM4 DNA
~ 50 kb, cos
T4 ligase
n-mer
pHC 79
6·4 kb
amp
tet
BamH1
Col E1 origin
λ cos
BamH1
Sau3A partial digest
T4 ligase
30-50 kb fragments
lambda
in vitro
packaging
Infect E.coli
select on amp
Isolate
CCC Cosmid
DNA
recombinant
(amp r) colonies
E. coli
recombinant phasmid library
Transfect into M. smegmatis
spheroplasts
recombinant
plaques
Infect into
BCG
BCG
M. smegmatis (mc2 6)

combines two important features: positive selection and cloning sites flanked by bidirectional transcription terminators that seem to improve the stability of hybrid plasmids. Further cloning studies in *Streptomyces* with this plasmid are planned and the results are eagerly awaited.

Jacobs et al. [70] developed a system of introducing foreign DNA into mycobacteria using a shuttle plasmid. This vector replicated as a plasmid in *E. coli* and as a bacteriophage in mycobacteria. The phage used was the ~50 kilobase TM4 from *M. avium,* which was known to replicate in *M. smegmatis,* BCG, and *M. tuberculosis.* Concatamers of the TM4 phage with cohesive ends were made, partially digested with *Sau*3A, and the fragments ligated to the cosmid pHC79 already digested with *Bam*HI. The ligation mixture was packaged into phage λ heads. *E. coli* was transduced with the phage, and *amp*$^r$ clones containing pHC79::TM4 DNA were selected and transfected into *M. smegmatis* (Fig. 4).

This technique was developed by Snapper et al. [71] in *M. smegmatis.* The mycobacteriophage L1 50 kilobase DNA was ligated to form concatamers and partially digested with *Sau*3A to generate fragments 30–50 kilobases long. The cosmid pHC79 was digested with *Bam*HI, ligated to the *Sau*3A fragments, and packaged into λ phage. The resultant cosmid was transduced into *E. coli* and ampicillin-resistant clones were isolated. These were shown to contain a piece of L1 genome containing pHC79; DNA from these clones could be used to generate shuttle phasmids replicating 1) as plasmids in *E. coli* and 2) as lysogenic phage in *M. smegmatis* (Fig. 5).

In another series of experiments, DNA of the pIJ666::pAL5000 DNA library (kan$^r$) was introduced into *M. smegmatis* cells by electroporation; 1 to 10 transformants carrying kan$^r$/μg library DNA were obtained. It was also possible to transform *M. bovis* BCG with the pIJ666::pAL5000 recombinant library.

Recently, Zainuddin et al. [72] have confirmed this strategy with the *E. coli* plasmid pIJ666 and *M. smegmatis* as the host.

**Fig. 5.** Gene expression in mycobacteria by using a temperate shuttle phasmid. Schematic representation of the use of temperate shuttle phasmids as cloning vectors to stably introduce genes of interest into the mycobacterial chromosome. Gene(s) of interest can be inserted into unique restriction sites in shuttle phasmid DNA by cosmid cloning in *E. coli*. Subsequently, the recombinants can be introduced into mycobacteria where the shuttle phasmid, carrying the gene(s) of interest, can lysogenize and be maintained stably as a prophage. The integration of the phasmid DNA into the mycobacterial chromosome is probably mediated by a site-specific recombination between a phage (Att) and bacterial (aTT) attachment site. Reproduced from Snapper et al. [71] with permission of the publishers of the Proceedings of the National Academy of Sciences U.S.A.

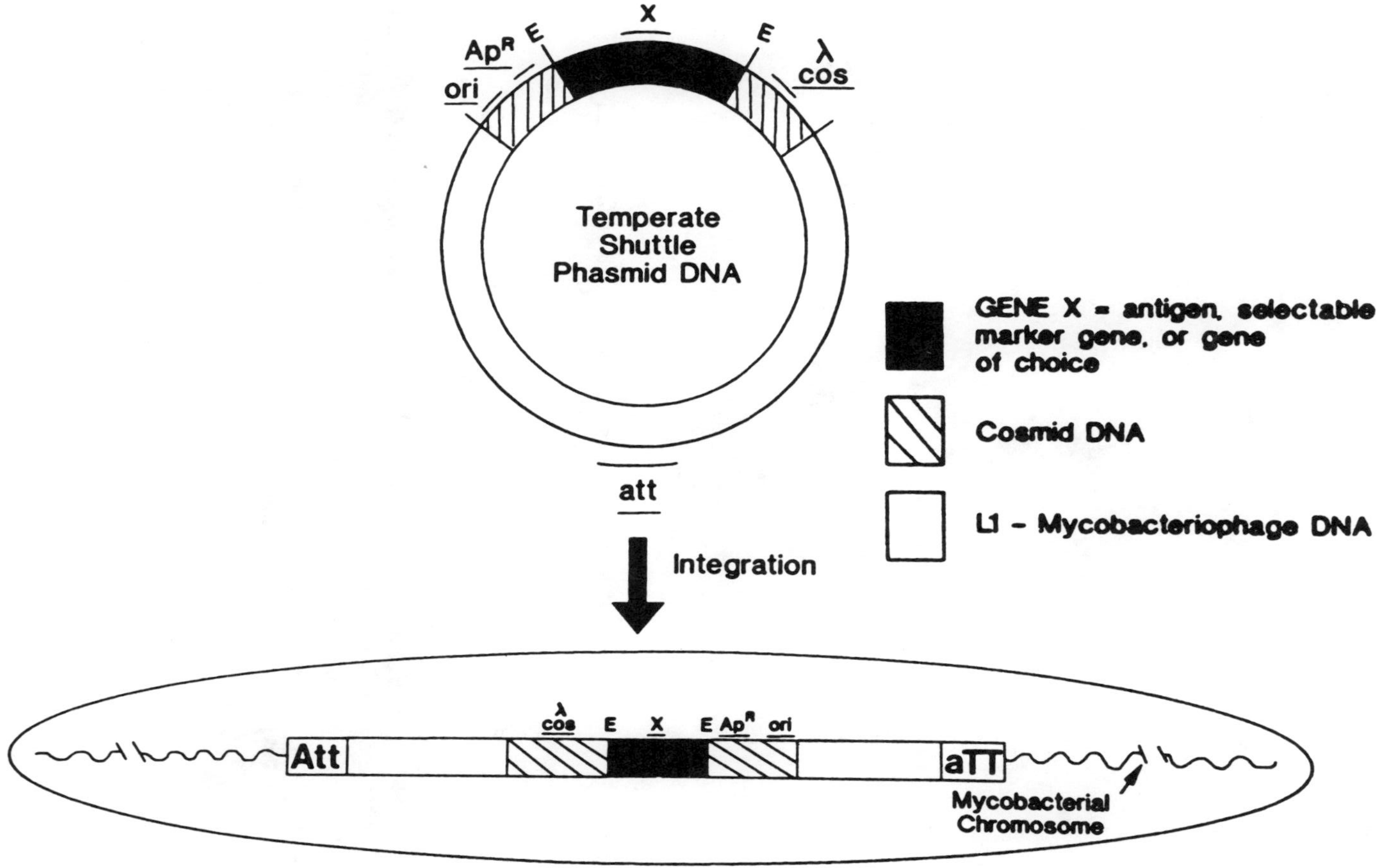

X
Ap^R
E
E
λ
cos
ori
Temperate
Shuttle
Phasmid DNA
att
Integration
GENE X = antigen, selectable marker gene, or gene of choice
Cosmid DNA
L1 – Mycobacteriophage DNA
Att
λ
cos
E
X
E
Ap^R
ori
aTT
Mycobacterial
Chromosome

**Fig. 6.** A relieved armadillo: no more experiments? (Reproduced with permission of Blackwell Scientific Publications Limited.)

The strategy of all these workers is to clone the genes for protective mycobacterial antigens into BCG vaccine strains using the shuttle plasmid. The TM4 phage will replicate in the *M. bovis* BCG Glaxo, Pasteur, and Copenhagen strains. These studies of the specific genetic determinants may ultimately lead to a recombinant mycobacterial vaccine that will contain the genes of protective epitopes of *M. leprae* [73,74]. It is essential that any such construct will stimulate cell-mediated responses, as antimycobacterial antibodies do not give protection in lepromatous leprosy. Immunological studies in leprosy are beyond the scope of this chapter and the reader may consult a number of useful reviews [75–77].

To conclude, it was pointed out by Mitchison [77] that second- generation vaccines would be useful because 1) distractor and suppressor epitopes could be deleted and 2) agretopes and their associated T epitopes could be optimally located in the construct. If the results of the various trials with the first-generation vaccines prove to be favorable, it would not seem to be overoptimistic to expect that a second-generation vaccine will be ready for trial by the middle of the next decade.

## REFERENCES


1. Plotkin SA, Mortimer EA: "Vaccines." Philadelphia: Saunders, 1988.
2. WHO Expert Committee on Leprosy: Sixth report. WHO Tech Rep Ser 768, 1988.
3. Stewart-Tull DES: Leprosy—In pursuit of a vaccine. Vaccine 2:238, 1984.
4. Denecke K: Die Züchtungsversuche der leprabakterien seit 1930. Zentralbl Bakteriol Parasitenk Infektionskr Hyg 187:17, 1963.
5. Pattyn SR: The problem of cultivation of *"Mycobacterium leprae."* Bull WHO 49:403, 1973.
6. Stewart-Tull DES: *Mycobacterium leprae:* The bacteriologists' enigma. In Ratledge C, Stanford J (eds): "The Biology of the Mycobacteria." London: Academic Press, 1982, Vol. 1, p 273.
7. Kinnear Brown JA, Stone MM: BCG vaccination of children against leprosy: First results of a trial in Uganda. Br Med J, p 7, 1966.
8. Kinnear Brown JA, Stone MM, Sutherland I: Trial of BCG vaccination against leprosy in Uganda. Lepr Rev 40:3, 1969.
9. Stanley SJ, Howland C, Stone MM, Sutherland I: BCG vaccination of children against leprosy in Uganda: Final results. J Hyg (Camb) 87:233, 1981.
10. Bechelli ML, Lwin K, Gallego Garbajosa P, Gyi MM, Uemura K, Sundaresan T, Tamondong C, Matejka M, Sansarricq H, Walter J: BCG vaccination of children against leprosy: Nine year findings of the controlled WHO trial in Burma. Bull WHO 51:93, 1974.
11. Scott GC, Russell DA, Boughton CR, Vincin DR: Untreated leprosy. Probability of shifts in Ridley-Jopling classification. Int J Lepr 44:110, 1978.
12. Fine PEM, Ponnighaus JM, Maine N, Clarkson JA, Bliss L: The protective efficacy of BCG against leprosy in northern Malawi. Lancet ii:499, 1986.
13. Tripathy: The case for BCG. Ann Natl Acad Med Sci (India) 19:12, 1983.
14. Shield MJ: The importance of immunologically effective contact with environmental mycobacteria. In Ratledge C, Stanford J (eds.): "The Biology of the Mycobacteria." London: Academic Press, 1983, Vol. 2, p 343.
15. Stanford JL: The BCG trials. Practitioner 227:x, 1983.
16. Lwin K, Sundaresan T, Gyi MM, Bechelli LM, Tamondong C, Gallego Garbajosa P, Sansarricq H, Noordeen SK: BCG vaccination of children against leprosy: Fourteen-year findings of the trial in Burma. Bull WHO 63:1069, 1985.
17. Mitchison DA: The virulence of tubercle bacilli from patients with pulmonary tuberculosis in India and other countries. Bull Int Union Tuberc 35:287, 1964.
18. Shepard CC: The experimental disease that follows the injection of human leprosy bacilli into footpads of mice. J Exp Med 112:445, 1960.
19. Rees RJW: Enhanced susceptibility of thymectomized and irradiated mice to infection with *Mycobacterium leprae*. Nature 211:657, 1966.
20. Colston MJ, Hilson GRF: Growth of *Mycobacterium leprae* and *M. marinum* in congenitally athymic (nude) mice. Nature 262:399, 1976.
21. Storrs E: The nine-banded armadillo: A model for leprosy and other biomedical research. Int J Lepr 39:703, 1971.
22. Kirchheimer WF, Storrs EE: Attempts to establish the armadillo (*Dasypus novemcinctus* Linn) as a model for the study of leprosy. I. Report of lepromatoid leprosy in an experimentally infected armadillo. Int J Lepr 39:693, 1971.

23. Storrs E: The armadillo as an experimental model for the study of human leprosy. Lepr Rev 45:8, 1974.
24. Kirchheimer WF, Storrs E, Binford CH: Attempts to establish the armadillo (*Dasypus novemcinctus* Linn) as a model for the study of leprosy. II. Histopathologic and bacteriologic post mortem findings in lepromatoid leprosy in the armadillo. Int J Lepr 40:229, 1972.
25. Rees RJW, Lowe C, McDougall AC: Preliminary bacteriological data on experimental *Mycobacterium leprae* infections in the hedgehog. J Med Microbiol 12:vi, 1979.
26. Convit J, Pinardi ME: Inoculation de *M. leprae* en dos especies de armadillo: *D. sabanicola, D. novemcinctus.* Acta Cient Venez 25:51, 1974.
27. Storrs EE, Walsh GP, Burchfield HP: Development of leprosy in another species of armadillo *Dasypus hybridus* (L): Genetic and immunological implications. J Trop Med Hyg 78:216, 1975.
28. Klingmüller G, Sobich E: Übertragung menschlicher Leprabackterien auf den Igel. Naturwissenschaften 64:645, 1977.
29. Donham KJ, Leininger JR: Spontaneous leprosy-like disease in a chimpanzee. J Infect Dis 136:132, 1977.
30. Meyers WM, Walsh GP, Brown HL, Binford CH, Imes GD, Hadfield TL, Schlagel CJ, Fukunishi Y, Gerone PJ, Wolf RH, Gormus BJ, Martin LN, Harboe M, Imaeda T: Leprosy in a Mangabey monkey—Naturally acquired infection. Int J Lepr 53:1, 1985.
31. Wolf RH, Martin LN, Gormus BJ, Baskin GB, Gerone PJ, Walsh GP, Meyers WM, Brown HL, Binford CH: Experimental transmission of leprosy in African green monkeys *(Cercopithecus aethiops)* and the rhesus monkey *(Macaca mulatta)*. Int J Lepr 51:664, 1983.
32. Smelt AHM, Rees RJW, Liew FY: Induction of delayed-type hypersensitivity to *Mycobacterium leprae* in healthy individuals. Clin Exp Immunol 44:501, 1981.
33. Gill HK, Mustafa AS, Godal T: Induction of delayed-type hypersensitivity in human volunteers immunized with a candidate leprosy vaccine consisting of killed *Mycobacterium leprae*. Bull WHO 64:121, 1986.
34. Harboe M, Closs O, Bjorvatu B, Kronvall G, Axelsen NH: Antibody response in rabbits to immunization with *Mycobacterium leprae*. Infect Immun 18:792, 1977.
35. Convit J, Aranzazu N, Pinardi M, Ulrich M: Immunological changes observed in indeterminate and lepromatous leprosy patients and Mitsuda -ve contacts after the inoculation of a mixture of *M. leprae* and BCG. Clin Exp Immunol 36:214, 1979.
36. White RG, Bernstock L, Johns RGS, Lederer E: The influence of components of *M. tuberculosis* and other mycobacteria upon antibody production to ovalbumin. Immunology 1:54, 1958.
37. Stewart-Tull DES, Shimono T, Kotani S, Kato M, Ogawa Y, Yamamura Y, Koga T, Pearson CM: The adjuvant activity of a non-toxic water-soluble glycopeptide present in large quantities in the culture filtrate of *M. tuberculosis* strain DT. Immunology 29:1, 1975.
38. Stewart-Tull DES, Davies M: Adjuvant activity of *Mycobacterium leprae*. Infect Immun 6:909, 1972.
39. Etemadi AH, Convit J: Mycolic acids from "noncultivable" mycobacteria. Infect Immun 10:236, 1974.
40. Devi S, Stewart-Tull DES: Detection of mycolic acid in leprous tissue. Int J Lepr 47:484, 1979.
41. Draper P, Dobson G, Minnikin DE, Minnikin SM: The mycolic acids of *Mycobacterium leprae* harvested from experimentally infected nine-banded armadillos. Ann Microbiol Inst Pasteur 133B:39, 1982.

42. Stewart-Tull DES: The immunological activities of bacterial peptidoglycans. Annu Rev Microbiol 34:311, 1980.
43. Stewart-Tull DES: Immunopotentiating activity of peptidoglycan and surface polymers. In Stewart-Tull DES, Davies M (eds): "Immunology of the Bacterial Cell Envelope." Chichester: John Wiley and Sons, 1985, p. 56.
44. Stewart-Tull DES: Immunostimulation with peptidoglycan or its synthetic derivatives. Prog Drug Res 32:305, 1988.
45. Convit J, Ulrich M, Aranzazu N, Castellanos PL, Pinardi ME, Reyes O: The development of a vaccination model using two microorganisms and its application in leprosy and leishmaniasis. Lepr Rev 57(Suppl 2):263, 1986.
46. Fine PEM: BCG vaccination against tuberculosis and leprosy. Br Med Bull 44:691, 1988.
47. Pönnighaus JM, Fine PEM: The Karonga Prevention Trial—Which BCG: Lepr Rev 57(Suppl 2):285, 1986.
48. Fine PEM, Pönnighaus JM: Leprosy in Malawi 2. Background, design and prospects of the Karonga prevention trial, a leprosy vaccine trial in northern Malawi. Trans R Soc Trop Med Hyg 82:810, 1988.
49. Jayaraman KS: India carries out large-scale tests of anti-leprosy vaccine. Nature 328:660, 1987.
50. Jayaraman KS: Leprosy-vaccine. Large-scale trials begin in India. Nature 317:665, 1985.
51. Draper P: Leprosy vaccine. Nature 319:94, 1986.
52. Rees RJW: Leprosy vaccine. Nature 330:690, 1987.
53. Bapat CV, Ranadive KJ, Khanolkar V: *In vitro* cultivation of an acid-fast mycobacterium isolated from human lepromatous leprosy. Indian J Pathol Bacteriol 1:156, 1958.
54. Deo MG, Bapat CV, Bhalerao V, Chaturvedi RM, Bhatki WS, Chulawala RG: Antileprosy potentials of ICRC vaccine. A study in patients and healthy volunteers. Int J Lepr 51:540, 1983.
55. Talwar GP, Fotedar A: Two candidate antileprosy vaccines—Current status of their development. Int J Lepr Myc Dis 51:550, 1983.
56. Jayaraman KS: Disputes delay development of leprosy vaccine for India. Nature 333:590, 1988.
57. Goodfellow M, Wayne LG: Taxonomy and nomenclature. In Ratledge C, Stanford J (eds): "The Biology of the Mycobacteria." London: Academic Press 1982, Vol. 1, p. 471.
58. Stanford JL: Immunologically important constituents of mycobacteria: Antigens. In Ratledge C, Stanford J (eds): "The Biology of the Mycobacteria." London: Academic Press, 1983, Vol. 2, p 85.
59. Stanford JL: Immunotherapy for mycobacterial disease. In Ratledge C, Stanford J, Grange JM (eds): "The Biology of the Mycobacteria." London: Academic Press, 1989, Vol. 3, p 567.
60. Clark-Curtiss JE, Jacobs WR, Docherty MA, Ritchie LR, Curtiss R: Molecular analysis of DNA and construction of genomic libraries of *Mycobacterium leprae*. J Bacteriol 161: 1093, 1985.
61. Jacobs WR, Docherty MA, Curtiss R, Clark-Curtiss JE: Expression of *Mycobacterium leprae* genes from a *Streptococcus mutans* promoter in *Escherichia coli* K-12. Proc Natl Acad Sci USA 83:1926, 1986.
62. Young RA, Mehra V, Sweetser D, Buchanan T, Clark-Curtiss J, Davis RW, Bloom BR: Genes for the major protein antigens of the leprosy parasite *Mycobacterium leprae*. Nature 316:450, 1985.
63. Young RA, Bloom BR, Grosskinsky CM, Ivanyi J, Thomas D, Davis RW: Dissection of *Mycobacterium tuberculosis* antigens using recombinant DNA. Proc Natl Acad Sci USA 82:2583, 1985.

64. Thole JER, Dauwerse HG, Das PK, Groothuis DG, Schouls LM, van Embden JDA: Cloning of *Mycobacterium bovis* BCG DNA and expression of antigens in *Escherichia coli*. Infect Immun 50:800, 1985.
65. Hopwood DA, Bibb MJ, Chater KF, Kieser T. Plasmid and phage vectors for gene cloning and analysis in *Streptomyces*. Methods Enzymol 153:116, 1987.
66. Hopwood DA, Bibb MJ, Chater KF, Janssen GR, Malpartida F, Smith CP: Regulation of gene expression in antibiotic-producing *Streptomyces*. In Booth I, Higgins CR (eds): "Regulation of Gene Expression." Cambridge: Cambridge University Press, 1986, p 251.
67. Kieser T, Moss MT, Dale JW, Hopwood DA: Cloning and expression of *Mycobacterium bovis* BCG DNA in *"Streptomyces lividans."* J Bacteriol 168:72, 1986.
68. Lamb FI, Colston MJ: Cloning of *Mycobacterium leprae* genes in *Streptomyces*. Lepr Rev 57(Suppl 2):225, 1986.
69. Kieser T, Melton RE: Plasmid pIJ699, a multi-copy positive-selection vector for *Streptomyces*. Gene 65:83, 1988.
70. Jacobs WR, Tuckman M, Bloom BR: Introduction of foreign DNA into mycobacteria using a shuttle phasmid. Nature 327:532, 1987.
71. Snapper SB, Lugosi L, Jekkel A, Melton RE, Kieser T, Bloom BR, Jacobs WR: Lysogeny and transformation in mycobacteria: Stable expression of foreign genes. Proc Natl Acad Sci USA 85:6987, 1988.
72. Zainuddin ZF, Kunze ZM, Dale JW: Transformation of *Mycobacterium smegmatis* with *Escherichia coli* plasmids carrying a selectable resistance marker. Mol Microbiol 3:29, 1989.
73. Bloom BR: Learning from leprosy: A perspective on immunology and the Third World. J Immunol 137:1, 1986.
74. Modlin RL, Rea TH: Leprosy: New insight into an ancient disease. Am Acad Dermatol 17:1, 1987.
75. Bloom BR, Godal T: Selective primary health care: Strategies for control of disease in the developing world V. Leprosy. Rev Infect Dis 5:765, 1983.
76. Bloom BR, Mehra V: Immunological unresponsiveness in leprosy. Immunol Rev 80:5, 1984.
77. Mitchison NA: The lessons to take home. Lepr Rev 57(Suppl 2):305, 1986.

Bacterial Vaccines, pages 223–254


# Vaccination Against the Cholera/Coli Family of Enterotoxins by Synthetic Peptides

**Chaim O. Jacob**

Department of Microbiology and Immunology, Stanford University School of Medicine, Stanford, California 94305



## I. INTRODUCTION AND BACKGROUND

### A. Introduction

The basis for the utilization of synthetic antigens for vaccination was established during studies for antigenicity of proteins [1] and by the finding that they are capable of eliciting antiprotein immune response [2]. It has also been shown that in the case of molecules with biological activity, only a small fraction of the antibodies they elicit possesses neutralizing capacity. As a corollary it was suggested that by adequate molecular engineering it should be possible to design synthetic material that would elicit immunity against a causative agent (bacteria or virus), leading eventually to multivalent synthetic vaccines [3]. Moreover, the synthetic approach should allow the choice of the carrier, as well as the adjuvant, the goal being the production of multivalent vaccines with built-in adjuvanticity.

The antigenic determinants of most globular proteins are conformation dependent, yet relatively short peptide segments that are exposed on surface or in ''corners'' of their spatial structure can constitute major antigenic determinants. Antibodies to such peptides react with intact protein, either because the peptide three-dimensional structure already resembles the structure they have within the protein or because that peptide stretch within the protein can transconform itself easily upon reacting with the combining site of the antipeptide antibody. This is by no means general, as many peptides provoke antibodies totally incapable of reacting with the parent protein. Most of the success thus far has been in synthesizing peptides derived from relevant regions of proteins, which, after covalent linking to appropriate carriers, are capable of producing antibodies that react with the respective virus or bacterial toxin.

Vaccination against both bacterial and viral diseases has been one of the major achievements in immunology since the beginning of the last century. However, while existing vaccines have certainly diminished the incidence, morbidity, and mortality of many infectious diseases, some infectious diseases are still very difficult to prevent and control by prophylactic means. The present mode of vaccination includes the use of killed or live attenuated microbial agents, purified viral proteins or their subunits, and bacterial toxoids. In many cases successful vaccines have been developed, but several major problems still await solution; for example, the lack of sufficient source material, since not all viruses can be cultured and developed into effective vaccines, or safety considerations due to the hazardous effects of some vaccines, caused by insufficient attenuation. Another problem, not yet overcome, is that many existing viruses constantly evolve into new strains, with different serological specificities, thus requiring the continuous development of new vaccines. The synthetic approach offers a potential solution to these problems. According to this approach, vaccines will consist of a synthetic material comprising only the relevant antigenic moiety leading to protective immunity. Such vaccines can be synthesized, either chemically or by genetic engineering, in large amounts and will not present safety hazards to the individual or the community [4,5].

The vaccines of the future should ideally contain within one molecule a specific synthetic antigenic determinant capable of provoking a protective immune response and a synthetic adjuvant, both covalently attached to a macromolecular synthetic carrier. If possible, several specificities characteristic of different diseases should be linked covalently within the same macromolecule. Ultimately, the choice of the carrier may depend—for optimal results—on the genetic background of the recipient. In order to obtain long-lasting immunity it will be necessary also to take into consideration the correct texture of the synthetic vaccines to be used for immunization.

Appropriate objectives for this type of investigation are the bacterial toxins, immunization against which (e.g., with tetanus, diphtheria, or pertussis toxoids) has been used in the past to provide neutralization of toxicity, and hence protection.

In the last several years, I have used the cholera toxin B subunit system as a model to study the feasibility and the constraints of the use of short synthetic peptides as the basis for vaccine development. The purpose of this chapter is to summarize our efforts to develop and evaluate the possibilities of using the synthetic approach for vaccination and to try to draw some conclusions that are relevant to the broader question of synthetic vaccine development in general.

### B. Cholera Toxin

The cholera toxin is an 84 kDa protein composed of two noncovalently bound subunits, denoted as A and B. The A subunit, although synthesized as a single polypeptide chain, is proteolytically nicked to form two polypeptide chains, $A_1$ and $A_2$, with molecular weights of approximately 23,000 and 6,000 respectively, linked by a single disulfide bond [6,7]. The $A_1$ polypeptide is the proenzyme form of an ADP ribosyltransferase that is responsible for the biological activity of the toxin through its ability to activate target-cell adenylate cyclase [8,9].

The B subunit, the immunologically dominant region, is composed of five identical, noncovalently associated B chains composed of 103 amino acids that are responsible for binding the toxin to target-cell membrane receptors containing the oligosaccharide of the $GM_1$ ganglioside [9–11]. Neutralizing antibodies raised against the holotoxin react mainly with the B subunit, and antibodies against the B subunit are capable of neutralizing the biological activities of the intact toxin [12].

These experiments suggest that most of the antigenic determinant in the toxin are within the B subunit, while subunit A might be in the center of the molecule protected from the outside. Thus, immunological experiments are consistent with the proposed structure and arrangement of the subunits in cholera toxin as a tight ring formed by the noncovalently bound B chains. The A subunit is linked to and partially inserted in the B ring through weak noncovalent interactions. Reduction of whole toxin releases the $A_1$ fragment from the $A_2$-5B complex, indicating that $A_1$ is attached to the B ring by its $A_2$ portion.

### C. Cholera/Coli Family of Enterotoxins

The cholera enterotoxin (CT), first purified from culture supernates of classical biotype, Inaba serotype, O group 1 vibrio cholerae strain 569B [13], is the prototype of an expanding family of structurally, functionally, and immunologically related diarrheagenic proteins produced by a number of species of enteric bacteria [14]. The family includes, at a minimum, at least three distinct CTs [15–17], two different heat-labile enterotoxins (LTs) from strains of *Escherichia coli* isolated from humans (H) [18–20], and LT isolated from porcine strains of *E. coli* (P-LT) [21,22]; heat-labile enterotoxin from *Salmonella* (S-LT) [23,24]; cholera-related enterotoxin from non-O1 *Vibrio cholerae* [25]; *Vibrio mimicus* toxin [26]; and most recently *Campylobacter jejuni* [27]. Similar activities reported with other enteropathogenic bacteria have not been sufficiently characterized to be included (see Fig. 1).

The best-known enterotoxins of this group are the LTs produced by human strains (H-LT) or porcine strains (P-LT) of enterotoxigenic strains of *E. coli* (ETEC) [20,28–30]. Acute diarrheal diseases caused by ETEC are a major health problem both for humans and animal husbandry. These organisms constitute a common cause of morbidity among persons from temperate zones who travel to the tropics and a significant cause of morbidity among infants in underdeveloped countries. A problem of great economical importance is mortality among newborn domestic animals worldwide, especially piglets, calves, and lambs [30].

## II. EPITOPE SELECTION

The aim of epitope analysis is to identify those fragments of the protein that will preferentially induce the expected type of immune response and to select those for chemical synthesis.

The question is, therefore, what are the principles underlying peptide antigenicity or are there any general methods that allow one to predict whether a given peptide segment of a protein will generate antibody that will react with the native protein?

Various concepts have been introduced in recent years based on the antigenic analysis of several proteins. The parameters surface accessibility, hydrophilicity, evolutionary variability, and segmental mobility have been introduced to predict the position of epitopes from these parameters.

Historically, the first concept was that amino acid sequences that form "corners" of the protein molecule are likely to be antigenic and thus anti-

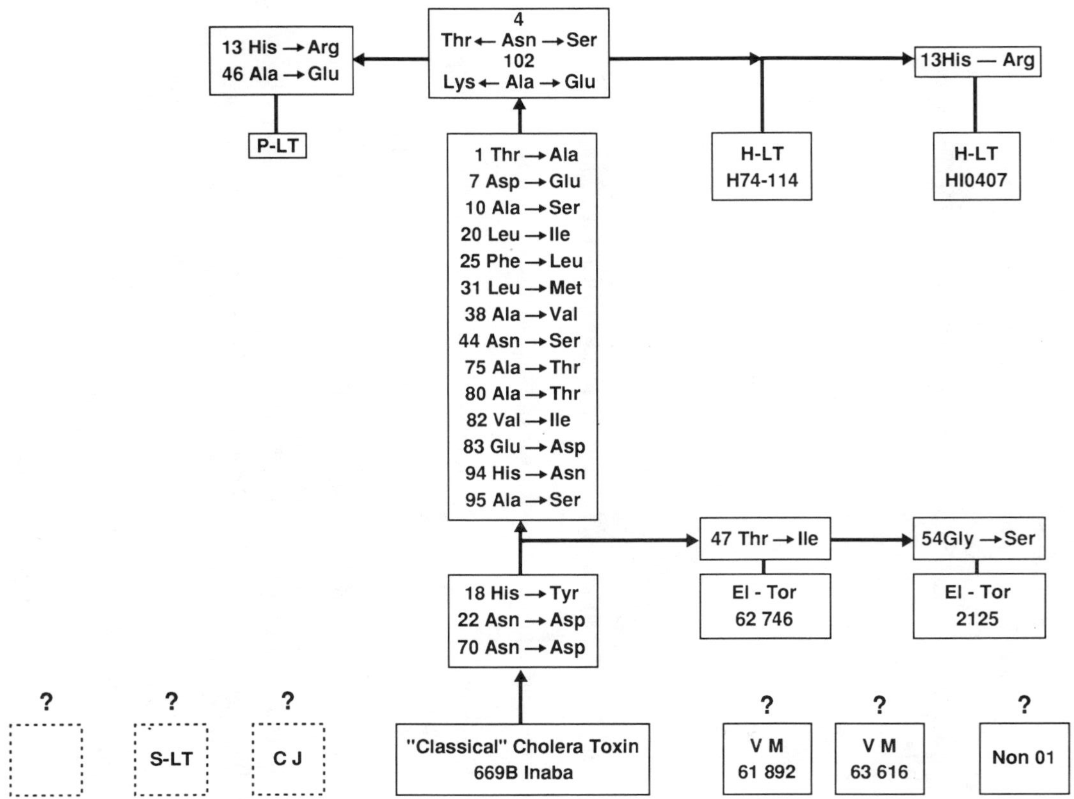


**Fig. 1.** The cholera/coli B-subunit family tree was devised according to Finkelstein et al. [14].

bodies raised to them would react with the native protein. A second concept was that a fragment of the protein composed of "hydrophilic" amino acids is likely to be exposed on the "surface" of the protein molecule and thus would be highly antigenic. Based on this concept, several prediction methods for local maxima in the hydrophilicity plots of proteins have been developed and used with reasonable success to predict antigenicity [31].

It was also shown, in some proteins, that amino acid sequences that form the protein chain termini correlated with antigenicity. Another concept introduced was that antibodies to the regions of the protein with high mobility react with peptide analogues because those peptides can assume in solution the conformation of the mobile segments [32,33]. It was suggested that it may be possible to predict segmental mobility from the primary structure of proteins [34].

Identification of antigenic determinants is possible also by chemical methodology, including fragmentation of the native protein either by chemical cleavage or by controlled proteolysis. The resultant fragments are then screened for immunologically active components that can bind to antibodies and interfere with their interaction with the intact antigen. Chemical modification of various amino acid residues or site-specific mutagenesis are two additional tools that can add significantly to identification of antigenic determinants. Crystallographic studies of the three-dimensional structure of proteins constitute a powerful tool to predict antigenic determinants.

To locate the regions relevant for antigenic determinants in the cholera toxin B subunit, we have considered many of the concepts and techniques delineated above. The following is a summary of the results obtained.

In view of its immunodominant role, the B subunit was the obvious candidate for our investigation (Fig. 2). To locate the regions relevant for antigenic determinants, the B subunit was cleaved by cyanogen-bromide (CNBr) to yield three fragments with the following sequences: 1–37 and 69–101 (linked by a disulfide bridge between cysteines at positions 9 and 86), 38–68, and 102–103. The fragments were separated on a Sephadex G-50 column, and the two large peptides were tested for reactivity with antisera prepared against CT. The largest peptide was partially cross-reactive with anti-CT sera, whereas peptide 38–68 did not bind directly to antisera but was capable of inhibiting binding of intact CT to homologous antisera. Considerations based on the results obtained with the CNBr fragments of the B subunit, as well as several previous studies on antigenically important regions of this protein [35,36], dictated the selection of peptides for chemical synthesis. In this context, a peptide surrounding the arginine at position 35, namely, residues 30–42 (CTP2), was synthesized, because it was suggested, based on chemical modification experiments, that arginine 35 or the region surrounding it may be involved in antibody and receptor binding activity

| | | | | | 5 | | | | | 10 | | | | | 15 | | | | | 20 |
|---|---|---|---|---|---|---|---|---|---|---|---|---|---|---|---|---|---|---|---|---|
| **H-LT** | Ala | — | — | Ser | — | — | Glu | — | — | Ser | — | — | — | — | — | — | — | Tyr | — | Ile |
| **CT** | Thr | Pro | Gln | Asn | Ile | Thr | Asp | Leu | Cys | Ala | Glu | Tyr | His | Asn | Thr | Gln | Ile | His | Thr | Leu |
| **P-LT** | Ala | — | — | Thr | — | — | Glu | — | — | Ser | — | — | Arg | — | — | — | — | Tyr | — | Ile |

| | | | | | 25 | | | | | 30 | | | | | 35 | | | | | 40 |
|---|---|---|---|---|---|---|---|---|---|---|---|---|---|---|---|---|---|---|---|---|
| **H-LT** | — | Asp | — | — | Leu | — | — | — | — | — | Met | — | — | — | — | — | — | Val | — | — |
| **CT** | Asn | Asn | Lys | Ile | Phe | Ser | Tyr | Thr | Glu | Ser | Leu | Ala | Gly | Lys | Arg | Glu | Met | Ala | Ile | Ile |
| **P-LT** | — | Asp | — | — | Leu | — | — | — | — | — | Met | — | — | — | — | — | — | Val | — | — |

| | | | | | 45 | | | | | 50 | | | | | 55 | | | | | 60 |
|---|---|---|---|---|---|---|---|---|---|---|---|---|---|---|---|---|---|---|---|---|
| **H-LT** | — | — | — | Ser | — | — | — | — | — | — | — | — | — | — | — | — | — | — | — | — |
| **CT** | Thr | Phe | Lys | Asn | Gly | Ala | Thr | Phe | Gln | Val | Glu | Val | Pro | Gly | Ser | Gln | His | Ile | Asp | Ser |
| **P-LT** | — | — | — | Ser | — | Glu | — | — | — | — | — | — | — | — | — | — | — | — | — | — |

| | | | | | 65 | | | | | 70 | | | | | 75 | | | | | 80 |
|---|---|---|---|---|---|---|---|---|---|---|---|---|---|---|---|---|---|---|---|---|
| **H-LT** | — | — | — | — | — | — | — | — | — | Asp | — | — | — | — | Thr | — | — | — | — | Thr |
| **CT** | Gln | Lys | Lys | Ala | Ile | Glu | Arg | Met | Lys | Asn | Thr | Leu | Arg | Ile | Ala | Tyr | Leu | Thr | Glu | Ala |
| **P-LT** | — | — | — | — | — | — | — | — | — | Asp | — | — | — | — | Thr | — | — | — | — | Thr |

| | | | | | 85 | | | | | 90 | | | | | 95 | | | | | 100 |
|---|---|---|---|---|---|---|---|---|---|---|---|---|---|---|---|---|---|---|---|---|
| **H-LT** | — | Ile | Asp | — | — | — | — | — | — | — | — | — | — | Asn | Ser | — | — | — | — | — |
| **CT** | Lys | Val | Glu | Lys | Leu | Cys | Val | Trp | Asn | Asn | Lys | Thr | Pro | His | Ala | Ile | Ala | Ala | Ile | Ser |
| **P-LT** | — | Ile | Asp | — | — | — | — | — | — | — | — | — | — | Asn | Ser | — | — | — | — | — |

| | | | |
|---|---|---|---|
| **H-LT** | — | Glu | — |
| **CT** | Met | Ala | Asn |
| **P-LT** | — | Lys | — |

**Fig. 2.** Amino acid sequences of B subunits of cholera toxin (CT) and heat-labile toxin of *Escherichia coli* of human origin (H-LT) and porcine origin (P-LT). Sequences were derived from Lai [11] for CT and from Leong et al. [19] for H-LT and P-LT.

[35]. Another peptide, residues 83–97 (CTP6), consisting of the region surrounding tryptophan at position 88, was synthesized, because results based on chemical modification of this residue suggest that the tryptophan is near or in the binding site of the B subunit [36]. Since it was claimed that the point of highest local average hydrophilicity is located in or immediately adjacent to an antigenic determinant [31], a peptide containing this area of the B subunit (residues 79–84) was chosen according to a statistical method of prediction. On the basis of the reported role of tyrosine residues for antigenicity [37], peptides containing tyrosine were included, namely, CTP1 (residues 8–20) and CTP5 (residues 75–85). Because the latter peptide contained only 11 residues, we prepared an additional peptide elongated by a few residues (residues 69–85, CTP4). Peptide 50–64 (CTP3) is a part of the

inhibitory CNBr fragment 38–68. Later on, CTP3 was elongated by five amino acid residues, resulting in CTP3a (residues 45–64), and two other peptides were synthesized: CTP2a (residues 27–36) and CTP2b (residues 21–36). We have thus studied nine peptides, some of them overlapping, covering about 80% of the B subunit of CT sequence.

The peptides were synthesized by solid-phase methodology, coupled originally to tetanus toxoid as a carrier and injected in complete Freund's adjuvant to achieve immune response. Of most interest among these peptides were CPT1 (residues 8–20) and CTP3 (residues 50–64). Antisera against these two peptides cross-reacted with the intact toxin to a level very similar to that of the homologous peptide–antipeptide reaction and also exerted significant inhibition of the biological activities of cholera toxin. Thus, they were able to inhibit the cholera toxin–induced cyclic AMP production in various cells to an extent of 60–70% and to neutralize significantly the biological activity of cholera toxin as followed by skin vascular permeability assay and by fluid accumulation in ligated small intestinal loops [38].

It is of interest that peptide CTP5 (residues 75–85) was incapable of inducing any cross-reactivity despite the fact that it contains the area of the B subunit (residues 79–84) that has the highest local average hydrophilicity and contains a tyrosine residue that was reported to have a significant role for antigenicity [37]. Similarly, CTP2 (residues 30–42), which was synthesized around arginine 35 (which was suggested to be involved in antibody and receptor binding activity [35]), induced antibodies that were unable to show any cross-reactivity with the native toxin and had no effect on the receptor binding activity [39,40].

CTP6 (residues 83–97), consisting of the region surrounding tryptophan at position 88 and synthesized because of suggestions that this residue is near or within the receptor binding site of B subunit [36], gave rise to antisera of the highest titer of homologous reaction, but showed only slight reactivity with the intact protein [39]. Moreover, antibodies to CTP6 had no effect on the GM1 binding of cholera toxin B subunit [40].

On the other hand, CTP3 induced antibodies that, though of lower (than CTP6) absolute titer, gave a very strong cross-reactivity with the intact toxin and neutralized its biological activities [38]. We have some evidence that CTP3 induces antibodies against the GM1 binding site, or near it, and exerts its effect by interaction with the B subunit and by blocking its attachment to the GM1 receptor [40] (see Fig. 3).

It is of interest that elongation of a peptide may affect the reactivity of the antibodies elicited by it in both positive and negative directions [41]. It may convert the peptide from one that is incapable of provoking antibodies cross-reactive with the parent protein to one that induces such cross-reactive antibodies. Thus, elongation of CTP2a (residues 27–36) by six amino acids

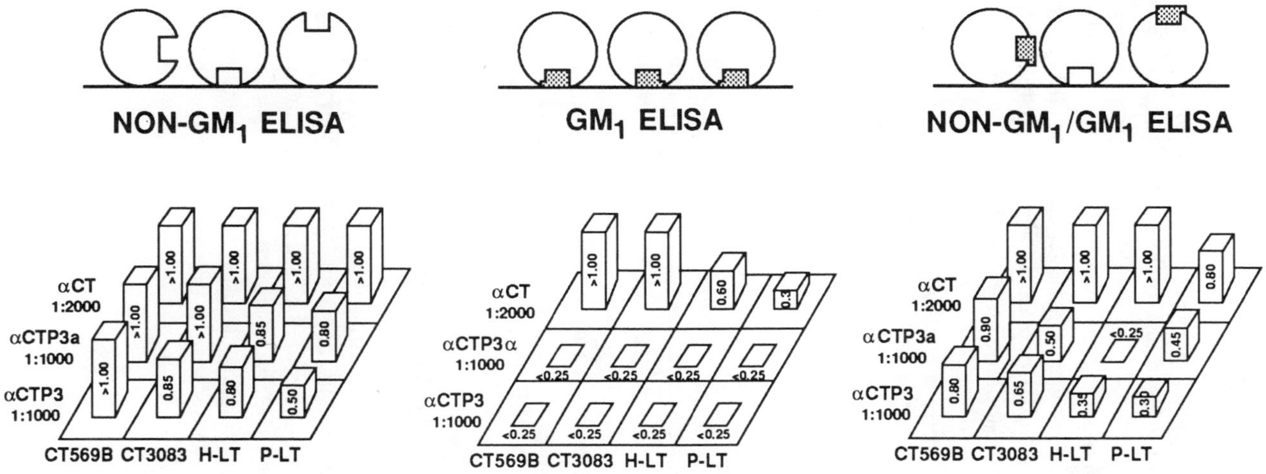


**Fig. 3.** Reactivity of antisera against CTP3 and against CT are compared in $GM_1^-$, $GM_1^+$, and $GM_1^-/GM_1^+$ ELISA with CT, H-LT, and P-LT. Top shows diagrammatically the different ways ELISA was performed: Non $GM_1^-$ ELISA—native proteins are bound directly to plate and then incubated with different antisera; $GM_1$ ELISA—first $GM_1$ ganglioside is immobilized on the plate, then native proteins are added; Non $GM_1/GM_1$ ELISA—$GM_1$ is added to the already immobilized proteins [for detailed discussions see 14 and 40].

converted it to one that elicited cross-reactive antibody, CTP2b (residues 21–36). Similarly, elongation of CTP5 by six amino acids resulted in CTP4 (residues 69–85), which elicited cross-reactive antibodies. Antibodies to a peptide will cross-react with the parent protein if they either possess in free solution a conformation similar to the one within the native protein or if they can easily transconform into such conformation. Therefore, if the protein segment is more flexible, the chance of cross-reaction is higher, even though there will be cases when a small peptide is already capable of having a relatively rigid conformation similar to the one it possesses within a native protein.

It is obvious for the above reasons that it is not realistic to expect that antibodies to any peptide corresponding to a sequence segment within a native protein will cross-react with that protein. On the other hand, elongation of a peptide, which already leads to cross-reactive antibodies by addition of amino acid residues, does not necessarily augment such cross-reactivity. As a matter of fact, judging from cross-reactions with toxins immunologically related to cholera toxin, such cross-reactivities may diminish. Consequently, elongation of CTP3 by five amino acids, CT3a (residues 45–64), induced antibodies that showed cross-reactivity of a definitely lower level with porcine labile toxin of *E. coli* (P-LT) and *Vibrio mimicus* toxin than antibodies to the shorter CTP3 peptide [41,42]. Even with whole CT or whole H-LT, the reactivity was not enhanced by the addition of five amino acid residues to CTP3 (Fig. 4). These results are in accord with the observation that antibodies against CTP3 reacted with human labile toxin (H-LT) of 30 strains of pathogenic *E. coli.* In the case of nine of these strains, anti-CTP3 showed even higher reactivity than anti-whole cholera toxin [43]. It is therefore possible that a shorter peptide that is still cross-reactive may give a better reactivity with a series of proteins related immunologically to the parent protein than the cross-reaction between the complete native proteins themselves. This argument may constitute an important consideration in the evaluation of synthetic peptides as the basis for potential vaccines.

As for the question of whether there are any general methods that allow one to predict if a given peptide segment will generate antibody that will react with the native protein, it seems that at present a single criterion for the identification of antigenic epitopes in native proteins is unlikely to be universally applicable [44].

## III. EPITOPE SYNTHESIS

The synthesis of linear peptides that correspond to various regions of the protein of interest is a crucial step in the development of a possible vaccine. It is not my intention here to give a thorough discussion of the methodologies

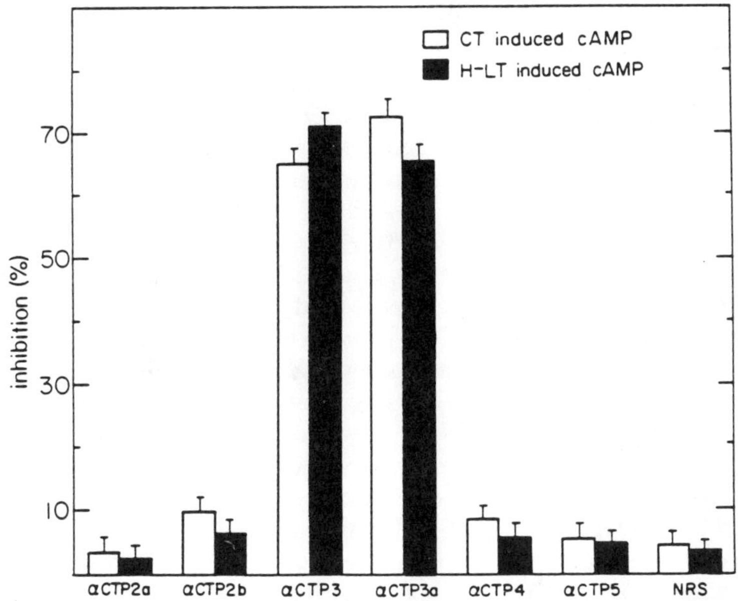


**Fig. 4.** Inhibition of adenylate cyclase activation by CT and H-LT with the various antipeptide antisera.

involved [for review see 45–47], but rather to briefly summarize some of the methodological concepts to be taken into consideration.

The solid-phase approach to peptide synthesis is the current method of choice in peptide chemistry. The characteristics of this method that make it attractive, particularly when a large number of peptides have to be prepared for screening, are its simplicity, speed, avoidance of intermediate isolation, and automatization.

The majority of syntheses based on Merrifield's methodology are carried out on polystyrene-derived supports and use the t-butyloxycarbonyl (t-Boc) group for N-α protection. For the simultaneous cleavage of the completed peptide from the resin and N-terminal and side-chain deprotection, acidolysis procedures are used. The major drawback in the Merrifield synthesis is that certain reactions in the growing peptide chain may not go to completion mainly because the peptide chain may be significantly different in its polar nature from the hydrophobic polystyrene support. This led to the development of peptide synthesis by the Fmoc-polyamide procedure [48,49]. The syntheses are carried out on polar polydimethylacrylamide-derived supports, with the use of base-sensitive α-amino protecting groups (fluorenylmethoxycarbonyl = Fmoc) and a polar solvent (DMF) throughout the synthesis. The major advantage of this method is the use of supports whose solvation properties are similar to those of the growing peptide chain.

Two recent innovative methods are relevant in the screening process of numerous peptides for properties of bioactivity or immunogenicity, as is the case for epitope selection for vaccine development purposes. First, Geysen and coworkers [50] developed a methodology that enables the simultaneous synthesis of many related peptides. The insoluble matrix is constructed from molded polyethylene rods grafted with acrylic acid to give a surface coating of polyacrylic acid. A diamine linker is used to convert free carboxylic into amino functions. Assembly of the polyethylene-derived rods allows mini-scale "conventional" solid-phase peptide synthesis at the tips of the rods within a 96-well Teflon tray. The antigenicity of the peptides produced by parallel synthesis is performed on the resulting peptide-carrier conjugate using the solid-phase enzyme-linked immunosorbent (ELISA) technique.

The second method for the simultaneous synthesis of families of peptides devised by Houghten [51] uses polymeric carriers in permeable polypropylene bags ("tea bags") that are insoluble and solvent resistant. The bags are combined in a reaction vessel and subjected to the various manipulations of the solid-phase synthesis. Bags are removed or added to the assembly for the purpose of different coupling steps. Cleavage is done simultaneously for up to 20 peptide-resin conjugates. Each peptide is then individually separated from the insoluble resin and assayed for its antigenicity. Unlike the previous method, here a purification step and analysis of the product is possible.

## IV. EPITOPE PRESENTATION

### A. Chemical Coupling of Peptide to Carrier

Several investigators have reported achieving antibody response upon immunization with synthetic peptides unattached to any carrier [52,53]. However, in most cases, the synthetic peptides, if not large enough, are not by themselves good immunogens. It is therefore the usual procedure to attach the synthetic peptide, which represents the desired epitope, to macromolecular carriers, such as proteins or synthetic polyamino acids. Various methods are available to bind or cross-link the epitope with the carrier. We will not discuss here the chemistry of these reactions; rather, we will emphasize that choosing a method of peptide-carrier binding may also play a significant role in the final molecular engineering of the synthetic vaccine.

The most commonly used reagents for the formation of peptide-protein conjugates—glutaraldehyde, soluble diimide (e.g., 1-ethyl-3-(3-dimethylaminopropyl)-carbodiimide hydrochloride) and bis-diazotized benzidine—are of bifunctional nature. Although the above methods have found wide application in vaccine development, they have certain drawbacks relevant to their function in endowing immungenicity to the synthetic peptides. Since the peptide and the protein have a multitude of functional groups such as amino, carboxy, aliphatic hydroxyls, and sulfhydryl groups, the linkage between the two occurs in a random fashion. The major problem resulting from this random interaction is the possibility of creating new antigenic determinants that can hamper the specificity of the immune response. Methods employing hetero-bifunctional reagents may lead to more specific couplings. A group of such compounds have an amino reactive, N-hydroxysuccinimide active ester moiety and a sulfhydryl reactive moiety. In the first of the two consecutive steps, free amino functions on the protein molecule (devoided of free cysteines) are acylated. Second, the synthetic peptide containing an added free sulfhydryl moiety is allowed to react with the modified protein. Alternatively, a free thiol function can be added to the protein, followed by reaction with a peptide containing a maleimide function [54].

Another coupling procedure uses para-aminophenylacetic acid (PAPA), which can generate diazonium moieties. The compound is initially added to a peptide through an amide bond, followed by diazotization, which enables direct coupling of the peptide to the protein carrier [55]. The advantage of binding via PAPA residue is the specificity of the conjugation that occurs only between the N-terminal amino group of the PAPA and histidine or tyrosine residues on the carrier.

In the cholera toxin system the purified peptides were conjugated to protein carriers either using a water-soluble carbodiimide as a coupling agent or through an azo bond when PAPA derivatives of the peptides were employed. Both methods of coupling yielded adequate conjugates [56].

### B. Selection of a Carrier

In the design of a synthetic vaccine, primary consideration should be given to the choice of the carrier. In our studies of cholera toxin–derived peptides, various approaches have been explored. We have used tetanus toxoid (TT) and bovine serum albumin (BSA) as protein carriers. The powerful immunogen KLH has been used successfully as a carrier more than any other protein. BSA has also been a popular carrier for coupling peptides. In our hands, tetanus toxoid as a carrier molecule was more effective than BSA.

One of the main putative advantages of the synthetic vaccines of the future is that they should provoke antibodies against relevant epitopes only. However, problems of hypersensitivity to the carrier may occur when repeated injections are necessary. From this point of view, the use of protein carriers like KLH and BSA is obviously undesirable for the immunization of humans. We have used TT as one effort in this direction because TT is already used for vaccination in humans. However, this approach as well is not without problems. Previous exposure to TT may lead to suppression of the immune response against synthetic epitopes conjugated to it. This phenomenon is called carrier-induced epitope suppression [57]. Such was the case for the CTP3 peptide of cholera toxin [56] and also for synthetic peptides derived from diphtheria toxin and streptococcal M protein [58]. To overcome this problem, it was suggested that purified protein derivative of the tuberculosis bacillus (PPD) be used as a carrier of choice, because it provides powerful T-cell help but gives rise to virtually no antibody response against itself [59].

An alternative approach is to use multichain synthetic polypeptides as carriers. When such a carrier is being used, the result may depend on the nature of the macromolecular carrier. Thus, the synthetic branched polymer multipoly (DL-alanyl)--poly (L-lysine)(A--L) was not an efficient carrier and did not lead to a significant antibody production against the cholera toxin peptides as compared to immunization with the respective free peptides [56]. However that same A--L carrier proved to be an efficient carrier for synthetic peptides in the lysozyme system [60], the bacteriophage MS2 [61], and the diphtheria toxin system [62]. On the other hand, another synthetic polymer (T,G)-A- -L was found to be an effective carrier for the CTP3 peptide [63]. These results indicate that the suitability of a carrier is correlated with the characteristics and specificity of the antigenic determinant coupled to it.

We have used palmitic acid coupling to CTP3 and CTP1 through an N-terminal lysine residue, which in turn was derived by linkage of palmitic acid at both the $\alpha$- and $\epsilon$-amino groups. In the case of CTP3, the conjugation of dipalmityl side chains had an effect comparable with TT conjugation on the level of either the homologous antipeptide response or the cross-reactivity against the intact cholera toxin [56]. Hopp [64] applied the conjugation of dipalmityl lysine for peptides of the hepatitis B surface antigen system. He reported that dipalmityl conjugates were even more effective than KLH conjugates. In our system, a higher efficacy of protein carriers like TT over dipalmityl lysine could be demonstrated (in the case of CTP1). Nevertheless, the addition of lipid moiety is sufficient for the induction of antibodies against the intact cholera toxin.

Peptide polymerization seems to be a decisive parameter in endowing a peptide with immunogenic properties. In our study, CTP3 and CTP6 were polymerized, each one separately. The results demonstrated that polymerization was indeed effective in augmentation of the antibody response; in the case of CTP3, the polymer was as effective as the conjugate with TT [56]. The advantage of polymerization is that it can be applied when there is more than one peptide. In such cases, the resultant polymer will lead to efficient immune response against both peptides, to an extent similar to that achievable by attaching each peptide to a macromolecular carrier. In this respect it is interesting that when two different peptides were attached to the same

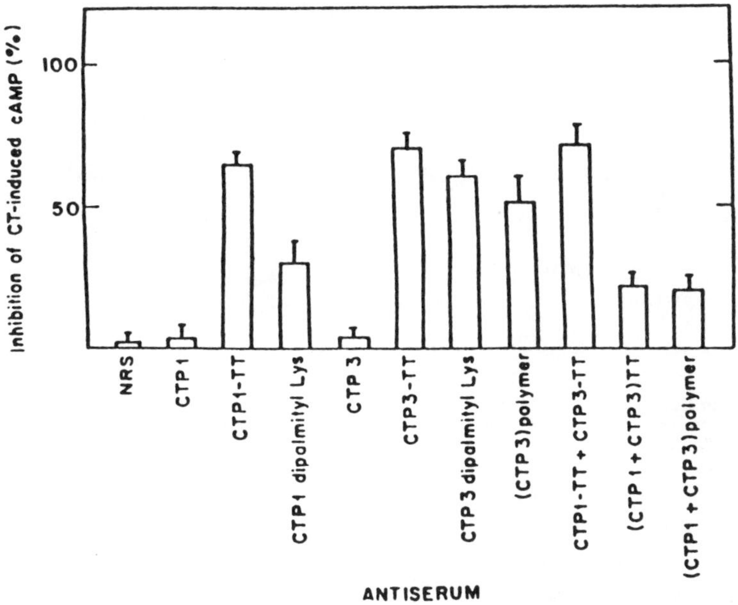


**Fig. 5.** Inhibition of cholera toxin–induced cyclic AMP by different antisera. All measurements were made after 3h incubation of 50ng CT with 1:10 dilution of the indicated antisera. NRS, normal rabbit serum. Values are mean ± SEM.

carrier molecule, the effectivity of the immune response was considerably lower than when each peptide was attached separately [56] (see Fig. 5).

## C. Selection of Adjuvant

An adjuvant can be defined as a substance that nonspecifically enhances the specific response to a given antigen. For experimental immunization in laboratory animals, Freund's complete and incomplete adjuvants are most commonly used. They contain a water-in-oil emulsion with or without killed mycobacteria. Obviously, these adjuvants are not suitable for human use because they induce local reactions, granuloma formation, inflammation, and fever.

At the present time, the adjuvants most commonly used are alum for humans and saponin or mineral oil in the veterinary field. Aluminum compounds may, however, produce local reactions and have the potential of favoring IgE responses.

We have studied the effectiveness of potential new adjuvants using the peptide CTP3 in rabbits and rats. These include the synthetic adjuvant N-

**TABLE I. Antibody Response Achieved by Immunization With CTP3 Conjugates**

| | Antibody Titer[a] | |
|---|---|---|
| Immunizing agent | vs. CTP3 | vs. CT |
| CTP3-TT in PBS | <10 | <10 |
| CTP3-TT + MDP | 10–60 | 10–40 |
| CTP3-TT + Thr-MDP | 10–40 | 10–25 |
| CTP3-TT-Lys-MDP | 180–260 | 90–120 |
| CTP3-(T,G)-A- -L | 10–30 | 10 |
| CTP3-(T,G)-A- -L-Lys-MDP | 160–270 | 90–120 |
| CTP3-TT + avridine | 10–30 | 10 |
| CTP3-(T,G)-A- -L + avridine | <10 | <10 |
| CTP3-dipalmityl-Lys + MDP | <10 | <10 |
| CTP3-dipalmityl-Lys + Thr-MDP | <10 | <10 |
| CTP3-dipalmityl-Lys + avridine | <10 | <10 |

[a]The serum dilution that gave 50% antibody binding.

acetyl-muramyl-L-alanyl-D-isoglutamine (MDP, muramyl dipeptide) and two derivatives, threonyl-MDP and MDP-lysine. Another adjuvant, avridine (CP-20, 961; (N,N-dioctadecyl-N′, N′-bis (2-hydroxyethyl)-propanediamine), a synthetic lipoidal amine [65], was also tested.

Conjugates of CTP3 with T,G-(A--L) or TT were administered with either MDP or threonyl-MDP in aqueous medium or given after covalent conjugation with MDP-Lys. Avridine was used with CPT3-TT and CTP3-T,G-(A--L). Dipalmityl-Lys-CTP3 conjugates were also administered to rabbits and rats with MDP, threonyl-MDP, and avridine. Groups of four rabbits and five SPD rats were used for these experiments; the results are summarized in Table I. Both MDP and avridine were found to be ineffective in inducing antipeptide or anti-CT antibodies when used in aqueous medium. No difference was observed between MDP or threonyl-MDP. In contrast to these results, when the adjuvant MDP-Lys was covalently conjugated to the peptide-carrier conjugate, highly significant antibody titers toward the homologous peptide as well as against native CT were obtained [63]. In this context no significant differences in the antibody level were obtained when TT or the synthetic T,G-(A--L) were used as the carrier (Fig. 6). Thus, a totally synthetic vaccine with built-in adjuvanticity, namely CTP3-T,G-(A--L)Lys-MDP, could be constructed and successfully used to induce antibody production when administered in physiological medium [63].

## V. VACCINE DESIGN

In designing an effective vaccine against cholera toxin we should ask, what should this vaccine achieve? This section discusses several possible

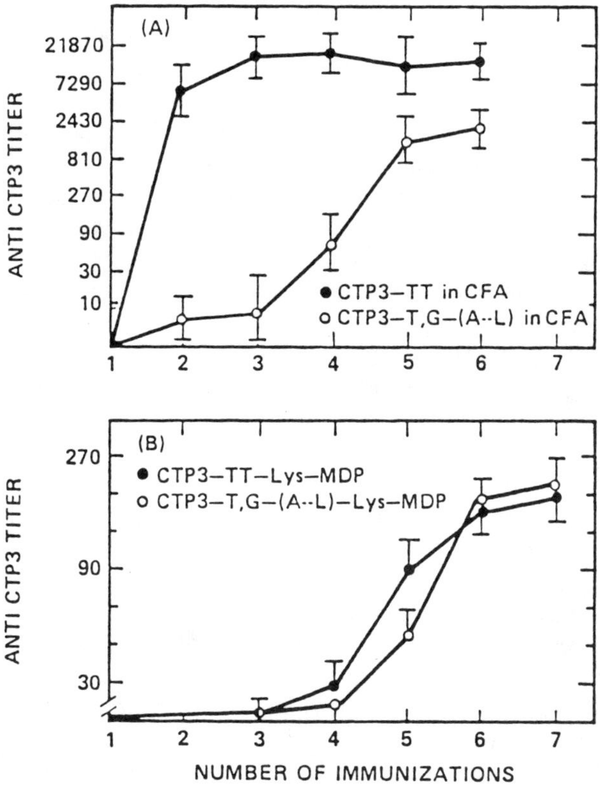


**Fig. 6.** Immune response toward CTP3 induced by immunization in complete Freund's adjuvant **(A)** and with the covalently conjugated MDP **(B).**

methods of vaccination using the synthetic approach: 1) immunization with synthetic peptides to provoke systemic, neutralizing antibodies toward cholera and related toxins; 2) priming immunization by the synthetic peptides; 3) promotion of mucosal immunity in the gut, presumably mediated by IgA; and 4) the possibility of developing a live vaccine based on the synthetic approach and combining recombinant DNA techniques with the synthetic peptide technology.

## A. Systemic Neutralizing Antibody Response by Synthetic Peptides

Of most interest among these peptides were CTP1 (residues 8–20) and CTP3 (residues 50–64). Antisera against these two peptides exerted significant inhibition of biological activity of cholera toxin [38,39]. The toxic effect of CT can be demonstrated by skin vascular permeation and by fluid accumulation in ligated small intestinal loops, as well as, on the biochemical level, by the induction of adenylate cyclase. The inhibitory effect of the antipeptide sera was manifested in all the assays of the biological activity of the toxin with very good correlation between the biochemical level and the end biological effect of the toxin [38]. In both cases inhibition values of 60 to 70% were obtained. Simultaneous immunization with the two peptides CTP1 and CTP3 was carried out in order to evaluate whether the combined immune response might lead to augmentation of the neutralizing effect achieved by each peptide separately. The peptides were administered to rabbits when bound together to the same tetanus toxoid carrier molecule, presented on separate carrier molecules, or in a polymerized form given without any carrier. However, when compared to the results obtained with the individual peptides, either as TT conjugate or in the polymerized form, there was no elevation in the anticholera toxin binding or augmentation of the neutralizing efficiency of the antibodies obtained as a result of the combined immunization [38]. Similarly, simultaneous addition of the two antisera (anti-CTP1 and anti-CTP3) did not result in additivity of the inhibitory effect [38].

It is noteworthy that all synthetic peptides investigated as such do not induce any activation of adenylate cyclase as measured by cAMP production or fluid secretion into ligated ileal loops [38]. This implies a low probability of undesirable biological effects as a result of their administration, which is a crucial consideration for the possible use of synthetic material for vaccination.

In view of the immunological similarity between CT, H-LT, and P-LT, it is conceivable that these toxins may serve for interchangeable immunization against the different infections. It was also suggested that these three enterotoxins share common antigens and, at the same time, have their own unique antigenic determinants [40]. Using the synthetic peptides we could corroborate this assumption by demonstrating the different reactivity of antibodies to synthetic peptides with the different toxins. For example, we could demonstrate that CTP3 is most definitely a shared epitope among the investigated toxins. The antibodies it induced were highly cross-reactive and reacted to a very similar level with all three toxins, in contrast to the gradual reactivity of the antibodies against intact CT with H-LT and P-LT [42,43].

Furthermore, anti-CTP3 reacted in an identical fashion with all four genetically engineered B-subunit proteins in which substitutions were made in the B subunit of P-LT such that the resulting chimeras have the H-LT-B amino acid residues at positions 102, 46, 102, and 46, and at 13 and 4. In contrast, anti-CT was able to distinguish among them [42]. This further emphasizes the fact that this region (50–64) is a conserved antigenic determinant in the various toxins. On the other hand, CTP1 represents a specific determinant for CT with partial similarity to H-LT-B only.

As expected, antisera to both CTP3 and CTP1, which were found inhibitory toward CT, were equally effective in neutralizing the biological activity of H-LT as well [43]. This was manifested by inhibition of the adenylate cyclase activity (Fig. 4) and the fluid secretion into ligated ileal loops of rats.

Of special interest is the observation of cross-reactivity of the antipeptide sera with multiple strains of *E. coli* identified as LT positive by ELISA. The reactivity of the antipeptide sera with randomly selected crude LT preparations deriving from 31 different pathogenic *E. coli* strains isolated from humans with diarrhea was tested by radioimmunoassay (RIA) and compared with that of anti-LT serum and anti-CT. Anti-CTP3 was able to cross-react to a different extent with 30 of 31 LT preparations. Anti-CTP1 cross-reacted with 24 of the different LTs, while anti-CT serum cross-reacted with 26 of the LT preparations. Thus, some of the LT preparations are recognized by antibodies against the synthetic peptides but not by anti-CT [43]. Hence, when cross-reactivity is considered, a synthetic peptide like CTP3 may prove to be an even more efficient immunizing agent than the intact toxin.

### B. Priming Immunization

The ability of synthetic peptides to prime the immune system toward a secondary stimulus with whole organism or native protein is of great importance in vaccination. This approach for immunization might be of general practical value in endemic areas, where the population is probably constantly vulnerable to a low level of a particular infectious agent. This exposure has no influence on the unprimed immune system, but could serve as a booster in the case of individuals primed with an appropriate peptide, leading to secondary immunological reaction. The basis for priming effect lies within the heterogenic nature of the immune response. Antibodies that are formed against synthetic peptides most probably have various specificities, only some of which may be relevant to the native protein. This is due in part to the flexible conformation of peptides in solution [32]. A secondary stimulus with the native protein serves as a specific booster only to those cells that produce antibodies that recognize the native protein or causative agent, thus selectively enhancing the specific antibody population. Because the administration of the native protein serves as a booster to cells that were already sensitized, a very low dose should be sufficient for eliciting a secondary immune response against the intact protein. This phenomenon was indeed observed with several systems using synthetic peptides [66,67], although not in every case [68].

Rabbits were "primed" by one or two injections, at 4 week intervals, of 1 mg peptide-TT conjugate, followed by a single booster injection of a subeffective dose (1 μg) of CT. All cholera toxin peptides investigated were able to prime for a significant neutralizing anti-CT immune response as demonstrated by high levels of cross-reacting antibodies and by efficient inhibition of CT-induced cAMP [66]. The priming effect was achieved irrespective of whether or not the immunization with peptide conjugate alone can lead to neutralizing response. In fact, even conjugates of peptides that do not induce any CT–cross-reactive antibodies also led to a priming phenomenon, though to a lower level (e.g., CTP2 induced antibodies capable of inhibiting about 50% of CT-induced cAMP). The priming effect was even more pronounced after a single immunization with the carrier-linked peptides, which as such did not result in any immune response. After boosting, both the level of anti-CT antibodies and their neutralizing capacity were essentially the same as those obtained by boosting after two injections of the conjugate. Control experiments of a single administration of 1 μg CT to unprimed rabbits or to rabbits primed with a peptide unrelated to CT sequence gave no response whatsoever, namely, neither cross-reacting antibodies nor any neutralization of CT activity. Moreover, a single dose of CTP3, by itself nonimmunogenic, primed rabbits to a vigorous immune response toward subsequent administration of single small subimmunizing doses of any of three toxins (CT, H-LT, P-LT). We have further shown that the dose of the boosting agent, the CT, has a major role in the level of the priming effect, while the number of the priming injections does not seem to be important. It was also demonstrated that there is a definite requirement for the presence of a macromolecular carrier. Both native proteins (TT or BSA) and synthetic polypeptide (A- -L) were adequate for this purpose. Of equal importance is the role of adjuvants; injections of CTP3-TT without adjuvant, followed by CT boost, failed to give any immune response. An interesting observation is the specificity of the antibodies formed as a result of priming with the synthetic peptide conjugate followed by a booster with the intact CT. Intuitively, one would guess that most of the resultant antibodies will be specific toward the priming agent. However, as manifested by inhibition measurements, this was not the case. Unless a repeated immunization with the peptide conjugate was performed, which led to appreciable levels of antipeptide antibodies, the specificity of the antibodies elicited after the booster was directed mainly toward the whole CT molecule and only a small fraction were specific toward the peptide used for priming [66].

## C. Local Immunization

There is accumulated evidence concerning the role of intestinal sIgA antibodies in the defense mechanism against several bacterial and viral infections of the gastrointestinal tract in different species [69,70], including experimental cholera [71]. It was therefore of interest to induce sIgA anti-

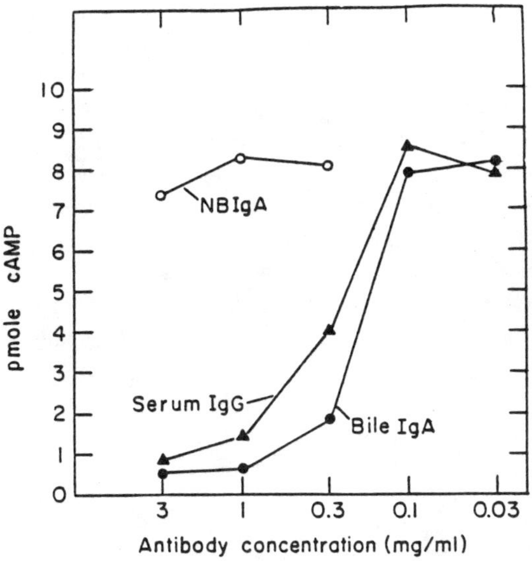


**Fig. 7.** Comparison of the reactivity of purified secretory IgA-anti-CT and purified IgG-anti-CT in inhibiting CT-induced cAMP production. CT-neutralizing sIgA were isolated from the bile of rats immunized with CT directly into their intestines. CT-neutralizing IgG were isolated from sera of rats immunized with CT intradermally in CFA. NBIgA, normal bile IgA.

bodies toward a synthetic peptide from CT-B subunit sequence. In the particular case of whole CT or its native B subunit, sIgA antibodies are consistently obtained by various protocols (Fig. 7), but this does not apply to other soluble proteins even if they have intestinal-wall binding properties [72]. In view of the availability of synthetic peptides that contain neutralizing epitopes, I pursued this line of research by inducing sIgA response toward the neutralizing epitope CTP3 (residues 50–64) of CT-B subunit. Moreover, CTP3 peptide can be used as a model system for induction of local immunity against synthetic peptides with relevance to other gastrointestinal infectious diseases.

The immunization protocols for sIgA responses to soluble proteins are ill defined, in contrast to those for systemic IgG responses. We used Peyer's patch immunization, which was shown to consistently induce strong primary and secondary sIgA responses, at least against particulate antigens [73]. Using this protocol we were able to demonstrate (for the first time, to my knowledge) the induction of sIgA antibodies against a synthetic peptide. These sIgA antibodies were cross-reactive with native CT and also partially neutralized its biological activity [74].

CTP3-TT conjugate was injected three times (2 weeks apart) mixed with

CFA into a dozen Peyer's patches (3–10 μg in 8–10 μl per patch) of each of adult OFA rats. Four days after the last dose, rat bile was accumulated for 3 to 5 days. Bile was collected on ice in the presence of EDTA, iodoacetamide, benzamidine, and azide to prevent protein degradation. Bile samples were filtered on AG1 × 8 resin columns to remove bile salts and pigments. Daily samples from individual rats were pooled and concentrated tenfold, and secretory IgA fractions were isolated by gel filtration on Ultragel Ac22 columns. The purified sIgA fractions were tested for antibody activity against the CTP3 peptide, the whole CT protein, and the TT carrier protein. Results of these experiments showed that the purified sIgA from CTP3-TT–immunized rats reacted with the homologous CTP3 peptide in solid-phase RIA and also with the whole CT, the two reactions being of comparable titers [74]. Immunoblotting experiments confirmed these results. Furthermore, these IgA samples also inhibited the biological activity of CT as demonstrated by inhibition of CT-induced cyclic AMP levels [74].

## VI. COMBINED USE OF SYNTHETIC PEPTIDES AND RECOMBINANT DNA TECHNOLOGIES

### A. CTP3-β-Galactosidase Fusion Proteins

We attempted to bridge the synthetic and recombinant DNA approaches. Our working hypothesis was that the expression of the CTP3 peptide by bacteria other than *V. cholerae* may provide an appropriate agent for induction of immunity toward cholera and *E. coli* toxins and that this could be approached by recombinant DNA techniques.

Two plasmids containing a synthetic gene coding for the region 50–64 in the amino acid sequence of cholera toxin, denoted pAM1 and pAM2, were inserted in phase into the gene coding for *E. coli* β-galactosidase (Fig. 8).

The vector used for constructing pAM1 is a derivative of pMC1403 with an *Eco*R1-*Bam*H1 fragment containing the $\lambda P_R$ promoter sequence joined to the lac operator and a synthetic ribosome binding sequence, followed by an initiation codon in frame with a truncated *lac*Z gene. The unique *Bam*H1 site overlapping the initiator ATG was used to introduce a 54-base-pair synthetic DNA duplex coding for CTP3. The design of the DNA complex includes single-stranded extensions allowing its ligation into a *Bam*H1 site. These termini extend the CTP3 coding sequence with codons for Met, Asp, and Pro at the 5′ end and with a Met codon at the 3′ end. The latter Met residue can be useful for the regeneration, if desired, of a CTP3-like peptide by CNBr cleavage of the hybrid protein. The synthetic DNA sequence is based on the preferred codon usage in *E. coli* and deviates from the corresponding sequences in the genes coding for the subunit of cholera and *E. coli* enterotoxins. A *Kpn*I site was incorporated into the sequence to facilitate the

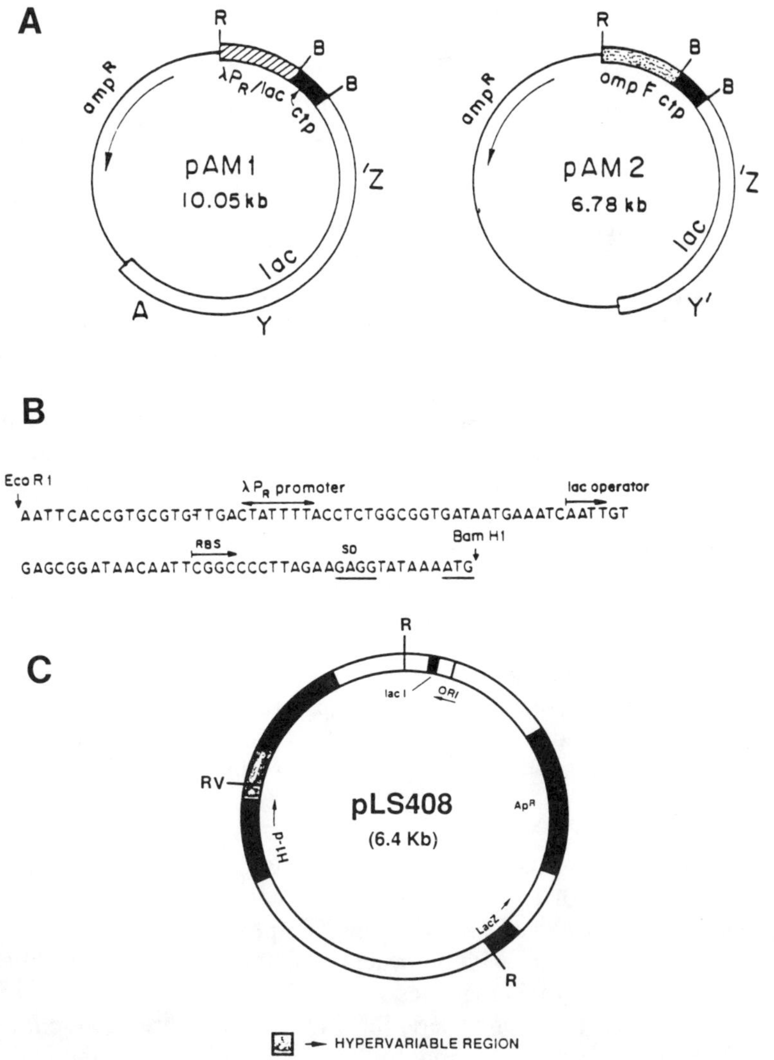


**Fig. 8.** Structure of vectors used for the expression of CTP3. **A:** General structure of vectors for the expression of CTP3-β-galactosidase. Hatched box, composite $P_R$/Lac promoter (in pAM1); dotted box, *ompF* promoter and amino-terminal coding sequence (in pAM2); filled boxes, CTP3 coding sequence (ctp); open boxes, parts of the *lac* operon; line, part of pBR322 sequence. Restriction sites: R, *Eco*R1; B, *Bam*H1. **B:** Sequence of the regulatory region in pAM1. Arrows designate cleavage sites by the indicated restriction enzymes. RBS, ribosome binding site; SD, Shine and Dalgarno sequence. **C:** General structure of the flagellin exposition vector.

analysis of recombinants. In addition, care was taken to avoid possible formation of secondary structures that could sequester the Shine-Dalgarno and ATG sequences. Although the synthetic DNA can be inserted into the vector in two orientations, only the correct orientation will yield a fused β-galactosidase. In the opposite orientation, the reading frame is terminated within the insert.

For the construction of pAM2, the synthetic DNA fragment replaced the sequence between the two closely located *Bam*H1 sites of pORF1 [75]. Introduced in this way, the sequence of the cloned fragment matches the reading frames of both the upstream *omp*F and the downstream *lac*Z genes. The presence of the *omp*F signal peptide at the amino terminus of the hybrid protein specified by pAM2 could make, in principle, the protein a component of the bacterial outer membrane [76].

The resulting fusion protein that was expressed by the two vectors was enzymatically active and reacted with antibodies against β-galactosidase. It also reacted with antibodies against CTP3 and, though to a much lesser extent, with antibodies against intact CT [77]. Immunization with the fusion protein did not lead to significant titer of antibodies recognizing CT. However, when followed by a booster injection of a minute amount (1 μg) of CT, too small to provide any immune response by itself, it led to a substantial level of neutralizing anti-CT antibodies. Similar results were obtained after boosting with the heat-labile toxin of *E. coli*. These findings demonstrate that when relevant antigenic determinants are defined, their corresponding synthetic oligonucleotides can be utilized, thus combining genetic engineering and peptide chemistry for future synthetic vaccines.

### B. CTP3-*Lam*B Fusion Protein

We reasoned that the fact that the CTP3-β-galactosidase fusion proteins did not induce high levels of antipeptide antibodies and only priming response could be achieved was because CTP3 peptide was not exposed on the surface of the bacteria. With the recent availability of some exposition vectors, we have tested this hypothesis.

The *Lam*B protein is a trimeric integral outer membrane protein from *E. coli* K12 that functions as a pore for permeation of maltose and maltotriose and is required for the entry of higher dextrins. Secondly, *Lam*B is functioning as a cell surface receptor for bacteriophages.

Recently, Maurice Hofnung's group in Paris have shown that it is possible to insert a foreign epitope into certain sites on the *Lam*B protein without affecting most of the properties of this protein [78]. The hybrid proteins are stable and nontoxic and therefore can be produced at a high level without affecting the exponential growth rate of the bacteria. They are able to yield trimers of *Lam*B protein that appear to be correctly inserted into the outer membrane of the bacteria. These hybrid *Lam*B protein–possessing bacteria are able to grow on dextrins and are sensitive to *Lam*B-specific phages [78].

Specifically, two sites seem to be suitable to insertion of foreign epitopes: at amino acid 153 and at amino acid 374. In these cases, the epitopes are exposed and recognized by specific antibodies at the cell surface.

We have used the *Lam*B "exposition vector" for the insertion of an oligonucleotide specifying the CTP3 peptide [79]. Recombinant plasmids that express hybrid CTP3-*Lam*B polypeptides in *E. coli* were constructed. Two distinct constructs were obtained with either one or two CTP3 peptides inserted in series within the *Lam*B primary sequence. Expression of the hybrid products was measured in *E. coli* K12 and in an *aroA Salmonella typhimurium* live vaccine strain. All strains express hybrid CTP3-*Lam*B polypeptides that are exported to the outer membrane. These hybrid polypeptides become major surface constituents upon induction with IPTG at about $10^6$ copies per cell. Antigenicity of the genetically fused peptides was demonstrated by immunoblot using polyclonal and monoclonal antibodies raised against synthetic CTP3. Immunofluorescence studies with these antibodies reveal that CTP3 is exposed at the cell surface. Immunization of rabbits with the gel-purified hybrid polypeptides elicited high titers of anti-CTP3 serum antibodies. This further demonstrates the usefulness of the *Lam*B vector for the highly efficient presentation of foreign epitopes at the cell surface. Experiments to evaluate the immune response of mice against CTP3 expressed in the *S. typhimurium* live vaccine strain are in progress.

### C. CTP3-Flagellin Hybrid Protein

The flagellar filament which is part of the structure responsible for motility in flagellated bacteria is composed of a single protein, the flagellin, that is polymerized to form a left-handed helix. The flagellar filament is a major bacterial immunogen and a high degree of variability in flagellar antigens in salmonella and other microorganisms is well known. The recognition of different flagellar serotypes in salmonella has been a very important tool for epidemiologists, since H antigens together with O antigens are useful for tracing outbreaks of food poisoning. Following the cloning and sequencing of several flagellin genes of salmonella in the last several years, it became clear that there is conservation of both the amino- and carboxy-terminal regions of the proteins, while the central portion of flagellin proteins is hypervariable, showing no more than 30% homology between different serotypes [80]. This suggests that the central region of the protein is the primary antigenic determinant, while the terminal regions are important for proper assembly of the protein for the formation of the filament. Because deletions of portions of the central hypervariable region of the flagellin gene caused little effect on the function of the protein, we reasoned that it is possible to insert and replace a portion of the hypervariable region by a foreign antigenic fragment and thus generate a neoflagellar antigen.

With Salete Newton and Bruce Stocker, we tested this hypothesis with the CTP3 peptide of cholera toxin B subunit [81]. The flagellin gene used, *H1-d,* determines the phase-1 flagellar antigen, *d,* of *Salmonella muenchen* by specifying a protein of 509 amino acids; its sequence includes a ''hyper-variable segment'' of approximately 350 base pairs with no more than 30% identity, at the amino acid level, with the corresponding sequences in three other salmonella flagellin genes [80]. Deletion of a 48-base-pair *Eco*Rv-*Eco*Rv fragment in the hypervariable region of the *H1-d* flagellin gene was produced and then transferred to a PUC19 plasmid to produce plasmid pLS408 (Fig. 8). This deletion reduced but did not abolish flagellar function. The remaining *Eco*RV site of pLS408 allows the blunt-end insertion of oligonucleotides.

A synthetic 45-base-pair double-stranded oligonucleotide specifying CTP3, with codon usage corresponding to that in sequenced salmonella flagellin genes and with a new *Cla*I site to facilitate analysis, was inserted at the unique *Eco*RV site of plasmid pLS408. The recombinant plasmid was introduced into CL447, a flagellin-deficient *E. coli* strain. The oligonucleotide insert improved flagellar function, as judged by the rate of spread in semisolid medium of clones with the insert.

To investigate the function of the recombinant flagellar gene in strains of salmonella, the plasmid containing CTP3 was used to transform various nonflagellated salmonella strains. The transductants obtained had functional flagella as shown both by rapid-spreading growth in semisolid medium and by translational motility of nearly all the bacteria of log-phase broth cultures examined by microscopy. Western blots of lysed bacteria or disaggregated flagella showed that a single band bound both rabbit anti-*d*-antibody and TE33 (a mouse monoclonal anti- CTP3 antibody) [81]. (The absence of amino acids 62–64 in one plasmid resulted in reduced binding of monoclonal antibody TE33, which suggests that these amino acids are part of the binding site for this monoclonal). Anti-CTP3 antibody immediately immobilized and rapidly agglutinated bacteria containing the plasmid with the CTP3 insert-bearing flagellin gene, but had no effect on the same strain made motile by plasmids with the wild-type gene or on that with the 48-base-pair deletion. Exposure of the CTP3 epitope at the surface of the flagella was also shown by immunofluorescent visualization and gold immunolabeling of the flagella of formalin-fixed bacteria [81].

Immunogenicity of the insert-bearing flagellin was subsequently shown in mice by intraperitoneal administration of live flagellated bacteria of a strain attenuated by nonreverting mutation. Such salmonella strains, made aromatic dependent by nonreverting *aroA* mutation, are nonvirulent and have been found to be safe and effective as live vaccine [82]. Mice were injected with $5 \times 10^6$ bacteria, either formalin killed or live; three intraperitoneal injections were given at 7 day intervals. One week after the last dose, the mice were bled and their sera tested by ELISA with CTP3 peptide or cholera toxin as coating antigen. Antibody to the inserted epitope was detected in all the sera, and they reacted as strongly with cholera toxin as with CTP3 peptide [81].

Thus the hypervariable region of flagellin gene *H1-d* can be used as a site for insertion of a synthetic oligonucleotide specifying a foreign epitope that is then expressed as part of the flagellar antigen in live vaccine strains of salmonella and will induce the production of antibody specific for the foreign epitope.

## VII. CONCLUDING REMARKS

The results that have been obtained with synthetic cholera toxin peptides share features with those obtained with peptides of other proteins. It is now quite clear that virtually any peptide can be made to elicit antibodies if appropriately engineered (carrier, adjuvant, etc.). For some peptides, antibodies that react with the native parent molecule (the toxin) may be elicited, while for relatively few peptides, neutralizing antibody can be elicited. As the complexity of an antigen is reduced with regard to size, charge, conformation, and "mobility," its capacity to induce neutralizing antibodies is affected. As there are no methods to predict discontinuous epitopes as yet, and since many, if not most, immunogenic epitopes are discontinuous, the design of efficacious synthetic vaccines is particularly challenging.

General considerations important for choosing peptides for synthesis are dealt with in this chapter. However, at present a single criterion for identification of antigenic epitopes in a given protein molecule is unlikely to be universally applicable. At the level of synthesis, the ability to test the immunoreactivity of a peptide in situ should prove useful in peptide evaluation and design [50].

The use of immunopotentiating conjugates along the line of (T,G)- A- -L with a backbone carrier and protective antigen can be envisaged in future vaccines. Peptide sequences of protective antigens will continue to be genetically engineered or chemically synthesized and these will require immunopotentiators or carriers to produce successful vaccines. Various possibilities that exist now have been described. As the synthetic peptide epitopes vary in amino acid composition and sequence, in size and in conformation, it is difficult to imagine that one type of carrier will be the solution in all cases. In the case of strong immunogenic carriers such as proteins, one must also weigh the advantages of a strong immunogen, leading to a big crop of antibodies (including antibodies toward the attached epitope), vs. a weak immunogen, which would lead to a smaller crop of total antibodies, but a much higher proportion against the desired epitope.

As cited in the chapter, a variety of carriers and adjuvants have been employed. No single combination stands out as being clearly superior to others. Improvements are needed for the duration of the response to vaccination, as well as its level, specificity, and isotype (for example, mucosal immune response for cholera synthetic vaccine).

I have also summarized our most recent work in which an innovative method was developed to insert foreign epitopes into flagellin in a manner that permits assembly of functional flagella [81]. The fact that flagella protrude out from the cell is significant because they are not masked by lipopolysaccharide, which sometimes obscures outer-membrane-protein antigenic determinants or epitopes embedded within them from surveillance by the immune system. The method developed should thus afford an innovative way to present specific epitopes and determine the nature of the responses in the secretory, humoral, and cellular immune systems.

Obviously, it was not my intention in this chapter to review all recent advances in vaccine development. Although the flagellin system was emphasized (for obvious reasons), it should be mentioned that most attention on carrier vaccines has focused in recent years on vaccinia virus systems, not discussed here.

A further note of caution: More and more published studies indicate that single epitopes often do not elicit an immune response in certain individuals of an outbred population. In fact, immunization with an antigen with multiple epitopes may be needed in order to elicit a protective immune response in the largest fraction of a population that is very heterogeneous with regard to MHC antigens. This point was not discussed in this chapter. The papers by Benjamini et al. [83] and Good et al. [84] are instructive in this regard.

Although problems still exist and many improvements have to be introduced, there are obvious advantages to the synthetic approach to vaccine development. Synthetic vaccines will have much-reduced side effects; they will provide multivalent vaccines with built-in adjuvanticity; and, as has been demonstrated above, their high potential as general vaccines against the cholera/coli family of toxins emphasizes their future prospects.

## REFERENCES

1. Sela M: Antigenicity: Some molecular aspects. Science 166:1635, 1969.
2. Arnon R, Maron R, Sela M, Anfinsen CB: Antibodies reactive with a native lysozyme elicited by a completely synthetic antigen. Proc Natl Acad Sci USA 68:1450, 1971.
3. Arnon R: Synthetic vaccines—dream or reality? In Kohn A, Klinberg AM (eds): ''Immunity of Viral and Rickettsial Diseases.'' New York: Plenum Press, 1972, p. 209.

4. Arnon R, Shapira M, Jacob CO: Synthetic vaccines. J Immunol Methods 61:261, 1983.
5. Sela M, Arnon R, Jacob CO: Synthetic peptides with antigenic specificity for bacterial toxins. In: "Ciba Foundation Symposium 19: Synthetic Peptides as Antigens." Chichester: John Wiley and Sons, 1986, p. 184.
6. Gill DM: The arrangement of subunits in cholera toxin. Biochemistry 15:1242, 1976.
7. Kurosky A, Markel DE, Touchstone B, Peterson JW: Chemical characterization of the structure of cholera toxin and its natural toxoids. J Infect Dis 133:S14, 1976.
8. Gill DM, Meren R: ADP-ribosylation of membrane proteins catalyzed by cholera toxin: Basis of the activation of adenylate cyclase. Proc Natl Acad Sci USA 75:3050, 1978.
9. Lay CY: The chemistry and biology of cholera toxin. CRC Crit Rev Biochem 9:171, 1980.
10. Kurosky A, Markel DE, Peterson JW: Covalent structure of the β chain of cholera enterotoxin. J Biol Chem 252:7257, 1977.
11. Lai CY: Determination of the primary structure of cholera toxin B subunits. J Biol Chem 252:7249, 1977.
12. Finkelstein RA: Antitoxic immunity in experimental cholera: Observations with purified antigens and the ligated ileal loop model. Infect Immun 1:464, 1970.
13. Finkelstein RA, LoSpalluto JJ: Pathogenesis of experimental cholera: Preparation and isolation of choleragen and choleragenoid. J Exp Med 130:185, 1969.
14. Finkelstein RA, Burks ME, Zupan A, Dallas WS, Jacob CO, Ludwig DS: Antigenic determinants of the cholera/coli family of enterotoxins. Rev Infect Dis 9:S490, 1987.
15. Finkelstein RA, Vasil ML, Holmes RK: Studies on toxinogenesis in *Vibrio cholerae.* I. Isolation of mutants with altered toxinogenicity. J Infect Dis 129:117, 1974.
16. Lockman H, Kaper JB: Nucleotide sequence analysis of the A2 and B subunits of *Vibrio cholerae* enterotoxin. J Biol Chem 258:13722, 1983.
17. Mekalanos JJ, Swartz DJ, Pearson GDN, Harford N, Groyne F, de Wilde M: Cholera toxin genes: Nucleotide sequence, deletion analysis and vaccine development. Nature 306:551, 1983.
18. Geary SJ, Marchlewicz BA, Finkelstein RA: Comparisons of heat-labile enterotoxins from porcine and human strains of Escherichia coli. Infect Immun 36:215, 1982.
19. Leong J, Vinal AC, Dallas WS: Nucleotide sequence comparison between heat-labile toxin B-subunit cistrons from *Escherichia coli* of human and porcine origin. Infect Immun 48:73, 1985.
20. Yamamoto T, Yokota T: Sequence of heat-labile enterotoxin of *Escherichia coli* pathogenic for humans. J Bacteriol 155:728, 1983.
21. Clements JD, Finkelstein RA: Isolation and characterization of homogeneous heat-labile enterotoxins with high specific activity from *Escherichia coli* cultures. Infect Immun 24:760, 1970.
22. Dallas WS, Falkow S: Amino acid sequence homology between cholera toxin and Escherichia heat-labile toxin. Nature 28:499, 1980.
23. Peterson JW: Salmonella toxin. Pharmacol Ther 11:719, 1980.
24. Finkelstein RA, Marchlewicz BA, McDeonald RJ, Boesman-Finkelstein M: Isolation and characterization of a cholera-related enterotoxin from Salmonella typhimurium. FEMS Microbiol Lett 17:239, 1983.
25. Yamamoto K, Takeda Y, Miwatoni T, Craig JP: Evidence that a non-O1 *Vibrio cholerae* produces enterotoxins that is similar but not identical to cholera enterotoxin. Infect Immun 41:896, 1983.
26. Spira WM, Fedorka-Cray PJ: Purification of enterotoxins from *Vibrio mimicus* that appear to be identical to cholera toxin. Infect Immun 45:679, 1984.
27. Walker RI, Caldwell MB, Lee EC, Guerry P, Trust TJ, Ruiz-Palacios GM: Pathophysiology of Compylobacter enteritis. Microbiol Rev 50:81, 1986.

28. Kunkel SL, Robertson DC: Purification and chemical characterization of the heat-labile enterotoxin produced by enterotoxigenic *Escherichia coli.* Infect Immun 25:586, 1979.
29. Gyles CL, Barnum DA: A heat-labile enterotoxin from strains of *Escherichia coli* enteropathogenic for pigs. J Infect Dis 120:419, 1969.
30. Levine MM: *Escherichia coli* infections. In Germanier R (ed): "Bacterial Vaccines." New York: Academic Press, 1984, p. 187.
31. Hopp TP, Woods KR: Prediction of protein antigenic determinants from amino acid sequences. Proc Natl Acad Sci USA 78:3824, 1981.
32. Westhof E, Altschuk D, Moras D, Bloomer AC, Mondragon A, Klug A, Van Regenmortel MHV: Correlation between segmental mobility and the location of antigenic determinants in proteins. Nature 311:123, 1984.
33. Tainer JA, Getzoff ED, Peterson Y, Olson AJ, Lerner RA: The atomic mobility component of protein antigenicity. Annu Rev Immunol 3:501, 1985.
34. Karplus PA, Schultz GE: Prediction of chain flexibility in proteins. A tool for the selection of peptide antigens. Naturwissenschaften 72:S212, 1985.
35. Duffy LK, Lai CY: Involvement of arginine residues in binding site of cholera toxin subunit B. Biochem Biophys Res Commun 91:1005, 1979.
36. DeWolf MJS, Fridkin M, Epstein M, Kohn LD: Structure-function studies of cholera toxin and its A and B promoters. Modification of tryptophan residues. J Biol Chem 256:5481, 1981.
37. Arnon R, Sela M: Studies on the chemical basis of the antigenicity of proteins. II. Antigenic specificity of polytyrosyl gelatins. Biochem J 75:103, 1960.
38. Jacob CO, Sela M, Pines M, Hurwitz S, Arnon R: Both cholera toxin-induced adenylate cyclase activation and cholera toxin biological activity are inhibited by antibodies against related synthetic peptides. Proc Natl Acad Sci USA 81:7893, 1984.
39. Jacob CO, Sela M, Arnon R: Antibodies against synthetic peptides of the B subunit of cholera toxin: Crossreaction and neutralization of the toxin. Proc Natl Acad Sci USA 80:7611, 1983.
40. Finkelstein RA, Burks MF, Zupan A, Dallas WS, Jacob CO, Ludwig DS: Epitopes of the cholera family of enterotoxins. Rev Infect Dis 9:544, 1987.
41. Jacob CO, Harari I, Arnon R, Sela M: Antibodies to cholera toxin enteric peptides of increasing size and their reactivity with related toxins. Vaccines 4:95, 1986.
42. Jacob CO, Arnon R, Finkelstein RA: Immunity towards heat labile enterotoxins of porcine and human *Escherichia coli.* Infect Immun 52:562, 1986.
43. Jacob CO, Pines M, Arnon R: Neutralization of heat-labile toxin of *E. coli* by antibodies to synthetic peptides derived from the B subunit of cholera toxin. EMBO J 3:2889, 1984.
44. Skehel JJ, Ada GL: Are peptides good antigens? Nature 316:764, 1985.
45. Bodanszky M, Bodanszky A: "The Practice of Peptide Synthesis." Berlin: Springer-Verlag, 1984.
46. Barany C, Merrifield RB: Solid-phase peptide synthesis. In Gross FE, Meienhofer J (eds): "The Peptides." New York: Academic Press, 1979, Vol. 2, p 1.
47. Fridkin M, Buchta R: Chemistry of immunogenic peptide synthesis. In Arnon R (ed): "Synthetic Vaccines." Boca Raton, FL: CRC Press, 1979, Vol. I, p 7.
48. Arshady R, Atherton E, Clive DLJ, Sheppard RC: Peptide synthesis. I. Preparation and use of polar supports based on poly (dimethylacrylamide). J Chem Soc Perkin 1:529, 1981.
49. Atherton E, Logan CJ, Sheppard RC: Peptide synthesis. II. Procedures for solid-phase synthesis using Nα-fluroenylmethoxycarbonylamino acids on polyamide supports. J Chem Soc Perkin 1:538, 1981.

50. Geysen HM, Meleon RH, Barteling SJ: Use of peptide synthesis to probe viral antigens for epitopes to a resolution of a single amino acid. Proc Natl Acad Sci USA 81:3998, 1984.
51. Houghten RA: General method for the rapid solid-phase synthesis of large numbers of peptides: Specificity of antigen-antibody interaction at the level of individual amino acids. Proc Natl Acad Sci USA 82:5131, 1985.
52. Dreesman GR, Sanchez Y, Ionescu-Matiu I, Sparrow JT, Six HR, Peterson DL, Hollinger FB, Melnick JL: Antibody to hepatitis B surface antigen after a single inoculation of uncoupled synthetic HBsAg peptide. Nature 295:158, 1982.
53. Chow M, Yabrow R, Bittle J, Hogle J, Baltimore D: Synthetic peptides from four separate regions of the poliovirus type I capsid protein VP1 induce neutralizing antibodies. Proc Natl Acad Sci USA 82:910, 1985.
54. Lin T, Zinnecker M, Hamaoka T, Katz DH: New procedures for preparation and isolation of conjugates of proteins and a synthetic co-polymer of D-amino acids and immunochemical characterization of such conjugates. Biochemistry 18:690, 1979.
55. Koch Y, Wilchek M, Fridkin M, Chobsieng P, Zor U, Lindmer HR: Production and characterization of an antiserum to synthetic gonadropin releasing hormone. Biochem Biophys Res Commun 55:616, 1973.
56. Jacob CO, Arnon R, Sela M: Effect of carrier on the immunogenic capacity of synthetic cholera vaccine. Mol Immunol 22:1333, 1985.
57. Herzenberg LA, Tokuhisa T, Parks DR, Herzenberg LA: Epitope-specific regulation. J Exp Med 155:1741, 1982.
58. Schutze MP, Leclerc C, Jolivet M, Audibert F, Chedid L: Carrier-induced epitopic suppression, a major issue for future synthetic vaccines. J Immunol 135:2319, 1985.
59. Lachmann PJ, Strangeways L, Vyakarnam A, Evan G: Raising antibodies by coupling peptides to PPD and immunizing BCG-sensitized animals. In Porter R, Whelan J (eds): ''Ciba Symposium on Synthetic Peptides as Antigens.'' Chichester: John Wiley and Sons, 1986, p 25.
60. Arnon R, Maron E, Sela M, Anfinsen CB: Antibodies reactive with a native lysozyme elicited by a completely synthetic antigen. Proc Natl Acad Sci USA 68:1450, 1971.
61. Langbeheim H, Arnon R, Sela M: Antiviral effect on MS-2 coliphage obtained with a synthetic antigen. Proc Natl Acad Sci USA 74:4636, 1976.
62. Audibert F, Jolivet M, Chedid L, Arnon R, Sela M: Successful immunization with a totally synthetic diphtheria vaccine. Proc Natl Acad Sci USA 79:5042, 1982.
63. Jacob CO, Arnon R, Sela M: Anti-cholera response elicited in aqueous solution by a completely synthetic antigen with built-in adjuvanticity. Immunol Lett 14:43, 1986.
64. Hopp TP: Immunogenicity of a synthetic HBsAg peptide: Enhancement by conjugation to a fatty acid carrier. Mol Immunol 21:13, 1984.
65. Siddiqui WA, Kan SC, Kramer K, Case S, Palmer K: Use of a synthetic adjuvant in an effective vaccination of monkeys against malaria. Nature 289:64, 1981.
66. Jacob CO, Grossfeld S, Sela M, Arnon R: Priming immune response to cholera toxin induced by synthetic peptides. Eur J Immunol 16:1057, 1986.
67. Emini EA, Jameson BA, Wimmer E: Priming for an induction of anti-poliovirus neutralizing antibodies by synthetic peptides. Nature 304:699, 1983.
68. Francis MJ, Fry CM, Rolands DJ, Brown F, Bittle JL, Houghton RA, Lerner RA: Priming with peptides of foot and mouth disease virus. In Lerner RA, Brown F (eds): ''Vaccines 1985: Molecular and Chemical Basis of Resistance to Parasitic Bacterial and Viral Diseases.'' Cold Spring Harbor, NY: Cold Spring Harbor Laboratory, 1985, p 203.
69. LaBrooy JT, Shearman DJC, Rowley D: Antibodies in serum and secretion one year after Salmonella gastroenteritis. Clin Exp Immunol 48:551, 1982.

70. McNabb PC, Tomasi TB: Host defense mechanisms at mucosal surfaces. Annu Rev Microbial 35:459, 1981.
71. Svennerholm AM, Gothefors L, Sack DA, Bardhan PK, Holmgren J: Local and systemic antibody responses, and immunological memory in humans after immunization with cholera B subunit by different routes. Bull WHO 62:909, 1984.
72. Elson CO, Ealding W: Generalized systemic and mucosal immunity in mice after mucosal stimulation with cholera toxin. J Immunol 132:2636, 1984.
73. Andrew E, Hall JG: IgA antibodies in the bile of rats. I. Some characteristics of the primary response. Immunology 45:169, 1982.
74. Jacob CO, Vaerman JP: Induction of rat secretory IgA antibodies against cholera toxin by a synthetic peptide. Immunology 59:129, 1986.
75. Weinstock GM, Aphhys C, Berman ML, Hampar B, Jackson D, Silhavi TJ, Weissman J, Zweig M: Open reading frame expression vector: A general method for antigen production in *Escherichia coli* using protein fusion to β-galactosidase. Proc Natl Acad Sci USA 80:4432, 1983.
76. Weinstock GM: Vectors for expression open reading frame DNA in *Escherichia coli* using *lac*Z gene fusions. In Setlow JK, Hollaender A (eds): ''Genetic Engineering Principles and Methods.'' New York: Plenum Press, 1984, Vol. 6, p 31.
77. Jacob CO, Leitner M, Zamir A, Solomon D, Arnon R: Priming immunization against cholera toxin and *E. coli* heat labile toxin by a cholera toxin short peptide β-galactosidase hybrid synthesized in *E. coli.* EMBO J 4:3339, 1985.
78. Charbit A, Boulain JC, Ryter A, Hofnung M: Probing the topology of a bacterial membrane protein by genetic insertion of a foreign epitope; expression at the cell surface. EMBO J 5:3029, 1986.
79. Bavoil P, Newton SMC, Stocker BAD, Jacob CO, Falkow S: Presentation of a cholera toxin epitope at the surface of a Salmonella typhimurium live vaccine strain (abstract). 1988 Albany Conference on Vector Based Vaccines.
80. Joys TM: The covalent structure of the phase-1 flagellar filament protein of Salmonella typhimurium and its comparison with other flagellins. J Biol Chem 260:15758, 1985.
81. Newton SMC, Jacob CO, Stocker BAD: Immune response to cholera toxin epitope inserted in Salmonella flagellin. Science 1989 244:70, 1989.
82. Hoiseth SK, Stocker BAD: Aromatic-dependent Salmonella typhimurium are non-virulent and effective as live vaccines. Nature 219:238, 1981.
83. Benjamini E, Andria ML, Estin CD, Norton FL, Leung CY: Studies on the clonality of the response to an epitope of a protein antigen: Randomness of activation of epitope-recognizing clones and the development of clonal dominance. J Immunol 141:55, 1988.
84. Good MF, Miller LH, Kumar S, Quakyi IA, Keister D, Adams JH, Moss B, Berzofsky JA, Carter R: Limited immunological recognition of critical malaria vaccine candidate antigens. Science 242:574, 1988.

Bacterial Vaccines, pages 255–279


# Liposomes as Carrier and Immunoadjuvant of Vaccine Antigens

**Nico van Rooijen**

Department of Histology, Medical Faculty, Free University, Amsterdam, The Netherlands

# I. INTRODUCTION

## A. Liposomes

Liposomes are artificially prepared spheres, consisting of concentric phospholipid bilayers separated by aqueous compartments. They are formed when phospholipids are confronted with water. The phospholipid molecules prefer to find a conformation in which their hydrophobic fatty acid chains are prevented from contact with water. For that reason, the formation of phospholipid bilayers in which the relatively hydrophilic head groups make up both of the outer parts of each bilayer, while the hydrophobic fatty acid groups are located directly opposite each other as the inner part of the bilayer, is not surprising (see Fig. 1).

Part of the aqueous solution, together with hydrophilic molecules dissolved in it, will be encapsulated during the formation of the liposomes, whereas lipophilic molecules may be associated with the phospholipid bilayers. Amphipathic molecules, as the phospholipids themselves, attempt to find a conformation with their hydrophilic parts extended in the aqueous compartments and their hydrophobic parts inserted in the bilayers. Liposomes may differ in their dimensions, composition (different phospholipids and cholesterol contents), charge (resulting from the charges of the composing phospholipids), and structure (multilamellar liposomes consisting of two or more concentric bilayers, separated by aqueous compartments or unilamellar liposomes consisting of only one phospholipid bilayer surrounding one aqueous compartment). The distribution of intravenously administered liposomes in the body is strongly influenced by these liposomal character-

**Fig. 1.** Schematic representation of a liposome with entrapped and (partly) surface-exposed antigen molecules. Part of the antigenic determinants (epitopes) of a surface-exposed antigen molecule may be masked depending on the interaction between the hydrophobic part of the antigen molecule and liposome. All antigenic determinants of an entrapped antigen molecule are masked. Without the help of macrophages only the exposed parts of the antigen can be recognized.

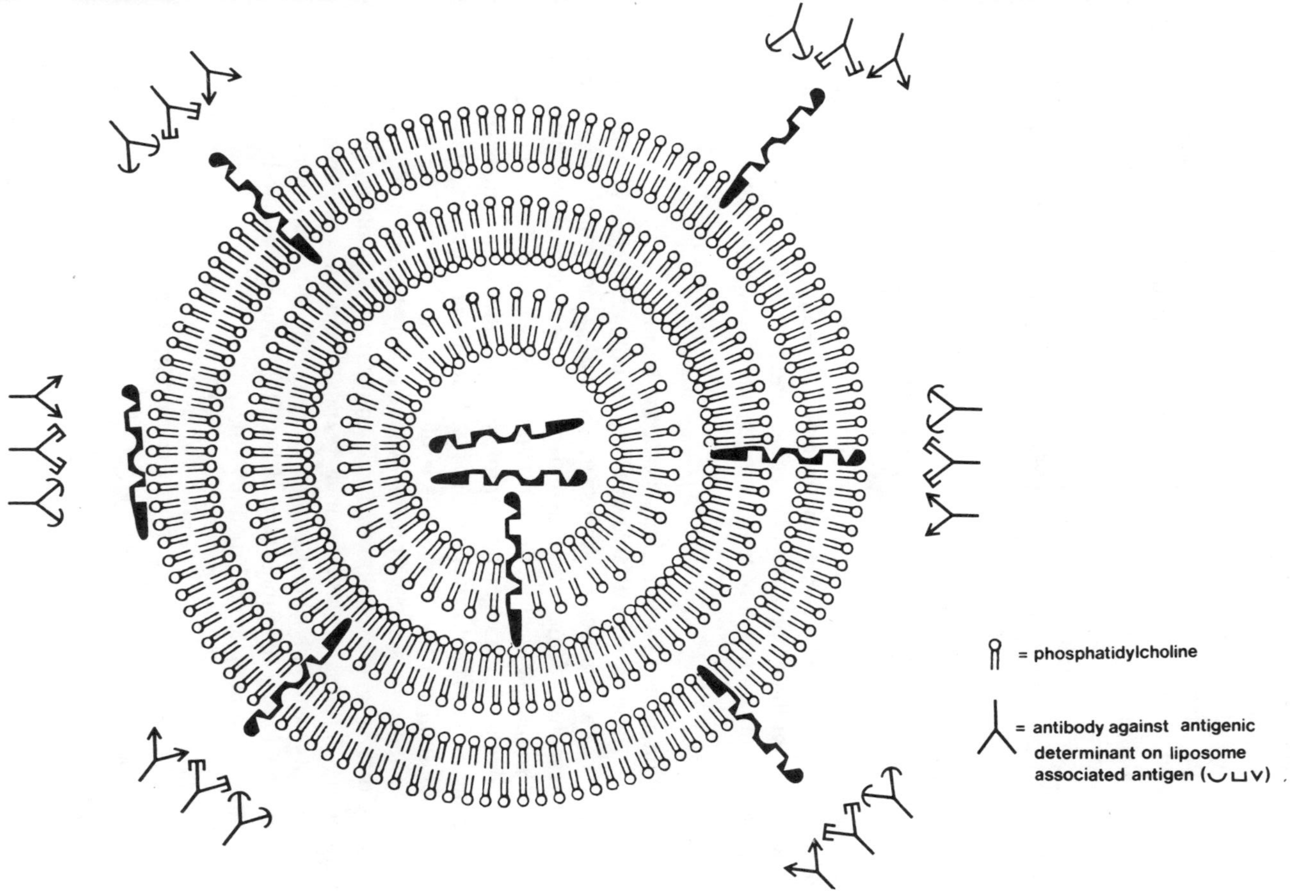
= phosphatidylcholine
= antibody against antigenic determinant on liposome associated antigen (◡⊔∨).

istics [1]. A more purposeful in vivo distribution of liposomes may be achieved when (monoclonal) antibodies directed against surface determinants of the intended target cells are attached to the liposomes in order to target the liposomes to these cells [1]. Inclusion of cholesterol in liposomes increases the stability of their phospholipid bilayers and makes them less susceptible to destruction by various serum components [2,3]. Cholesterol inclusion also greatly influences the interaction of liposomes with lymphoid and nonlymphoid cells involved in immune reactions [4]. One of the main advantages of liposomes as an immunoadjuvant is that they are biodegradable and may be composed of nontoxic and immunologically inert phospholipids [4].

### B. Immune Reactions Against Liposome-Associated Antigens or Haptens

The first report on the immunoadjuvant activity of liposomes in the immune response against diphtheria toxoid [5] has been followed by many studies confirming their capability to enhance immune responses to various antigens [4,6]. Interest centered on the possible applicability of liposomes as an adjuvant for the preparation of vaccines and for the presentation of tumor antigens to the immune system. Antigen presentation by liposomes may concern:

1. Presentation of haptens. In this case the liposome serves as a carrier for the hapten. Haptens are not able to elicit an immune response when given without a carrier molecule. As usual, the carrier determines the type of immune response, being a thymus-independent IgM response without generation of immunological memory for liposomal carriers [7–9] (Fig. 2).
2. Presentation of antigen (e.g., protein) molecules with various antigenic determinants. These antigens may also elicit an immune response when given free in solution, but the response is strongly enhanced when the antigen is given in a liposome-associated form. Antigens may be associated with liposomes either by encapsulation (masked antigenic determinants) or by coupling to the surfaces of the liposomes (exposed antigenic determinants; see Fig. 1). Liposome-associated antigens elicit a thymus-dependent immune response, including the production of both IgM and IgG antibodies and the generation of immunological memory [6].

## II. CHARACTERISTICS OF THE LIPOSOME-MEDIATED IMMUNE RESPONSE

### A. Thymus-Independent Response to Haptenated Liposomes

Haptenated liposomes evoke a hapten-specific humoral IgM immune response. An IgG response was not detected after primary or secondary im-

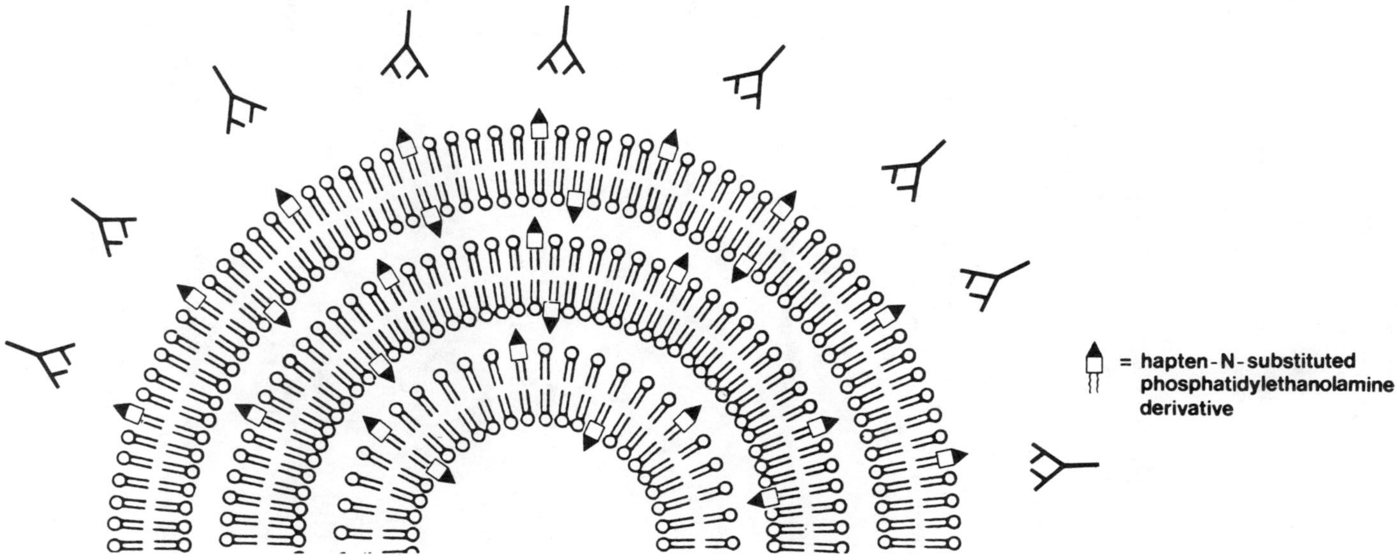


**Fig. 2.** Schematic representation of a liposome serving as a carrier for haptens. The hapten was coupled to phosphatidylethanolamine that was consecutively incorporated in liposomes.

munization with haptenated liposomes; in addition, mice depleted of, or deficient in, T lymphocytes appeared to respond to haptenated liposomes. For that reason, haptenated liposomes were concluded to belong to the group of TI antigens [8]. TI antigens are further divided into two types: TI-1 and TI-2. TI-2 antigens activate B lymphocytes from all the normal mouse strains except those of CBA/N mice, which carry an X-linked immune defect, and the B cells from neonatal mice of all strains. TI-1 antigens, having intrinsic mitogenic properties, stimulate B cells of all mouse strains, including those with the X-linked immune defect. Antibody responses to TI-2 antigens like trinitrophenyl (TNP)-Ficoll obviously require the presence of mature B cells. Accessory cell–derived factors appeared not essential for responses to TI-1 antigens such as TNP-lipopolysaccharide (LPS) and TNP-*Brucella abortus,* although these factors could enhance the responses; TI-2 responses require such factors [see, e.g., 10]. Studies by Snippe et al. [11] showed that CBA/N mice respond to haptenated liposomes, and for this reason, these antigens have to be considered TI-1 antigens. However, contrary to other known TI-1 antigens they appear to be nonmitogenic in vivo. Their immunogenicity can be completely blocked by prior injection of Ficoll bearing the same hapten. The TI-2 character of haptenated liposomes is also supported by a similar in vivo behavior of the response against such antigens and that against TNP-Ficoll [9,12]. For the above reasons Snippe et al. [11] suggested that haptenated liposomes could represent a distinct subclass of TI (-3?) antigens, partly resembling TI-1 antigens and partly resembling TI-2 antigens.

### B. Thymus-Dependent Response to Liposome-Associated (Protein) Antigens

Contrary to haptenated liposomes, liposome-associated (protein) antigens elicit an IgG response in addition to the preceding IgM response [13,14], start the generation of immunological memory [15], and appear to behave as real T-cell–dependent (TD) antigens [16,17]. Recent studies have focused on the antigen-specific interleukin-2 (IL-2) production, which appeared to be enhanced when the rabies antigens used for priming were given either specifically incorporated in liposomal bilayers [18] or nonspecifically associated with liposomes [19]. Production of IL-2, mainly by T-helper lymphocytes, is one of the early events, indicating that a thymus-dependent (TD) immune response has been initiated [20,21].

### C. Association of (Protein) Antigens With Liposomes: Entrapped and Surface-Exposed Antigens

To enhance the antibody response to antigens, liposomes can be associated with antigens in various ways (Fig. 2). Surface exposition can be achieved by a number of methods for covalent attachment of the antigen to one of the phospholipids to be used for preparation of the liposomes, by nonspecific adsorption of (partly) hydrophobic antigens with the phospholipid bilayers, or by "hydrophobization" of hydrophilic antigens followed by nonspecific adsorption. Antigens can also be entrapped in the liposomes when dissolved in the aqueous solution used for preparation of the liposomes. For many antigens the latter procedure will result in liposomes with antigen entrapped in their aqueous compartments and also exposed on their outer surfaces. The ratio between entrapped and exposed antigen may vary to a large extent and will depend on the structure of the antigen, e.g., the presence of hydrophobic groups. This may have influenced the results of studies comparing the effectiveness of liposome-entrapped and surface-exposed antigens in immunopotentiation [22–24].

There is now much evidence that liposomes act as an adjuvant for both entrapped and surface-exposed antigens [25–27]. However, the ratio between entrapped and surface-exposed antigen may strongly influence the isotype of antibody production. An important difference between liposome-entrapped antigens and surface-exposed antigens is that the antigenic determinants (epitopes) of the latter antigens are exposed and are directly accessible to cells of the immune system, whereas those of the entrapped antigen are masked (Fig. 1). For that reason it seems difficult to escape the conclusion that macrophages are required for the processing of liposome-entrapped antigens, because only these cells are able to release the antigens after digestion of the liposomal bilayers by phospholipases in their lysosomal compartments.

### D. Influence of Liposomes on the Isotype of Immunoglobulins Produced During the Immune Response Against Associated Antigens

After immunization with haptenated liposomes, IgM, but no IgG, responses were elicited [7,8]. However, no information was available about the relative participation of IgM and IgG antibodies in the response against liposome-associated protein antigens. We studied the IgM–anti-human serum albumin (HSA) response and the IgG–anti-HSA response after intravenous administration of liposome-associated HSA to rabbits [13,29]. Blood samples, obtained from the rabbits by cardiac puncture at different time intervals after injection of liposome-associated or free antigens, were used for two different antibody determinations:

1. A hemagglutination technique was used for measurement of the IgM–anti-HSA antibody titers.
2. For measurement of IgG–anti-HSA antibody titers, HSA-coupled Sepharose beads were incubated with doubling dilutions of the rabbit sera

and consecutively with sheep antirabbit IgG conjugated with horseradish peroxidase (HRP). After the enzymatic reaction with 3,3′-diaminobenzidine-tetrahydrochloride/$H_2O_2$, the beads treated with IgG–anti-HSA stained brown, whereas the beads treated with control serum or too high dilutions of the sera containing IgG–anti-HSA remained colorless.

The IgM response against 0.24 mg of liposome-associated HSA started at day 4 after injection and a mean peak titer of about 18 was reached at 10 days. The IgM response against a similar amount of free HSA started at day 10 after injection and a mean peak titer of about 10 was reached at 14 days. No such marked differences were found when the IgG responses against liposome-associated and free HSA were compared, although 0.24 mg of liposome-associated HSA produced higher peak titers than a similar amount of free HSA. When we compared the immune response against 0.24 mg of liposome-associated HSA with that against 10 mg of free HSA, the IgM–anti-HSA response against the liposome-associated antigen appeared earlier than that against the free antigen. In contrast, the IgG–anti-HSA response against the liposome-associated antigen started well after that against the free antigen.

From the above results [13] obtained with antigen exposed on the surfaces of liposomes, we concluded that the immunoadjuvant effect of liposomes during the primary immune response against an antigen is mainly due to an enhanced IgM antibody production. Contrary to our results on the relative effectiveness of liposome-entrapped and surface-exposed antigens in immunopotentiation [22], Shek and Sabiston [23] concluded that an antigen entrapped in liposomes was more immunogenic than the same antigen when it was surface adsorbed. Because an albumin antigen was used in both the studies of these authors and in our studies, the discrepancy between the results of both studies deserves attention. Shek and Sabiston used an antigen-specific, plaque-forming cell technique, by which they could determine the IgG response against the antigen [16]. In our studies we determined both the IgM response and the IgG response against the liposome-associated antigen [13,29]. As mentioned above, we demonstrated that the adjuvant effect of liposomes with respect to surface-exposed antigens concerned mainly the IgM response, whereas we found only a moderate effect on the IgG response. So it may well be that there is no real discrepancy between the results of our studies and those of Shek and Sabiston.

We postulated recently that liposome-entrapped antigens may especially stimulate the IgG response against antigens, whereas surface-exposed antigens could stimulate the IgM response against the same antigens [6]. Liposomes could then be used both to enhance the immune response against associated antigens and to modulate it (isotype selection). Although Latif and

**TABLE I. Comparative Studies on the IgM and IgG Antibody Production Against Liposome-Associated Albumin Antigen**

| | | Association of antigen and production of IgM and IgG anti-albumin antibodies | | | | |
|---|---|---|---|---|---|---|
| | | Antigen entrapped in liposomes | | Antigen exposed on surfaces of liposomes | | |
| Source of albumin | Immune response studied in | IgM anti-albumin production | IgG anti-albumin production | IgM anti-albumin production | IgG anti-albumin production | References |
| Bovine serum | mouse | N.D. | $++^2$ | N.D. | $+^2$ | 23 |
| human serum | rabbit | N.D. | N.D. | $++^1$ | $+^1$ | 13,29 |
| bovine serum | mouse | $\pm^2$ | $++^2$ | $+^2$ | $+^2$ | 31 |
| human serum | mouse | $+^1$ | $++^1$ | $++^1$ | $+^1$ | 32 |

ND = not determined
+ + = high antibody production
+ = moderate antibody production
± = low antibody production
$^1$ = antibody production measured by determination of serum antibody titers
$^2$ = antibody production measured by determination of number of antibody-forming cells

Bachhawat [14] reported a strong adjuvant effect of liposomes with respect to the IgG response against surface-exposed lysozyme and Davis et al. [30] found that adjuvanticity of liposomes is reflected in most antibody subclasses and that there is no shift in subclasses compared to the response obtained with the free antigen, strong support for our hypothesis came from recent studies [31,32] (Table I). Shahum and Therien [31] compared the immune response against different bovine serum albumin (BSA) preparations in mice. The antigen was given free in solution, encapsulated in liposomes, or covalently linked to the liposomal surfaces. The crucial finding was that, although both types of liposome association were equally efficient in potentiating the humoral immune response as a whole, encapsulation mainly enhanced the IgG isotype production, with little or no effect on the IgM response, while covalent linkage stimulated the production of IgM at least as well as IgG.

Further modulation of the immune response with respect to IgM and IgG production could be achieved by simultaneous incorporation of antigen and a polyclonal T-cell activator in liposomes [33]. We compared the immune response in mice against liposome-entrapped HSA with that against HSA exposed on the surfaces of liposomes [32]. These results also confirm that antigens coated on liposomes primarily enhance the IgM response, whereas liposome-entrapped antigens primarily enhance the IgG response. Considering the IgM response, HSA exposed on the outer surfaces of liposomes elicited an eightfold higher serum antibody concentration at day 5 than HSA entrapped in liposomes. To the contrary, at the same time interval after antigen administration, liposome-entrapped HSA produced a 32-fold higher IgG–anti-HSA antibody concentration than did surface-exposed HSA. Considering the normal kinetics of IgM and IgG production during the primary immune response against TD antigens, it is not surprising that these differences were less pronounced at day 7 after antigen injection [32].

### E. Efficiency of Liposomes as Carriers of Antigen in Priming and Booster Injections

In order to study the specific effect of liposomes on the generation of immunologic memory, booster injections were given as antigen free in solution in all animals, whereas priming injections were given with either free antigen or liposome-associated antigen. The results of these experiments showed that liposomes not only enhance the primary immune response against associated antigens, but also induce an increased immunological memory [15]. We also investigated the adjuvant effect of liposomes when used as carriers for antigens administered both in the priming and in the booster injection [34]. From these results, we concluded that after a priming injection with a liposome-associated albumin antigen, liposomes do not further enhance the secondary immune response when they are also used for the booster injection. With respect to strong antigens such as serum albumins, the immunopotentiating effect of liposomes obviously concerns only the primary antigen administration, but not any further injections. This is not surprising in the light of our recent findings that macrophages are the primary target cell for liposome-associated antigens [32]. Small protein antigens such as albumins will not be ingested by macrophages unless they are aggregated, complexed with antibodies, or associated with liposomes [35]. However, during the secondary immune responses, the antigens entering in the circulation meet an excess of antibodies, and the formed immune complexes will be rapidly ingested by macrophages.

## III. CELLS INVOLVED IN THE IMMUNE RESPONSE AGAINST LIPOSOME-ASSOCIATED ANTIGENS

### A. In Situ Immune Response in the Spleen After Intravenous Injection of Liposome-Associated Antigens

Compared to the fate of free (protein) antigens during the primary immune response, the bulk of liposome-associated antigens will be ingested by macrophages in the liver and in the marginal zone and red pulp of the spleen [36]. Moreover, small (protein) antigens will penetrate much easier, e.g., in the lymphoid cell compartments of the white pulp of the spleen, than their larger liposome-associated counterparts [37]; they also have to be given in

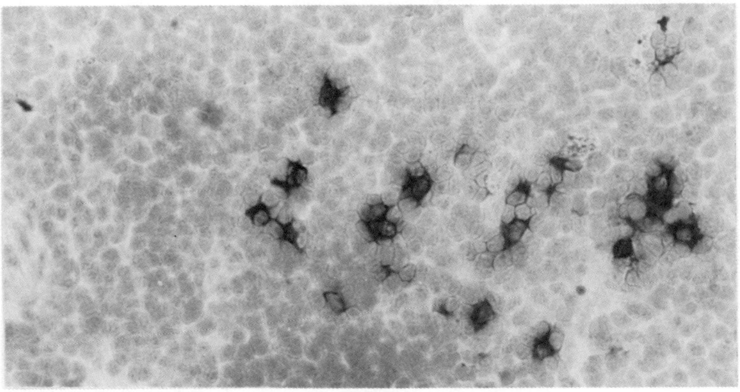

**Fig. 3.** Anti-human serum albumin (HSA) antibody producing cells are demonstrated in spleen tissue by incubation of cryostat sections with a HSA-Horseradish peroxidase (HRP) conjugate followed by enzymehistochemical demonstration of the HRP. Anti-HSA producing cells are present in the periphery of a periarteriolar lymphocyte sheath in the spleen of a rabbit at 6 days after intravenous injection of 50 μg liposome-associated HSA.

considerably larger amounts in order to elicit a comparable strong immune response [4]. In spite of the different distribution patterns of liposome-associated and free antigens, the localization pattern of specific antibody-forming cells developing in the spleen after intravenous administration of both antigen preparations did not differ for either the primary [29] or secondary responses [34]. Anti-HSA antibody-forming cells could be detected in the spleen of rabbits from about 4 days after an intravenous injection with either free HSA (10 mg) or liposome-associated HSA (0.05 mg). The bulk of these cells developed in the periphery of the periarteriolar lymphocyte sheaths and in the small sheaths of lymphoid tissue around the terminal arterioles (Fig. 3). Their differentiation pathway in the spleen is not different from that followed by developing antibody-forming cells against other TD and even TI antigens [37].

Recent lines of research have demonstrated that B-cell memory is short-lived in the absence of antigen [38,39]. There is a lot of evidence that immune complexes formed from the antigen administered, and specific antibody evoked in response to it, play an important role in the generation of immunological memory [40] and/or in the maintenance of a long-term antibody production [41]. This is believed to be mediated by a small part of these immune complexes, which escape degradation by macrophages and are retained for long periods on the cell processes of follicular dendritic cells in lymphoid follicles of spleen and lymph nodes [40,42,43]. From a viewpoint of application of liposomes for vaccination (generation of immunological

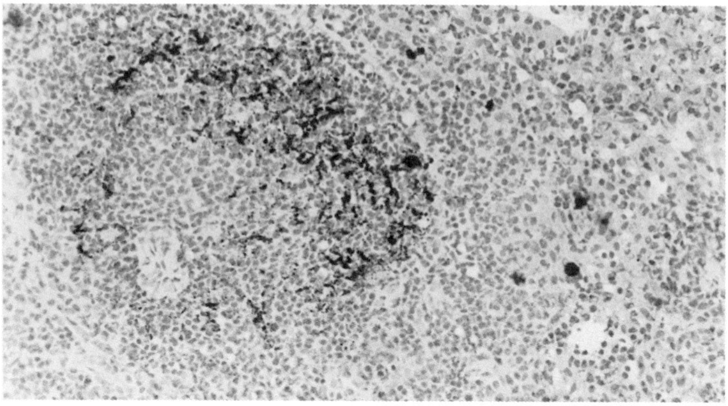

**Fig. 4.** Extracellular anti-HSA antibodies in a follicle center in the spleen of a rabbit treated as described in the legend of Fig. 3. Extracellular anti-HSA antibody is trapped in the follicles as an HSA-anti-HSA immune complex, and could be demonstrated due to its free antigen binding sites, using the same technique as described in the legend of Fig. 3. Reproduced by permission of Alan R. Liss Inc. from Van Rooijen et al., 1983. Anatomical Record, 206:189–198.

memory) it may thus be important that immune complexes are formed during the response. There is no reason to believe that association of antigens with liposomes completely prevents the formation of immune complexes. Apart from leakage of the antigen from the liposomes and antigen released from instable liposomes or liposomal bilayers affected by serum components (e.g., albumins) [44], part of the antigen may be released from the liposomes by complement-dependent immune lysis of the liposomes [see, e.g., 45], and macrophages may release (partly degraded) antigens after disruption of liposomal bilayers [35,46]. Indeed, trapping of immune complexes in follicles of the spleen was found during the primary immune response against liposome-associated HSA intravenously given to rabbits [29] (Fig. 4).

### B. Role of Macrophages in the Immune Response Against Liposome-Associated Antigens

Antigens that are exposed on the outer surfaces of liposomes may be recognized by receptors on immunocompetent cells. On the other hand, it is difficult to see how antigens that are entrapped in the aqueous compartments of liposomes (see Fig. 1) can elicit an immune response without the help of macrophages. The macrophages could ingest the liposomes, digest the lipo somal phospholipid bilayers with the help of lysosomal phospholipases, and process the released antigens as reviewed by Unanue [46].

We investigated the role of macrophages in mediating the adjuvant activity of liposomes using a new in vivo macrophage elimination technique. This approach, in which liposomes with entrapped dichloromethylene diphosphonate (Cl2MDP) were used for physical elimination of macrophages in spleen and liver, has been described in detail elsewhere [36]. Since liposome-entrapped drugs and liposome-entrapped antigens will be targeted to macrophages in the same compartments of the same organs, problems related to a different distribution pattern of macrophage-eliminating drugs and antigens are avoided. The primary antibody response to intravenously administered and liposome-associated HSA in mice was studied under conditions where no response could be detected against the non–liposome-associated form of the antigen. The positive response against the antigen, entrapped in and/or exposed on the surfaces of liposomes, thus resulted from the adjuvant action of the liposomes. Elimination of all red pulp macrophages, marginal metallophilic macrophages, and marginal zone macrophages in the spleen and Kupffer cells in the liver was obtained by intravenous injection of liposome-encapsulated Cl2MDP 2 days before antigen administration [36]. Selective elimination of marginal zone macrophages in the spleen was obtained by intravenous injection of liposome-encapsulated Cl2MDP 22 days before antigen [47]. Elimination of all red pulp macrophages, marginal metallophilic macrophages, and marginal zone macrophages in the spleen resulted in markedly reduced IgM and IgG anti-HSA antibody titers [32]. On average, the IgG–anti-HSA concentration in the blood serum of the mice at 7 days after antigen administration was reduced to 5–10% of the normal values. Selective elimination of marginal zone macrophages in the spleen did not result in reduced anti-HSA antibody concentrations at 7 days after antigen administration. It is concluded that macrophages in the spleen and/or in the liver are involved in the processing of liposome-associated HSA. Marginal zone macrophages, however, which are located close to the open ends of the white pulp capillaries and thus are the first macrophages to meet antigens entering into the spleen [37,47], are not required. A surprising finding was that the reduction in antibody production caused by elimination of macrophages was more impressive for antigen exposed on the outer surfaces of liposomes than for antigen entrapped in the aqueous compartments of the liposomes, a phenomenon not yet explained [32].

## IV. PREPARATION AND CHARACTERISTICS OF LIPOSOMES AND ASSOCIATION OF ANTIGENS

### A. Preparation of Liposomes and Entrapment of Antigens

Multilamellar liposomes can be prepared by different methods; however, in all methods the phospholipids are confronted with an aqueous solution, resulting in the formation of spheres consisting of concentric phospholipid bilayers separated by compartments containing the (entrapped) aqueous solution (see Fig. 1).

Multilamellar liposomes can be prepared from one single phospholipid, e.g., phosphatidylcholine (lecithin), or from mixtures of different lipids. By adding to the mixtures negatively charged phospholipids such as phosphatidic acid or dicetylphosphate or positively charged phospholipids such as stearylamine, the charge of the resulting liposomes can be influenced correspondingly. For a comparative study on adjuvant activity of various liposomes, we prepared neutral liposomes composed of phosphatidylcholine only, positive liposomes composed of phosphatidylcholine and stearylamine, and negative liposomes composed of phosphatidylcholine, cholesterol, and phosphatidic acid. The latter liposomes showed an immunoadjuvant action exceeding that of positive or neutral liposomes [28,48]. A detailed description of the preparation of liposomes composed of phosphatidylcholine, cholesterol, and phosphatidic acid follows.

1. Dissolve 75 mg of phosphatidylcholine, 11 mg of cholesterol, and 10.6 mg of phosphatidic acid (or dicethylphosphate) in 20 ml methanol/chloroform (1:1) in a 500 ml round-bottom flask.
2. Remove organic phase by low-vacuum rotary evaporation (at 37°C).
3. Dissolve lipid film in 10 ml chloroform.
4. Remove chloroform by low-vacuum rotary evaporation (at 37°C).
5. Disperse the thin film that forms on the interior of the flask in 10 ml phosphate-buffered saline (PBS) (for empty liposomes) by gentle rotation for 15 min at room temperature. If antigen encapsulation is intended, the antigen must be dissolved in the PBS solution used in this step. Keep in mind that (partly) hydrophobic antigens will be exposed on the outer surfaces of the liposomes in addition to their encapsulation [22]. If required, these surface-exposed protein antigens can be removed by trypsin treatment in step 10.
6. After complete removal of the lipid film from the interior of the flask, keep the milky white suspension for 2 h at room temperature.
7. Sonicate for 3 min at room temperature in a waterbath sonicator.
8. Keep suspension for another 2 h at room temperature (for liposome swelling).
9. Centrifuge at 100,000*g* for 30 min at 16°C.
10. Resuspend in saline or PBS, and wash one or two times if free antigen has to be removed, e.g., if the free antigen may have harmful effects on cells and tissues of the body. If no entrapment but only surface association of the partly hydrophobic antigen with liposomes is intended, the empty liposomes have to be resuspended in a strongly concentrated antigen solution (saline or PBS) for about 15 min and then washed [22,48]. On the other hand, if only

entrapment of the antigen is intended, trypsin can be used to remove surface-exposed antigen.

11. In order to check the character of liposomes, a sample can be stained with fluorochromes and studied with the fluorescence microscope [4]. After incubation of the multilamellar liposomes in a 1% aqueous solution of the fluorochromes acridine orange or eosin Y for several minutes, and washing to remove the free fluorochrome, a drop of the liposome suspension can be observed under the fluorescence microscope (1,000× magnification).

12. The antigen-associated liposomes can be stored at 4°C under nitrogen atmosphere for some time (depending on antigen and phospholipids used).

Another way to prepare multilamellar liposomes is by the ether evaporation method. Ether solutions of phospholipids or phospholipid mixtures are injected in warm aqueous solutions. After vaporization of the ether, liposomes are formed. When vacuum is drawn over the aqueous solution, rapid vaporization of the ether may be obtained at a lower temperature in order to prevent degradation of temperature-sensitive molecules [49]. Organic solvent vaporization methods for preparation of liposomes cannot be recommended for liposomal entrapment of antigen, because the organic solvents may alter the (protein) antigens. Liposomes must be associated with the antigens after their preparation. In addition, various other methods for liposome preparation have been developed. So-called dehydration–rehydration vesicles have been described [50], and these were successfully associated with antigen [26]. It is beyond the scope of this chapter to review extensively all methods for preparation of liposomes; the reader is referred to Gregoriadis [51].

### B. Immunogenicity of Liposomal Phospholipids

Apart from the phospholipid phosphatidylcholine mostly forming the backbone of liposomes, the negatively charged phospholipid phosphatidic acid, and the positively charged phospholipid stearylamine that were mentioned above, numerous other natural and artificially prepared phospholipids can be used for preparation of liposomes. The phospholipid composition of liposomes will influence to a large extent their in vivo behavior and their immunoadjuvant activity. It is not surprising that phospholipids, occurring in all cell membranes in high concentrations, generally are not very immunogenic themselves [52]. Although liposomes may serve as carriers for haptens and as a strong adjuvant for protein antigens, antibodies against their main component phosphatidylcholine were not elicited unless lipid A was inserted in the same bilayers [53]. Nevertheless, naturally occurring antibodies that react with liposomes of varied phospholipid composition have been found in normal rabbit [54] and human [55] sera and in the course of experimental trypanosomal infections in rabbits [56]. It may well be that these antibodies are produced by a distinct lineage of B cells, occurring in high concentrations in the peritoneum, but only in small numbers in the typical lymphoid organs such as spleen and lymph nodes. In the mouse, these Ly-1–positive B cells form a self-renewing population replenished from progenitors largely in the peritoneal cavity and preferentially develop in older normal and young autoimmune animals [57]. A high proportion of these Ly-1–positive B cells have surface IgM recognizing the common membrane phospholipid phosphatidylcholine [58].

An obvious question arising is, Why don't all cell membranes become damaged by the immunological attack from the antiphospholipid antibodies? Alving [55] postulated that the phospholipids of normal cells are shielded by adjacent glycolipids and proteins. This hypothesis is supported by experimental results of the author and his colleagues that antiliposome antibodies can be completely inhibited from binding to liposomes by including certain glycolipids in the liposomal bilayers at the glycolipid concentrations normally found in cell membranes. So, although the antibodies against the lipid bilayer cannot bind to normal (young) cells, it is suggested that they will attach to cells that have an exposed phospholipid bilayer, either due to damage (e.g., inflammation) or to age-related changes in the cell membrane (''senescence'' of the cell) and that they may selectively eliminate such cells [55].

It is also postulated that antiphospholipid antibodies may be produced as a reaction against LPS or lipid A derived from gram-negative bacteria and adsorbed to various cell types [55]. It has been established that the immunoadjuvant activity of liposome-inserted antigens can be greatly enhanced by the simultaneous incorporation of lipid-A [53] or LPS [59]. However, in addition, antibodies against the phospholipids themselves are also produced after incorporation of lipid A in liposomes [53]. So, although liposomes and lipid A can be used in combination to obtain a synergistic strong immunopotentiation, we should be aware of the potential negative side effects of the antiphospholipid (and autoimmune) antibodies that were not observed when lipid A was omitted from the antigen-associated liposomes. Finally, it should be mentioned that liposomes of certain lipid composition, prepared by the detergent removal method, induced the proliferation of spleen cells from different mouse strains in a way resembling the stimulation by LPS [60].

### C. Toxicity of Liposomal Phospholipids

One of the advantages of liposomes as an immunoadjuvant is that they are biodegradable. They are ingested by macrophages that consecutively can digest the phospholipid bilayers using the lysosomal phospholipases [61]. Indeed, ingestion and digestion of liposomes by macrophages, followed by intracellular release of encapsulated compounds, has been shown to occur [36], and it may well be that liposome-mediated targeting of antigens to macrophages foms part of the mechanism of the immunoadjuvant activity of liposomes [32]. Another important advantage of liposomes as an immunoadjuvant is that they may be composed of phospholipids, e.g., phosphatidylcholine, that are nontoxic in the doses to be used [62]. To the contrary, it has been demonstrated that after attachment of cholera toxin to liposomes containing ganglioside $G_{MI}$, toxicity of the toxin was completely eliminated and antigenicity was enhanced [63]. It has been reported that stearylamine- and cardiolipin-containing liposomes are toxic for cultured human cells at 200 μM liposomal lipid concentrations or less. Similarly, phosphatidylglycerol- and phosphatidylserine-containing liposomes were toxic at slightly higher concentrations [62]. However, the above phospholipids are not required for the preparation of liposomes with strong immunoadjuvant activity. Also, sphingomyelin liposomes that were shown to be toxic for certain cells in the spleen [64] have only moderate adjuvant activity [48]. In fact, simple, inexpensive, and easily obtainable phosphatidylcholine cholesterol liposomes meet the requirements of a good immunoadjuvant.

### D. Effects of Cholesterol as a Membrane Component of Liposomes

It has been discussed previously that cholesterol is likely to play an important role in the interaction between lymphocytes and liposomes [4]. Some new results with respect to the influence of cholesterol on the interaction between macrophages and liposomes deserve attention, in light of our recent finding that macrophages play a role in the immunoadjuvant action of liposomes [32]. Inclusion of cholesterol in liposomes is known to increase their half-life in the circulation. A dual mechanism is responsible for this phenomenon: 1) cholesterol stabilizes the liposomal membrane and makes it less susceptible to destruction by various serum components [2,3]; 2) cholesterol makes liposomes unfavorable for phagocytosis by hepatic Kupffer cells [65–67].

Patel and coworkers have found that macrophages in liver and spleen respond differently to cholesterol-containing liposomes [66,68]. Their results show that, whether in vivo or in vitro in the presence of serum, Kupffer cells avidly ingest cholesterol-poor but not cholesterol-rich liposomes, whereas splenic phagocytic cells preferentially take up cholesterol-rich rather than cholesterol-poor liposomes. The authors postulated that serum contains opsonins specific for hepatic and splenic macrophages and that these opsonins have different affinities for cholesterol-rich and cholesterol-poor liposomes. The implication is that the cholesterol contents of liposomes intended for use as immunoadjuvants may be important, because the cholesterol-rich liposomes will target their antigen primarily to macrophages in the spleen and cholesterol-poor liposomes will target their antigen primarily to macrophages in the liver. It has been shown that general approaches for improvement of liposome targeting to macrophages, e.g., the incorporation of mannosylated phospholipids [69], also improved the adjuvant activity of liposomes [70].

### E. Association of Antigens With Outer Surfaces of Liposomes

As mentioned above, evidence has been presented that antigens encapsulated in liposomes tend to elicit primarily an IgG antibody response, whereas antigens exposed on the outer surfaces of liposomes prefer to induce an IgM antibody response. The entrapment of antigens in the aqueous compartments of liposomes mainly depends on the concentration of antigens used and on the type of liposome to be prepared (e.g., unilamellar or multilamellar liposomes). The attachment of antigens to the outer surfaces of liposomes, however, may be manipulated to a large extent. It is beyond the scope of this chapter to describe in great detail the various procedures available for association of antigens with the surfaces of liposomes, but their principles will be briefly summarized [for more details see 28].

It is not surprising that hydrophobic proteins show more tendency to bind to liposomes than hydrophilic proteins. Hydrophobic proteins or proteins with hydrophobic parts will associate with the surfaces of liposomes in a nonspecific way, and need no modification for that reason. Covalent modification of proteins with hydrophobic compounds, which serve as an anchor to the phospholipid membranes, seems to be an effective approach for hydrophilic proteins [71,72]. Phosphatidylethanolamine may be an important component of liposomes, since this phospholipid contains a free amino group that can be used for covalent attachment of antigens. Covalent attachment of proteins to liposomes can also be performed using glycosphingolipids [73] or glutaraldehyde [14,74]. Alving et al. [75] studied the immune response against liposome-associated cholera toxin. Attachment of cholera toxin to the liposomes was performed by incorporation of a receptor, i.e., ganglioside $G_{MI}$, in the liposomal bilayers. The method of choice for association of any particular antigen with liposomes will depend on the chemical structure of the antigen and on the availability and costs of the antigen. The more expensive the antigen, the more efficient the coupling procedure should be.

### F. Synergistic Effects of Liposomes and Other Adjuvants

If the immunopotentiating effect of all adjuvants had the same mechanism of action, it would be rather easy to compare the effectiveness of different adjuvants. However, because different antigens elicit different types of immune reactivity and different adjuvants influence different steps in immune reactions, both the immunological goal of vaccination and the effectiveness of various adjuvants to be used in these vaccines differ to a large extent. For that reason, different adjuvants have been proven to be the best choice for different vaccines [76]. It is not surprising that different adjuvants influencing different steps in an immune response may have additive or synergistic effects when used in combination. We have already mentioned the synergistic effects of liposomes and lipid A [63,75] or LPS [59]. Other lipid substances that might be incorporated into liposomes to enhance the immunogenicity of liposome-associated antigens include acylated derivatives of muramyl dipeptide [77]. A very powerful adjuvant results if antigen-containing liposomes are combined with Freund's complete adjuvant [78]. On the other hand, lysophosphatidylcholine, which shows adjuvant activity when given alone (with an antigen), impairs the adjuvant activity of liposomes if incorporated in their bilayers [79]. Recently, Richards et al. [80] showed that the adjuvants liposomes, lipid A, and aluminum hydroxide were strongly additive or synergistic when used in combination to enhance the immune response against a synthetic malaria sporozoite antigen. It should be taken into account, however, that such a combination may produce undesired and autoimmune antibodies in addition to the intended response (see IV.B).

### G. Use of Liposomes as Adjuvant in Vaccines and for Presentation of Tumor Antigens

One of the most promising applications of liposomes at the moment seems to be their use as an adjuvant in vaccines [4,81,82]. The establishment of a strong immunological memory with respect to the antigen should be an important property of a vaccine adjuvant for long-term protection. For short-term protection and especially for postexposure treatment, e.g., with rabies immunosomes, which were found to be effective in protection of laboratory animals that had been experimentally injected with a lethal dose of a rabies wild strain [83], the primary immune response may be important.

Apart from the rabies immunosomes just mentioned, liposomes have proved to be efficient antigen carriers in the responses against cholera toxin [63], malaria sporozoite antigen [80,84], tetanus toxoid [26], influenza virus surface glycoproteins [85,86], *Entamoeba histolytica* protein antigens [87,88], *Plasmodium yoelii* antigens [89], and *Streptococcus mutans* antigens [90]. The list of antigens that may give some form of protection when administered after association with liposomes is growing continuously. Increasing interest in an alternative method of contraception has intensified the search for a suitable carrier of spermatozoal antigens to be used for development of a vaccine for immunoregulation of fertility. For this purpose, liposomes seem to be promising [91]. The immunogenicity of a tumor cell antigen is closely related to its presentation. So it is not surprising that liposomes have been proposed as carriers for tumor antigens in order to enhance the immune reactions against the tumor. Some successful results have already been reported [92–96].

## V. CONCLUSIONS

Liposomes may be considered a promising vaccine adjuvant. Like most other adjuvants, their precise mechanism of action remains to be clarified. Macrophages seem to play a crucial role in the processing of liposome-associated antigens. There are a number of important advantages of liposomes when used as a vaccine adjuvant:

1. Liposomes are biodegradable (they are digested by macrophages).
2. Liposomes are harmless if composed of nontoxic phospholipids and cholesterol.
3. Liposomes are immunologically inert if composed of, e.g., phosphatidylcholine and cholesterol.
4. Preparation of liposomes is simple and components may be inexpensive (egg lecithin).
5. Liposomes may be applied for intravenous administration to stimulate immune reactions in the spleen.
6. Phospholipid composition of liposomes may be varied in order to obtain the most efficient antigen-liposome preparation.
7. Combination with other immunoadjuvants may produce an extremely strong immune response if required (e.g., lipid A, aluminum hydroxide, muramyl dipeptide).
8. Liposomes may eliminate the toxicity of antigens in addition to their adjuvant activity (e.g., cholera toxin).
9. Liposomes can be used to modulate the immune response. Entrapped antigens preferentially elicit IgG antibodies. Antigens exposed on the surfaces preferentially elicit IgM antibodies.

## REFERENCES


1. Gregoriadis G (ed): "Liposomes as Drug carriers." Chichester: John Wiley and Sons, 1988.
2. Scherphof G, Roerdink F, Waite M, Parks J: Disintegration of phosphatidylcholine liposomes in plasma as a result of interaction with high-density lipoproteins. Biochim Biophys Acta 542:296, 1978.
3. Kirby C, Clarke J, Gregoriadis G: Effect of the cholesterol content of small unilamellar liposomes on their stability in vivo and in vitro. Biochem J 186:591, 1980.
4. Van Rooijen N, Van Nieuwmegen R: Immunoadjuvant properties of liposomes. In Gregoriadis G, Senior J, Trouet A (eds): "Targeting of Drugs." New York: Plenum Press, 1982, p 301.

5. Allison AC, Gregoriadis G: Liposomes as immunological adjuvants. Nature 252:252, 1974.
6. Van Rooijen N: Liposomes as immunological adjuvants: Recent developments. In Gregoriadis G (ed): "Liposomes as Drug Carriers." Chichester: John Wiley and Sons, 1988, p 159.
7. Yasuda T, Dancey GF, Kinsky SC: Immunologenic properties of liposomal model membranes in mice. J Immunol 119:1863, 1977.
8. Van Houte, AJ, Snippe H, Willers JMN: Characterization of immunogenic properties of haptenated liposomal model membranes in mice. I. Thymus independence of the antigen. Immunology 37:505, 1979.
9. Claassen E, Kors N, Van Rooijen N: Immunomodulation with liposomes: The immune response elicited by liposomes with entrapped dichloromethylene-diphosphonate and surface-associated antigen or hapten. Immunology 60:509, 1987.
10. Goud SN, Muthusamy N, Subbarao B: Differential responses of B cells from the spleen and lymph node to TNP-Ficoll. J Immunol 140:2925, 1988.
11. Snippe H, Van Houte AJ, Lizzio EF, Willers JMN, Merchant B: Characterization of immunogenic properties of haptenated liposomal model membranes in mice. VI. Response in B cell-defective CBA/N mice. Immunology 45:545, 1982.
12. Claassen E, Kors N, Van Rooijen N: Influence of carriers on the development and localization of anti-trinitrophenyl–antibody-forming cells in the murine spleen. Eur J Immunol 16:271, 1986.
13. Van Rooijen N, Van Nieuwmegen R: Association of an albumin antigen with phosphatidylcholine liposomes alters the nature of immunoglobulins produced during the immune response against the antigen. Biochim Biophys Acta 755:434, 1983.
14. Latif NA, Bachhawat BK: The effect of surface-coupled antigen of liposomes in immunopotentiation. Immunol Lett 15:45, 1987.
15. Van Rooijen N, Van Nieuwmegen R, Kors N: The secondary immune response against liposome associated antigens. Immunol Commun 10:59, 1981.
16. Shek PN, Sabiston BH: Immune response mediated by liposome-associated protein antigens. I. Potentiation of the plaque-forming cell response. Immunology 45:349, 1982.
17. Beatty JD, Beatty GB, Paraskevas F, Froese E: Liposomes as immune adjuvants: T cell dependence. Surgery 96:345, 1984.
18. Oth D, Mercier G, Perrin P, Joffret ML, Sureau P, Thibodeau L: The association of the rabies glycoprotein with liposome induces an *in vitro* specific release of interleukin 2. Cell Immunol 108:220, 1987.
19. Mansour S, Thibodeau L, Perrin P, Sureau P, Mercier G, Joffret ML, Oth D: Enhancement of antigen-specific interleukin 2 production by adding liposomes to rabies antigens for priming. Immunol Lett 18:33, 1988.
20. Robb RJ: Interleukin 2: The molecule and its function. Immunol Today 5:203, 1984.
21. Spitz M, Gearing A, Callus M, Spitz L, Thorpe R: Interleukin-2 in vivo: Production of and response to interleukin-2 in lymphoid organs undergoing a primary immune response to erythrocytes. Immunology 54:527, 1985.
22. Van Rooijen N, Van Nieuwmegen R: Liposomes in immunology: Evidence that their adjuvant effect results from surface exposition of the antigens. Cell Immunol 49:402, 1980.
23. Shek PN, Sabiston BH: Immune response mediated by liposome-associated protein antigens. II. Comparison of the effectiveness of vesicle-entrapped and surface-associated antigen in immunopotentiation. Immunology 47:627, 1982.
24. Snyder SL, Vannier WE: Immunologic response to protein immobilized on the surface of liposomes via covalent azo-binding. Biochim Biophys Acta 772:288, 1984.

25. Davis D, Gregoriadis G: Liposomes as adjuvants with immunopurified tetanus toxoid: Influence of liposomal characteristics. Immunology 61:229, 1987.
26. Gregoriadis G, Davis D, Davies A: Liposomes as immunological adjuvants: Antigen incorporation studies. Vaccine 5:145, 1987.
27. Vannier WE, Snyder SL: Antibody responses to liposome-associated antigen. Immunol Lett 19:59, 1988.
28. Van Rooijen N, Van Nieuwmegen R: Use of liposomes as biodegradable and harmless adjuvants. Methods Enzymol 93:83, 1983.
29. Van Rooijen N, Kors N, Van Nieuwmegen R, Eikelenboom P: The development of specific antibody-containing cells and the localization of extracellular antibody in the follicles of the spleen of rabbits after administration of free or liposome associated albumin antigen. Anat Rec 206:189, 1983.
30. Davis D, Davies A, Gregoriadis G: Liposomes as adjuvants with immunopurified tetanus toxoid: The immune response. Immunol Lett 14:341, 1987.
31. Shahum E, Therien HM: Immunopotentiation of the humoral response by liposomes: Encapsulation versus covalent linkage. Immunology 65:315, 1988.
32. Su D, Van Rooijen N: The role of macrophages in the immunoadjuvant action of liposomes: Effects of elimination of splenic macrophages on the immune response against intravenously injected liposome associated albumin antigen. Immunology 66:466, 1989.
33. Therien HM, Shahum E: Immunopotentiation of the humoral response by liposomes: Effect of a T cell polyclonal activator. Cell Immunol 116:320, 1988.
34. Van Rooijen N, Van Nieuwmegen R, Kors N, Eikelenboom P: The development of specific antibody-containing cells in the spleen of rabbits during the secondary immune response against free or liposome associated albumin antigen. Anat Rec 208:579, 1984.
35. Weissman G, Dukor P: The role of lysosomes in immune responses. Adv Immunol 12:283, 1970.
36. Van Rooijen N, Claassen E: In vivo elimination of macrophages in spleen and liver, using liposome-encapsulated drugs: Methods and applications. In Gregoriadis G (ed): "Liposomes as Drug Carriers." Chichester: John Wiley and Sons, 1988, p 131.
37. Van Rooijen N, Claasen E, Eikelenboom P: Is there a single differentiation pathway for all antibody-forming cells in the spleen? Immunol Today 7:193, 1986.
38. Collé JH, Truffa-Bachi P, Freitas AA: Secondary antibody responses of thymus-independent antigens. Decline and lifespan of memory. Eur J Immunol 18:307, 1988.
39. Gray D, Skarvall H: B-cell memory is short-lived in the absence of antigen. Nature 336:70, 1988.
40. Klaus GGB, Humphrey JH, Kunkl A, Dongworth DW: The follicular dendritic cell: Its role in antigen presentation in the generation of immunological memory. Immunol Rev 53:3, 1980.
41. Tew JG, Phipps RP, Mandel TE: The maintenance and regulation of the humoral immune response: Persisting antigen and the role of follicular antigen-binding dendritic cells as accessory cells. Immunol Rev 53:175, 1980.
42. Mandel TE, Phipps RP, Abbot A, Tew JG: The follicular dendritic cell: Long-term antigen retention during immunity. Immunol Rev 53:29, 1980.
43. Van Rooijen N: Immune complex trapping in lymphoid follicles: A discussion on possible functional implications. In Manning J (ed): "Phylogeny of Immunological Memory." Amsterdam: Elsevier, 1980, p 281.
44. Zborowski J, Roerdink F, Scherphof G: Leakage of sucrose from phosphatidylcholine liposomes induced by interaction with serum albumin. Biochim Biophys Acta 497:183, 1977.
45. Legros F, Cashillo M, Praet M, Vandenbranden M, Lemoine O, Cabiaux V, Van Vooren

JP, Nyabenda J, Dierckx P, Turneer M, Ruysschaert JM: Detection of lytic antimycobacterial antibodies by immune lysis of liposomes sensitized to tuberculin. J Immunol Methods 108:223, 1988.

46. Unanue ER: Antigen-presenting function of the macrophage. Annu Rev Immunol 2:395, 1984.
47. Van Rooijen N, Kors N, Kraal G: Macrophage subset repopulation in the spleen: Differential kinetics after liposome-mediated elimination. J Leukocyte Biol 45:97, 1989.
48. Van Rooijen N, Van Nieuwmegen R: Liposomes in immunology: Multilamellar phosphatidylcholine liposomes as a simple, biodegradable and harmless adjuvant without any immunogenic activity of its own. Immunol Commun 9:243, 1980.
49. Deamer D, Bangham AD: Large volume liposomes by an ether vaporization method. Biochim Biophys Acta 443:629, 1976.
50. Kirby C, Gregoriadis G: Dehydration-rehydration vesicles: A simple method for high yield drug entrapment in liposomes. Biotechnology 2:979, 1984.
51. Gregoriadis G. (ed): "Liposome Technology." Boca Raton, FL: CRC Press, 1984, Vols. 1–3.
52. Alving CR: Immune reactions of lipids and lipid model membranes. In Sela M (ed): "The Antigens." New York: Academic Press, 1977.
53. Schuster BG, Neidig M, Alving BM, Alving CR: Production of antibodies against phosphocholine, phosphatidylcholine, sphingomyelin and lipid A by injection of liposomes containing lipid A. J Immunol 122:900, 1979.
54. Strejan GH, Essani K, Surlan D: Natural occurring antibodies to liposomes. II. Specificity and electrophoretic pattern of rabbit antibodies reacting with sphingomyelin-containing liposomes. J Immunol 127:160, 1981.
55. Alving CR: Antibodies against lipids, lipid bilayers and liposomes: A theory of aging. In Bangham AD (ed): "Liposome Letters." London: Academic Press, 1983, p 269.
56. Richards RL, Aronson J, Schoenbechler M, Diggs CL, Alving CR: Antibodies reactive with liposomal phospholipids are produced during experimental *Trypanosoma rhodesiense* infections in rabbits. J Immunol 130:1390, 1983.
57. Stall AM, Farinas MC, Tarlinton DM, Labor PA, Herzenberg LA, Strober S, Herzenberg LA: Ly-1 B cell clones similar to human chronic lymphocytic leukemias routinely develop in older normal mice and young autoimmune animals. Proc Natl Acad Sci USA 85:7312, 1988.
58. Mercolino TJ, Arnold LW, Hawkins LA, Haughton G: Normal mouse peritoneum contains a large population of Ly-1 + (CD5) B cells that recognize phosphatidylcholine. J Exp Med 168:687, 1988.
59. Van Rooijen N, Van Nieuwmegen R: Endotoxin enhanced adjuvant effect of liposomes, particularly when antigen and endotoxin are incorporated within the same liposome. Immunol Commun 9:747, 1980.
60. Grover A, Sundharadas G: Effect of liposomes on lymphocytes: Induction of proliferation of B lymphocytes and potentiation of the cytotoxic response of T lymphocytes to alloantigens. Eur J Immunol 16:665, 1986.
61. Barrett AJ: Proteolytic and other metabolic pathways in lysosomes. Biochem Soc Trans 12:899, 1984.
62. Mayhew E, Ito M, Lazo R: Toxicity of non-drug-containing liposomes for cultured human cells. Exp Cell Res 171:195, 1987.
63. Alving CR, Richards RL, Moss J, Alving LI, Clements JD, Shiba T, Kotani S, Wirtz RA, Hockmeyer WT: Effectiveness of liposomes as potential carriers of vaccines: Applications to cholera toxin and human malaria sporozoite antigen. Vaccine 4:166, 1986.
64. Claassen E, Westerhof Y, Versluis B, Kors N, Schellekens M, Van Rooijen N: Effects of

chronic injection of sphingomyelin containing liposomes on lymphoid and non-lymphoid cells in the spleen. Transient suppression of marginal zone macrophages. Br J Exp Pathol, 69:865, 1988.
65. Patel HM, Tuzel NS, Ryman BE: Inhibitory effect of cholesterol on the uptake of liposomes by liver and spleen. Biochim Biophys Acta 761:142, 1983.
66. Dave J, Patel HM: Differentiation in hepatic and splenic phagocytic activity during reticuloendothelial blockade with cholesterol free and cholesterol rich liposomes. Biochim Biophys Acta 888:184, 1986.
67. Claassen E, Van Rooijen N: The effect of elimination of macrophages on the tissue distribution of liposomes containing 3H-metrotrexate. Biochim Biophys Acta 802:428, 1984.
68. Moghimi SS, Patel HM: Tissue specific opsonins for phagocytic cells and their different affinity for cholesterol-rich liposomes. FEBS Lett 233:143, 1988.
69. Barratt G, Tenu JP, Yapo A, Petit JF: Preparation and characterization of liposomes containing mannosylated phospholipids capable of targeting drugs to macrophages. Biochim Biophys Acta 862:153, 1986.
70. Garcon N, Gregoriadis G, Taylor M, Summerfield J: Mannose-mediated targeted immunoadjuvant action of liposomes. Immunology 64:743, 1988.
71. Torchilin VP, Klibanov AL: Immobilization of proteins on liposome surface. Enzyme Microb Technol 3:297, 1981.
72. Weissig V, Lasch J, Klibanov AL, Torchilin VP: A new hydrophobic anchor for the attachment of proteins to liposomal membranes. FEBS Lett 202:86, 1986.
73. Heath TD, Macher BA, Papahadjopoulos D: Covalent attachment of immunoglobulins to liposomes via glycosphingolipids. Biochim Biophys Acta 640:66, 1981.
74. Strejan GH, Gilbert JJ, St. Louis J: Effect of treatment with glutaraldehyde fixed myelin basic protein-liposomes on active induction and passive transfer of experimental allergic encephalomyelitis in Lewis rats. Cell Immunol 116:250, 1988.
75. Alving CR, Banerji B, Shiba T, Kotani S, Clements JD, Richards RL: Liposomes as vehicles for vaccines. Prog Clin Biol Res 47:399, 1980.
76. Gregoriadis G, Allison AC (eds): ''Immunological Adjuvants and Vaccines.'' New York: NATO ASI Plenum Press, 1989 (in press).
77. Siddiqui WA, Taylor DW, Kan SC, Kramer K, Richmond-Crum SM, Kotani S, Shiba T, Kusumoto S: Vaccination of experimental monkeys against *Plasmodium falciparum:* A possible safe adjuvant. Science 201:1237, 1978.
78. Gerlier D, Doré JF: Adjuvant effect of liposome presentation of soluble tumour associated antigen. In Gregoriadis G, Senior J, Trouet A (eds): ''Targeting of Drugs.'' New York: Plenum Press, 1982, p 323.
79. Van Rooijen N, Van Nieuwmegen R: Liposomes in immunology: Impairment of the adjuvant effect of liposomes by incorporation of the adjuvant lysolecithin and the role of macrophages. Immunol Commun 8:381, 1979.
80. Richards RL, Hayre MD, Hockmeyer WT, Alving CR: Liposomes, lipid A, and aluminum hydroxide enhance the immune response to ra synthetic malaria sporozoite antigen. Infect Immun 56:682, 1988.
81. Stewart-Tull DES: Immunopotentiating conjugates. Vaccine 3:40, 1985.
82. Gregoriadis G: Liposomes for drugs and vaccins. Trends Biotechnol 3:235, 1985.
83. Perrin P, Thibodeau L, Sureau P: Rabies immunosome (subunit vaccine) structure and immunogenicity. Pre- and post-exposure protection studies. Vaccine 3:325, 1985.
84. Richards RL, Wirtz RA, Hockmeyer WT, Alving CR: Development of liposomes as carriers for a human malaria peptide vaccine. In Gregoriadis G (ed): ''Liposomes as Drug Carriers.'' Chichester: John Wiley and Sons, 1988, p 235.

85. Boudreault A, Thibodeau L: Mouse response to influenze immunosomes. Vaccine 3 (Suppl) 1985, p. 231.
86. Torchilin VP, Burkhanov SA, Mazhul LA, Agava ON: Liposomal vaccine against influenza virus. In Gregoriadis G (ed): ''Liposomes as Drug Carriers.'' Chichester: John Wiley and Sons, 1988, p 229.
87. Vinayak VK, Nain PCK, Saxena A: Immunoprotective behaviour of plasma-membrane associated antigens of axenic Entamoeba histolytica. J Med Microbiol 24:297, 1987.
88. Nain PCK, Vinayak VK: Vaccination against experimental hepatic amoebic infection. An evaluation of phosphatidylcholine liposomes as immunopotentiator. J Parasitol 73:859, 1987.
89. Vinayak VK, Sharma R: Phosphatidylcholine liposomes as effective adjuvant for vaccination of mice against Plasmodium yoelii. Indian J Med Res 84:32, 1986.
90. Klein JP, Wachsmann D, Schöller M: Liposomes as immunological adjuvants for 'Streptococcus mutans' antigens. In Gregoriadis G (ed): ''Liposomes as Drug carriers.'' Chichester: John Wiley and Sons, 1988, p 261.
91. Mettler I, Czuppon AB: Induction of high titer antisera by liposome-incorporated spermatozoal antigens and reversible immunosuppression of fertility. In Gregoriadis G (ed): ''Liposomes as Drug Carriers.'' Chichester: John Wiley and Sons, 1988, p 207.
92. Gerlier D, Bakouche O, Doré JF: Liposomes as a tool to study the role of membrane presentation in the immunogenicity of a Mu-LV-related tumor antigen. J Immunol 131: 485, 1983.
93. Steele G, Ravikumar T, Ross D, King V, Wilson E, Dodson T: Specific active immunotherapy with butanol-extracted, tumor-associated antigens incorporated into liposomes. Surgery 96:352, 1984.
94. Le Grue S: Carrier and adjuvant properties of liposome-borne tumor-specific antigens. Cancer Immunol Immunother 17:135, 1984.
95. Raphael L, Tom BH: Liposome facilitated xenogeneic approach for studying human colon cancer immunity: Carrier and adjuvant effect of liposomes. Clin Exp Immunol 55:1, 1984.
96. Tom BH: Liposome-facilitated human colon cancer immunity: A proposed in vitro model. In Gregoriadis G (ed): ''Liposomes as Drug Carriers.'' Chichester: John Wiley and Sons, 1988, p 267.

Bacterial Vaccines, pages 281–306


# Surface Chemistry and Classification of Vaccine Adjuvants and Vehicles

**L.F. Woodard**

Department of Veterinary Sciences, University of Wyoming, Laramie, Wyoming 82070

## I. INTRODUCTION

It has been over 60 years since the discovery that addition of many foreign substances to toxoids could increase their immunogenicity [1]. Although tapioca, bread crumbs, starch oil, turpentine, toluol, and even pyogenic bacteria are no longer used, most commercial inactivated vaccines and toxoids produced today contain aluminates that were described in 1926 [2].

We are now entering an era where genetically engineered subunit proteins, viral vectors, and synthetic peptides will provide the antigenic stimuli of new vaccines. To provoke a protective immune response, advanced adjuvant and vehicle systems must be developed. The adjuvants and excipients must be nontoxic, noncarcinogenic, and not result in objectionable local or systemic reactions.

To develop new adjuvants, we must first understand their mechanisms of action at the cellular and molecular levels. Fresh approaches are needed. Although theories have been advanced to explain the properties that make substances adjuvant active, few unifying answers have been found. Confusion rather than order exists.

The purpose of this chapter is to detail adjuvant substances and the many aspects of surface science relating to adjuvant activity, emulsion technology, and vaccine design. As pointed out by Gall [3], most vaccine adjuvants are surface active. The principles of surface science are briefly outlined, especially as they relate to adjuvant activity and classification, in hope that a better understanding will promote accelerated research. An empirical classification of vaccine adjuvants is proposed.

The term "adjuvant," or "immunoadjuvant," has many meanings depending on the context in which it is used. For purposes of this review, only substances that are *physically included* in vaccines and have been shown to enhance antigenicity will be discussed. Immunostimulants and other biological response modifiers given before or after antigen may increase or modulate the immune response, but most will not be covered here. With a few exceptions, crude adjuvants such as certain bacteria (mycobacteria, pertussis, proprionibacteria, etc.) will not be reviewed, because defined components with adjuvanticity have been isolated. The term "vaccine" will be used in its broadest sense to include all antigenic preparations.

A review of the literature on a subject as broad as that of vaccine adjuvants will result in omissions. Most of these are deliberate. Previous review articles and recent papers that include extensive references will be cited where appropriate to avoid prolixity. Several current review articles and books are available [4–8].

## II. PRINCIPLES OF SURFACE SCIENCE AND EMULSION TECHNOLOGY

As mentioned in the introduction it has been noted that most vaccine adjuvants "are potentially or actually surface active" [3]. The significance of this thought became more apparent with studies of trehalose dimycolate and certain block polymer surfactants. These vaccine adjuvants have adhesive properties that concentrate antigen and certain immune-activating host proteins in a surface orientation conducive to enhancement of the immune response [9–13]. If one considers emulsions, such as the water-in-oil emulsion of Freund's incomplete adjuvant, that are not surface active per se, perhaps a better statement may be that "most vaccine adjuvants are surface active or have a special surface interface" [14]. Binding of antigen to a surface or presentation of a special type of surface for antigen adsorption appears to be critical for much of the biological activity of most agents reported as adjuvants. The biological activity may include facets of proper antigen presentation and recognition, antigen–antibody interaction, and activation of host proteins of inflammatory and immune system origin [9–11]. Proposed mechanisms of action of vaccine adjuvants have been reviewed [15–19].

For the immunologist, just what is meant by the terms "surface active" and "surface"? In the simplest terms, surface-active agents or surfactants are substances that have both polar and nonpolar moieties. In the case of living systems where water is the principal solvent, the polar groups have an affinity for water (hydrophilic), while the nonpolar groups are repelled by water (hydrophobic). Thus, surface-active agents concentrate at the surface formed by the interface of water and a nonpolar substance such as a lipid or lipid membrane. Most adjuvants are water-insoluble surfactants.

### A. Emulsions And Vaccine Vehicles

Mixing of oil and water is an excellent example to illustrate many of the principles of surface and colloidal science. The technology involved is also critical to the design of vaccine vehicles or excipients. As already stated, most adjuvants are water-insoluble surfactants, so lipoidal vehicles are necessary for proper delivery. These vehicles are generally oil emulsions or liposomes and provide the surface interphase necessary for solubilization and proper orientation of the adjuvant active material.

As explained by Garrett [20], "there is no force of repulsion between oil molecules and water molecules," but "oil does not mix with water because it is squeezed out by powerful forces between water molecules." The attraction between water molecules is simply greater than the attraction between water and hydrocarbon molecules or hydrocarbon and hydrocarbon molecules.

If one vigorously shakes oil and water, a liquid-in-liquid dispersion is temporarily formed. Likewise, solid-in-liquid, gas-in-liquid, solid-in-gas, etc., dispersions are possible. Such dispersions are classified by particle size: molecular ($<1$ nm), colloidal (1 nm to 0.1 μm), and coarse ($>0.1$ μm) [21]. An emulsion of oil and water is a coarse dispersion. Colloidal dispersions are also commonly used in vaccines and will be discussed later. Molecular dispersions like solutions need no further explanation. For now, the focus will be on emulsions to explain most principles of surface-active agents.

To disperse a small amount of oil into a larger amount of water requires the formation of a large surface area. For example, 10 ml of oil on top of a water column with a cross-sectional area of 1 $cm^2$ would result in 1 $cm^2$ of surface interface. If emulsified to an oil droplet diameter of 0.2 μm, the emulsion would contain $2.4 \times 10^{15}$ oil droplets and the surface area of these droplets would be about 300 square meters [21].

Generation of this amount of surface requires considerable kinetic energy. The resultant potential or stored energy gives the emulsion a high degree of instability. The main effect of surfactants known as emulsifiers is to reduce interfacial tension and give stability to the system.

## B. Emulsifiers

Emulsifiers are generally classified into three groups (realizing that overlap is common): the finely divided solids, the naturally occurring materials, and the synthetics [22]. Structurally, they all have polar (hydrophilic) and nonpolar (hydrophobic) regions. The polar regions align with the water phase, while the nonpolar regions align with the oil phase of an oil emulsion. This alignment can be thought of as a film that forms at the interface.

Emulsifiers with a proportionally larger hydrophilic moiety would tend to promote formation of an oil-in-water (O/W) emulsion. Those with a larger hydrophobic moiety would aid in formation of water droplets and, thus, a water-in-oil (W/O) emulsion. The droplet phase would be the *disperse* or *internal phase,* which is dispersed in the *continuous* or *external phase*. The latter is also known as the *bulk phase*. These terms should be kept in mind when discussing oil-emulsion vaccines.

## C. Classification of Emulsifiers

**1. Finely divided solids.** Many metal and mineralized substances have the tendency to concentrate at surface interfaces. Powdered quartz (silica), alumina, carbon black, and bentonite clay are examples, and all have been used as adjuvants. Depending on whether the surface is wetted by water (hydrophilic) or oil (hydrophobic), they can be used to stabilize O/W or W/O emulsions respectively.

**2. Naturally occurring materials.** Among the surfactants that occur naturally are phospholipids such as lecithin, sterols such as cholesterol, fatty acid esters such as lanolin, water-soluble gums, alginates from kelp, carrageenan from seaweed, cellulose derivatives such as methylcellulose, and even proteins. They are often used as auxiliary surfactants (e.g., emulsion stabilizers) because of their lack of efficiency [22].

Most proteins exhibit surface activity because of the presence of hydrophobic peptide sequences and side chains. As a result, they can form interfacial "skins" or films. They have been used in emulsions to prevent oil-droplet coalescence and in cleaners to increase detergent activity. Proteins are concentrated at the interface of surfaces, which often results in unfolding or denaturization [23]. As will be discussed, both concentration and unfolding often have important biological significance and may explain some aspects of antigenicity and adjuvanticity.

**3. Synthetic surfactants.** The synthetic group of surface-active agents is widely used as detergents, emulsifiers, wetting agents, spreading agents, antisplattering agents, etc. Although they may contain natural starting materials, such as the tallow in soap, their composition is generally well defined. This class can be further divided into anionic, cationic, or nonionic surfactants according to the charge of the ion species attached to the nonpolar portion of the molecule. For example, with sodium oleate soaps, the sodium cation is free to move in aqueous media, whereas the oleate anion remains at the interface. Thus, sodium oleate would be classified as an anionic surfactant. A basic surfactant like hexadecylamine would attract hydrogen ions from water and carry a net positive charge, resulting in a cationic classification.

The most widely used anionic surfactants are the salts of long-chain fatty acids (soaps), sulfated oils or alcohols, and aliphatic or aromatic sulfonates. Aliphatic nitrogenous and sulfonium bases constitute the main cationic surfactants. Of course, ampholytic agents can be synthesized.

Because of their insensitivity to pH and water hardness, the nonionic surfactants are widely used. Fatty-acid esters of polyhydric alcohols such as the Tween and Span series are frequently used in the laboratory and by commercial firms. They result from the dehydration of sorbitol (Spans) or polymerization of partially esterified Spans with ethylene oxide (Tweens). Ether-linked polymers of polyoxyethylene (hydrophilic) and polyoxypropylene (hydrophobic) are known as block copolymers and have many uses, including that of adjuvant. The nonionics also include the fatty-acid amides and other ester-linked polymers of alcohols or other hydroxyl-containing compounds.

**TABLE I. HLB Ranges and Functional Applications of Nonionic Surfactants**

| HLB range | Application |
|---|---|
| 0–2 | Spreading agent |
| 3–6 | Water-in-oil emulsifier |
| 7–9 | Wetting agent |
| 8–15 | Oil-in-water emulsifier |
| 13–15 | Detergent |
| 15–18 | Solubilizer |

From Hunter et al. [11] with permission of the *Journal of Immunology*.

## D. Selecting Nonionic Surfactants: The HLB System

A particularly useful numeric system for predicting functional activity of defined nonionic surface-active agents is known as the hydrophile–lipophile balance (HLB) system [22]. Positive values are assigned to each hydrophilic group and negative values are assigned to each hydrophobic group. The sum of these values is the HLB value for a given surfactant. Those with low values are strongly hydrophobic, whereas those with high values are strongly hydrophilic. Within general ranges of values, functional uses of surfactants can be predicted (Table I) [11]. Most commercial firms publish HLB values for their nonionic surfactants. A word of caution: HLB values assigned to ionic surfactants can be very misleading and should generally be ignored.

The HLB system is very useful in formulating emulsions and, as will be discussed later, has significant value in selecting nonionic surfactant adjuvants. By knowing the approximate HLB value required to emulsify a given oil, one can select emulsifiers that have this value or (preferably) blend emulsifiers to yield the desired value. Blends generally result in more stable emulsions than can be expected from a single emulsifier. A particularly useful guide to the HLB values required to emulsify various paraffinic and vegetable oils and methods of blending surfactants is available from ICI Americas [24].

Returning to the example of water and oil, an O/W emulsion of light mineral oil would require emulsifiers with an HLB value of 10 to 11 [24]. If one blended six parts of hydrophilic emulsifier with an HLB value of 15 and four parts of hydrophobic emulsifier with an HLB value of 4, the resultant emulsifier blend would have an HLB value of 10.6 and should be capable of emulsifying the oil. Several blends may have to be formulated before an acceptable stable emulsion results. Considerable art is involved. By combining the HLB system with spectrophotometric analysis of the emulsions, stable O/W emulsions suitable for vaccine and adjuvant carriers can be formulated for metabolizable n-alkanes and mineral oils [25]. Vegetable oils generally require an auxiliary surfactant for stability.

## III. PROPOSED CLASSIFICATION AND REVIEW OF VACCINE ADJUVANTS

Perusal of the literature on surface-active agents reveals thousands of compounds with dozens of uses. Examples of these uses are lubricants, glues, adhesives, soaps, detergents, emulsifiers, antisplattering agents, slipping agents, gelling agents, wetting agents, lather promotants, flocculating agents, dispersing agents, spreading agents, etc. To immunologists not familiar with surface science, choosing adjuvant-active substances without guidelines is nearly impossible.

The HLB classification of functional uses for nonionic surfactants (Table I) reveals the diversity to be found among so-called "emulsifiers." Indeed, surfactants include a multitude of agents not used in emulsions, but acting at the interphase between immiscible subtances.

The classification scheme already discussed for emulsifiers serves as a basis for classifying surfactant adjuvants: the finely divided solids, the naturally occurring materials, and the synthetics [14]. Keep in mind the overlap between groups. Modification of this classification scheme is well suited to categorizing surfactant adjuvants. It is broad and nonlimiting with regard to physical appearance, structure, and functional activity.

Classification of adjuvants should not be subject to constant change as we begin to understand their various modes of action at the cellular and molecular levels. Although it would be naive to think that a common trait, activity, or composition could be related to all known adjuvants and could explain their mode of action, most vaccine adjuvants are water-insoluble surfactants and/or provide a suitable surface for antigen binding. The notable exceptions are certain water-soluble substances such as nonlipidated muramyl dipeptides and certain cytokines.

### A. Surfactant Adjuvants

**1. Inorganic surfactant adjuvants.** The class of inorganic surfactant adjuvants could also be titled "metals and minerals" or "finely divided solids." It includes those basic metal salts, metal colloids, and mineralized substances capable of binding antigens or precipitating them as a complex. Foremost among these adjuvants are the aluminates. First described by Glenny et al. [2,26], sodium or potassium aluminum salts (alums) are still used extensively, especially in the production of bacterins. Not only are the protein antigens concentrated from culture broths, for example, but antibody production is greatly enhanced. Alternatively, aluminum hydroxide gels formed by a reaction of aluminum salts with ammonium hydroxide are used to adsorb antigens (e.g., toxoids) from solution and provide adjuvanticity.

At present, aluminum ''adsorbent'' adjuvants are the only adjuvants cleared for use in humans by the U.S. Food and Drug Administration. With exceptions, notably poultry vaccines, this is generally the case with U.S. Department of Agriculture–approved animal vaccines. Excellent review articles on aluminum adjuvants are available [7,27,28]. Manufacturers of aluminum hydroxide gels also can furnish selected references.

Antigen complexes with such metals as calcium, cerium, beryllium, iron, tungsten, mercury, platinum, and zinc have increased antigenicity of soluble antigens, although aluminates generally elicit the best antibody response [26,29–36]. It is interesting to note that aluminum stearate, an anionic surfactant, apparently does not have adjuvant activity [3]. However, it is used in vegetable-oil emulsions as a stabilizer.

Recent studies with beryllium have shown that macrophages in regional lymph nodes are damaged and immunoblastic proliferation occurs after injection [34,35]. Beryllium, lipopolysaccharide (LPS), and Freund's complete adjuvant have all been shown to increase Ia-positive macrophages, suggesting that some adjuvants may augment the immune response by increasing antigen processing and presentation [36].

Mineralized substances such as quartz (silica, silicon dioxide) [37,38] and bentonite [39] have been used to bind proteins and increase their antigenicity. Of interest are the metal lithium and the mineral selenium. Lithium is capable of reversing cAMP-mediated suppression of plaque-forming cells, enhancing E-rosette formation, and increasing antigen-induced or spontaneous in vitro antibody synthesis [40,41]. Selenium has been found to increase immune responsiveness when included in the diet [42,43], but, like lithium, it has apparently not been used as a vaccine adjuvant.

**2. Natural organic surfactant adjuvants and derivatives.** A wide variety of adjuvants have been derived from plant, animal, and microbial sources. As previously stated, whole bacteria such as the mycobacteria included in Freund's complete adjuvant will not be discussed, because defined components responsible for the adjuvant activity have been isolated. Certain analogues and isomers of natural products will be discussed here rather than under the synthetic subclass to avoid repetition.

The glycolipids known as ''cord factor'' are among the best-studied natural vaccine adjuvants [44]. This term designates a variety of trehalose mono- or dimycolates produced principally by the *Mycobacterium, Corynebacterium,* and *Nocardia* genera of bacteria. The trehalose dimycolates (TDM) of mycobacteria are the most-studied cord factor preparations and are capable of potentiating both cellular and humoral immune responses [10,44–52]. Synthetic trehalose diesters have also been found to have immunostimulatory activity [44,53,54].

The need of a lipophilic surface for the biological activities of TDM has been demonstrated [9,10,51,55]. Suspensions of TDM form micellar cylindricals incapable of binding proteins, while TDM (in a monolayer film) coating lipophilic beads or oil droplets binds protein and is adjuvant active [9,10]. Adjuvanticity may be related to fibrinogen binding, complement activation, and the resultant inflammatory response [10,56].

The LPS constituent of outer membranes of gram-negative bacteria has been extensively studied and shown to have a variety of immunomodulating and biological activities. Although the adjuvant property was noted over 30 years ago [57], native LPS is too toxic and pyrogenic for use in vaccines. Recent studies have shown the lipid A moiety of LPS to be the adjuvant active component [58,59].

The backbone structure of lipid A varies somewhat among gram-negative bacteria, but consists of a diphosphorylated glucosamine disaccharide [60]. Added to this backbone are a number of saturated fatty-acid esters or amide-linked acyl groups that give lipid A its amphipathic (polar and nonpolar) surfactant property. Structurally, lipid A resembles TDM [51].

Synthetic lipid A analogues have been shown to have much of the immunomodulatory activities of native lipid A [61,62]. Removal of the phosphoryl group from the reducing end of the molecule results in a nontoxic adjuvant [51,63].

Phospholipids such as techoic acids, sphingomyelin, lecithin, lysolecithin, lysolecithin analogues, and various other phosphocholines have limited adjuvanticity [1,64–68]. Endogenous lysolecithin formed by activation of phospholipase A in response to numerous adjuvants has been proposed as a mode of action [65,68]. The lysolecithins incorporated into liposomes may, in part, account for their adjuvant activity. Liposomes will be discussed later in this chapter.

Other emulsifying agents such as lanolin [69], methylcellulose [70], alginate [71], and saponins [1,6,70,72–76] have adjuvant activity. A matrix of Quil A saponin micelles and protein antigens known as iscom (immunostimulating complex) has been reported to be a potent method of enhancing antigenicity [77–79]. Several viral antigens have been incorporated onto Quil A micelles as subunit vaccines [78].

Vitamin A (retinol) and vitamin A analogues (retinoids) have been shown to enhance humoral and cell-mediated immunity [see 80 for citations], but in nearly all experiments they were not physically included with the antigen preparation. Vitamin E has been used as a successful vaccine adjuvant for at least two vaccines [81,82].

Glucan (zymosan), lentinan, and other glucopyranose polysaccharides of yeast and mushrooms have been shown to have general immunostimulatory [83–86] and adjuvant activities [87,88]. Similarly, the poly-N-acetylglucosamines and analogues of chitins have been shown to act as adjuvants [89,90].

Carrageenans (polyanionic sulfated algal polysaccharides) generally result in immunosuppression when given with antigens [91], although enhancement has been observed in some mouse strains [92]. Immunosuppression has been associated with increased prostaglandin E levels.

Fungal by-products such as amphotericin B [83,93] and monensin [94] are of peripheral interest to this review because of their surface and immunomodulating properties.

**3. Synthetic organic surfactant adjuvants.** The majority of compounds in this group of surface-active agents fall into two fairly distinct groups: the cationic aliphatic nitrogenous surfactants and the nonionic surfactants. Only a few examples of anionic synthetic adjuvants can be found.

Hexadecylamine was the first water-insoluble polar lipid adjuvant to be described [95–97]. Shortly thereafter, Gall [3,76] performed a series of classic experiments to determine structure–activity relationships of various surfactants. This was the first extensive attempt to discover new vaccine adjuvants with a systematic rather than empirical approach. He noted little adjuvanticity among the nonionic and anionic compounds tested, but aliphatic nitrogenous bases provided a number of adjuvant active compounds.

Adjuvant activity was found with aliphatic amine, guanidine, benzamidine, and thiouronium compounds, provided that the aliphatic chain (generally saturated alkyl substitution) was approximately C16 or greater. In some cases, dialkylation provided enhancement of adjuvanticity, suggesting that a decrease in polarity was beneficial. A simple decrease in water solubility could not explain all results, because basicity was also required. Highly cationic substances were more active than analogues at or near neutrality. Membrane activity, as evidenced by hemolytic activity in the absence of complement, was exhibited by most adjuvant active compounds, but could not explain the mode of action since many nonadjuvants were also hemolytic.

Noting the need for lipophilic vehicles for optimal activity of TDM [9,10], I have recently reexamined the structure activity of various aliphatic primary, secondary, tertiary, and quaternary aliphatic amine surfactants in an O/W emulsion [25]. Except for avridine (discussed below), none of the aliphatic amines were any more adjuvant active than the emulsion alone [98].

The most-studied alkyl amine is dimethyl dioctadecyl ammonium bromide (DDA). First noted in Gall's study [3,76], dozens of papers have appeared in which DDA has been used as an adjuvant. Recent papers provide most of the pertinent references [6,14,42,48,98–100].

Avridine (N,N-dioctadecyl-N′,N′-bis[2-hydroxyethyl] propanediamine; CP-20,961) is an alkyl diamine similar in activity do DDA [100,101]. It has been found to be more potent than DDA in O/W emulsions [98] and has numerous immunostimulatory activities by itself [100].

Octadecyl tyrosine hydrochloride is another alkyl amine that has been reported to be an effective adjuvant without the inherent immunostimulant properties of most alkyl amines [102].

The nonionic synthetic adjuvants are best represented by the block polymer surfactants. Repeating units of polyoxyethylene and polyoxypropylene give these synthetics their respective hydrophilic and hydrophobic characteristics. The adjuvant and inflammatory properties of selected block polymers have been extensively studied [5,6,11–13,98,103,104] Of these, copolymers L101, L121, and T1501 have received the most attention. The threonyl analogue of muramyl dipeptide has been incorporated into a squalane O/W emulsion with L121 to provide an especially useful adjuvant [105].

Surface studies of block polymers suggest that those that are poorly soluble in water but form an exposed hydrophilic surface with a large area are the best adjuvants [12,13]. This surface orientation interacts with antigen and host proteins in an effective manner without undue inflammatory responses. Reversed octablocks suspended in solution have been shown to be effective adjuvants for haptenated liposomes [106]. Also, adjuvant activity of aqueous suspensions of block polymers may differ from that seen when they are mixed in the oil phase of an O/W emulsion [107].

Among the nonionic synthetic glycolipids, I have recently described several with significant adjuvant activity when mixed with antigen in the oil phase of hexadecane-in-water emulsions. In terms of antibody response, two were equal to Freund's complete adjuvant: sorbitan trioleate (Span 85) and glycerol trioleate [98]. Acetylated sucrose dioctadecanate was also very active and propylene glycol octadecanate was only slightly less so. These glycolipids were initially selected solely on the basis of their low (<2.0) HLB values.

The anionic synthetic surfactant adjuvants are represented by considerably fewer substances than the cationics and nonionics. Among the anionics that have been reported as adjuvants, although not always physically included with the antigen, are dextran sulfate [108–111], sumarin [111,112], Evans blue dye [113], sodium dodecylbenzene sulfonate [114], sodium diethyldithiocarbamate [115], pyran [100], and styrene-maleic anhydride copolymer [116]. Dextran sulfate nanospheres with entrapped antigen have been described as an effective slow-release adjuvant [117]. Although Evans blue dye (related to sumarin) has been shown to enhance delayed hypersensitivity, immunity was decreased [118]. Collectively, although the anionic surfactants, both natural and synthetic, appear to be useful in studies concerned with immune mechanisms (especially cellular immunity), they are weak adjuvants as compared to those already discussed.

Small synthetic molecules with limited surface activity that are of peripheral interest include tilorone [100,119], hypoxanthine analogues [120], pyrimethamine [121], pyrimidinones [100], and polynucleotides [122].

## B. Vesicle Adjuvants

A second major category of vaccine adjuvants includes those preparations that provide a surface interphase for antigen adsorption. Although some solid-phase surface adjuvants have been described, most are liquid vesicles that provide an aqueous–lipid surface interface. For commercial purposes, liquid vaccines are probably the only acceptable form.

**1. Water-in-oil adjuvants.** The development of W/O emulsion adjuvants has been reviewed by Jules Freund, the man whose name is commonly associated with these mineral-oil preparations [123,124]. Freund's incomplete adjuvant (FIA or IFA) denotes an approximately 50% water:50% oil emulsion and Freund's complete adjuvant (FCA or CFA) denotes a similar preparation with inclusion of killed mycobacterial bacteria. Several other excellent review articles are available [5–7,125,126].

Over a 20 year period starting in 1937, Dr. Freund and colleagues determined that water droplets containing antigen (internal phase) emulsified in light mineral oil (continuous phase) could effectively boost the antibody response to foreign antigen. Inclusion of killed TB organisms usually enhanced this response and sensitized the subject to subsequent skin test. Delayed hypersensitivity occurred only when mycobacteria were included; this allergic state is delayed in time after skin test and has been used to detect cell-mediated immune responsiveness. For purposes of this review, the effect of mycobacteria will not be discussed. Defined components of these bacteria, especially muramyl dipeptide and trehalose dimycolate, are discussed separately. Inclusion of mycobacteria has been limited to experimental purposes. Use of FCA and FIA is coming under increased criticism because of granulomatous inflammation and abscessation [4].

Emulsification of mineral or paraffin oils to form W/O emulsions is accomplished with the aid of emulsifying surfactants. Emulsion formation requires the surfactant to be associated mainly with the bulk or continuous phase. In W/O emulsions, this is the oil phase and thus lipophilic surfactants are required (Table I). Lanolin, aquafor, and Arlacel A (principally mannide monooleate) are among the emulsifiers that have been used and refined Arlacel A is the principal emulsifier in use today [126]. Oil emulsions made with a blend of Span 85 and Tween 85 as emulsifiers were reported to be very stable and more adjuvant active than traditional FIA [127]. The authors speculated that their emulsions contained a "strong stimulating factor." In light of the finding that Span 85 is a good adjuvant [98], their thoughts may have been well founded.

The oil phase of W/O emulsion adjuvants is generally a light mineral (paraffinic) oil such as Drakeol 6VR, Bayol F, or Marcol 52 [126]. These oils contain normal, branched, and cyclic hydrocarbons plus other aromatics, but have been refined to reduce or eliminate substances such as polycyclic aromatics that are potential carcinogens [7]. Aliphatic and branched saturated paraffinic hydrocarbons have been shown to be effective replacements for refined mineral oils [128,129]. Vegetable oils such as peanut oil [130] have also been used, although an auxiliary surfactant such as aluminum stearate is necessary to stabilize such emulsions. Esters of fatty acids have been used as the oil phase, although the results must be interpreted with the knowledge that L122 block polymer, a moderately potent adjuvant, was used as the emulsifier [131].

To be effective, antigens must be incorporated into the internal phase of oil emulsions, i.e., the water phase of W/O emulsions [123,124] and the oil phase of O/W emulsions (discussed below) [25,46]. Apparently, the amphipathic nature of protein antigens results in maximal surface concentration (with subsequent epitope exposure due to denaturization) only when they are concentrated in the internal phase.

**2. Oil-in-water adjuvants.** In contrast to W/O emulsions, O/W emulsions have only limited adjuvant activity and have generally been used as low-viscosity vehicles for natural or synthetic lipoidal adjuvants. For example, mean log 2 titers against bovine serum albumin were 4.12 when injected in saline, 5.88 when given in an O/W emulsion, and 10.61 when given in FIA [98]. Inclusion of surfactant adjuvants such as sorbitan trioleate or glycerol trioleate into these O/W emulsions resulted in titers statistically equivalent to those induced with FIA or FCA (9.75 and 9.44 respectively).

Use of O/W emulsions as vehicles for delivery of lipoidal immunopotentiators became widespread with the finding that oil droplets were necessary for the activity of TDM [51,55]. These studies employed 1% Drakeol in water with 0.2% Tween 80 as the emulsifier. Subsequent research defined the role of surface in TDM's biological activity [9,10].

Preparation of stable O/W emulsions with a 70:30 blend of Tween 80 and Span 80 as emulsifiers and hexadecane as the metabolizable oil has been recently described [25]. These 5% hexadecane emulsions were more effective than a commercial peanut-oil emulsion (Lipovant) [132] and similar to a commercial soybean-oil emulsion (Intralipid) [133] when avridine adjuvant was incorporated.

Squalane, the hydrogenated form of squalene oil, has been shown to be an effective hydrocarbon vehicle for O/W emulsions with muramyl dipeptides, L121 block polymer, TDM, and synthetic lipid A adjuvants [51,105,134].

**3. Liposomes.** Like O/W emulsions, liposomes—artificial lipid bilayer vesicles—have limited adjuvant activity unless more adjuvant-active compounds are entrapped within the liposome or integrated into the bilayer. The surface-active organics used to form the laminar membranes include most of the natural surfactants also discussed as having adjuvanticity, including lecithin, stearylamine, and various phosphatidylcholines. Most of the phosphocholines now used have aliphatic hydrocarbon esters to increase their nonpolar moiety. Several excellent review articles are available that detail liposomal compositions, preparation, and uses [5,135–139].

Liposomes are attractive vaccine vehicles because antigens and adjuvants can be entrapped within or presented on the surface. Use of a potent adjuvant such as lipid A incorporated into the bilayer results in enhancement of immune responses [140]. However, immune responses against constituent phospholipids could be a serious problem if they are derived from natural body sources [141].

**4. Proteosomes.** Recently, lipophilic outer-membrane porin vesicles from meningococci have been described as a useful vaccine vehicle and adjuvant [142,143]. Unlike most carrier proteins, direct conjugation is not required. Synthetic peptides made lipophilic have been shown to be antigenic when hydrophobically bound to these "proteosomes" [143]. The proteosomes are B-cell mitogens and apparently provide good adjuvanticity without addition of extra adjuvant compounds.

**5. Miscellaneous.** Solid matrix implants of ethylene-vinyl acetate and antigen have been described [144]. These implants provide a sustained release of antigen with resultant antibody titers comparable to FCA. The polymer matrix is inert.

A special type of emulsion termed "microemulsions" has not to my knowledge been used as a vaccine vehicle, but would appear to be an item of interest in vaccine technology. Microemulsions are stable, isotrophic, low-viscosity dispersions of oil or water domains stabilized with surfactants and cosurfactants [145,146]. Their vesicle size is about 10–100 nm, so a large amount of surface area is available as compared to the macroemulsions previously discussed.

### C. Water-Soluble Adjuvants and Derivatives

Very few water-soluble vaccine adjuvants have been described. Most notable are the nonlipidated muramyl dipeptides and certain cytokines.

**1. Peptidoglycans.** Before the discovery that N-acetylmuramyl-L-alanyl-D-isoglutamine (MDP) was the minimal adjuvant active principle of bacterial cell walls, a number of water-soluble preparations were described. These peptidoglycans have been reviewed and adjuvant activities of various fractions detailed [147]. Depending on the organism, extraction procedure, and residue linkages, a wide spectrum of immunological activities were seen.

**2. Muramyl dipeptides.** Since the discovery that MDP was the minimal peptidoglycan component capable of replacing mycobacteria in FCA [148], hundreds of MDP derivatives and analogues have been synthesized. Several excellent review articles are available [5,8,149–151].

Development of acquired resistance to an intracellular parasite such as brucella requires that the MDP be in FIA or the oil phase of O/W emulsions [48,152,153]. Likewise, MDP must be incorporated into the oil phase of O/W emulsions for optimal antibody response to a soluble antigen such as bovine serum albumin [98]. Incorporation of MDP into liposomes also markedly increases cell-mediated immune responses [154] as does substitution with lipophilic moieties [155,156].

Two MDP derivatives show the most promise at this time for use in human and veterinary vaccines. Both lack the pyrogenicity of native MDP. The threonyl-MDP analogue has been reported to give excellent antibody and cell-mediated immune responses, especially when incorporated into an O/W emulsion of squalane with L121 block polymer [105,157]. Another derivative, murabutide (N-acetylmuramyl-L-Ala-D-Gln-n-butyl ester), is also receiving considerable attention in clinical trials [5,8,149–151]. Both of these MDP derivates stimulate interleukin-1 production [157].

**3. Cytokines.** Products from immune reactions and inflammation have immunomodulatory effects. The term ''autocoids'' has been suggested for these local hormonelike substances [100].

Interleukin-1 (IL-1) has received attention as a product of adjuvant-activated monocytes and as a natural adjuvant [5]. When given at the proper time after vaccination, increased antibody responses are seen [158]. The multiple roles of IL-1 as a mediator of fever, inflammation, and immunity have been reviewed [159–161]. It seems likely that recombinant technology could lead to an IL-1 fragment that would be useful as a vaccine adjuvant but without adverse side effects [162].

Interleukin-2 (IL-2) production is also enhanced by adjuvants such as lithium and MDP [163]. Like IL-1, it seems probable that recombinant IL-2 or fragments will have use in vaccine adjuvants. The use of IL-2 to restore responsiveness when given with antigen has been reported [164]. More recently, recombinant IL-2 incorporated with vaccine antigen has been shown to enhance cellular cytotoxicity and protection against genital herpes in guinea pigs [165].

Many of the vaccine adjuvants discussed have the ability to induce interferons [100]. Recent studies with recombinant gamma interferon have shown it to be comparable with saponin as measured by immune responses and protection [166]. Like IL-1 and IL-2, it is obvious that gamma interferon may have use as a natural body adjuvant.

## IV. ROLE OF SURFACE IN ANTIGEN RECOGNITION AND ADJUVANTICITY

With a few exceptions, nearly all of the vaccine adjuvants that have been described are either surface active or provide a surface interface. To some degree, the surface formed by these substances has been characterized [9–13,55,103]. Hunter et al. proposed that the ability of surfactants to act as adjuvants "is dependent on their ability to concentrate adjuvant, immunogen, and host proteins on hydrophobic surfaces, where they are more effectively presented to cells of the immune system" [11]. Even the muramyl dipeptides are more adjuvant active with addition of lipoidal moieties or when incorporated into vaccine vesicles.

Formation of a special surface—either by adhesive surfactant substances concentrating on a surface interface or by mixing of certain hydrocarbons with water to form an appropriate aqueous–nonaqueous surface interface—has several consequences of importance to antigen recognition and immune enhancement. Activation of chemotactic peptides such as complement fragments or fibrinopeptide B results in cell recruitment and interactions that could play a significant role in adjuvanticity. Perhaps even more importantly, proper antigen orientation and epitope exposure results when antigens are concentrated on surfaces.

Protein antigens, and probably most other complex antigens, concentrate at surface interfaces because of their amphipathic nature [23]. Change of protein structure results as nonpolar amino acid regions of the protein segregate to the nonaqueous surface interface and polar domains segregate to the aqueous interface. With some proteins this change could result in exposure of epitopes not exposed when proteins are in aqueous solution; with others, the partitioning may result in a stable secondary structure similar to that of proteins in a membrane [167].

It is known that hydrophilic domains of proteins or peptides represent the antigenic determinants. Thus, a good vaccine should present these domains to antigen-processing cells in such a manner that they, in turn, present them to lymphocytes in an optimal manner. Surface immunoglobin on B cells can recognize native solubilized antigen, whereas T cells can recognize only processed antigen on the surface of antigen-processing cells in association with histocompatibility antigens. In regard to T-cell antigenic sites, it has been proposed that they are amphipathic structures "induced and stabilized by hydrophobic interactions with structures on the surface of the presenting cell" [168]. T-cell antigens are characterized by conformational amphipathic antigenic sites (mainly α-helices) rather than segmental amphipathic domains [168–170].

## V. FUTURE DIRECTIONS

The future of vaccine adjuvant research lies not in the empirical approach of the last 60 years, i.e., measurement of immune responses or protection following injection of antigen, adjuvant, and vehicle. To make real progress, we must first understand antigen presentation, antigen processing, and the structural requirements of antigenic determinants. Only by understanding the interaction of surface interfaces with antigen can we begin to understand the mechanism of action of the majority of vaccine adjuvants. Once these interactions are known, we can eliminate much of the trial-and-error experiments and design the most effective adjuvants and vehicles. Although a number of modes of action of adjuvants have been proposed [15–19], we still do not know if these mechanisms are responsible for adjuvanticity or are just a consequence of intertwined immunological and inflammatory events.

With the exception of the muramyl dipeptides and certain cytokines, it appears that a special surface interface or alteration of a surface to give it special adhesive properties is a common trait. In some cases, such surfaces increase cell recruitment to the antigen depot by the generation of chemotactic peptides; in others, this does not occur. Many adjuvants have general immunomodulatory effects while others do not. Perhaps by studying the interaction of adjuvant surfaces with antigen, we can find some common properties of interest.

Because most of the antigens of interest for human and veterinary biologics are T dependent, the interaction of adjuvant surface formation on antigen processing and eventual presentation to T-cell receptors would appear to be central to the understanding of surfactant adjuvants. The fact that T-cell receptors recognize amphipathic epitopes on the surface of antigen-presenting cells (in conjunction with histocompatibility antigens), but not the native solubilized antigen, suggests that denaturization and amphipathicity as a result of protein concentration at a surface interface should be beneficial. Perhaps surface-bearing or surface-active adjuvants abbreviate configuration changes that would normally occur in vivo. Antigens in association with surfactant adjuvants and vesicles may be better positioned for processing and presentation. Epitopes that would normally be revealed only after processing might be readily available in higher densities at the time of initial contact by macrophages and dendritic cells.

Immunogenicity of antigens may be enhanced by surface adjuvants through their ability to add hydrophobicity. With small-molecular-weight substances such as synthetic peptides, the lack of amphipathic sequences in the native primary structure or the lack of amphipathic conformational secondary structure may prevent them from acting as antigens. In many cases, it seems likely that proper surface adjuvants could circumvent the need for large carrier proteins.

As we begin to understand fully the interaction of protein antigens with membranes of antigen-processing cells and the surface conformations of epitopes, the role of surface adjuvants should become clearer. Subunit and synthetic vaccines with defined synthetic adjuvants will provoke the desired immunologic response without adverse inflammatory reactions.

## REFERENCES


1. Ramon G: Sur l'augmentation abnormale de l'antitoxine chez les chevaux producteurs de serum antidiphtherique. Bull Soc Centr Med Vet 101:227, 1925.
2. Glenny AT, Pope CG, Waddington H, Wallace U: Immunological notes. XXIII. The antigenic value of toxoid precipitated by potassium alum. J Pathol 29:38, 1926.
3. Gall D: The adjuvant activity of aliphatic nitrogenous bases. Immunology 11:366, 1966.
4. Nervig RM, Gough PM, Kaeberle ML, Whetstone CA: "Advances in Carriers and Adjuvants for Veterinary Biologics." Ames: Iowa State University Press, 1986.
5. Warren HS, Vogel FR, Chedid LA: Current status of immunological adjuvants. In Paul WE (ed): "Annual Review of Immunology." Palo Alto, CA: Annual Reviews Inc., 1986, p 369.
6. Dalsgaard K: Adjuvants. Vet Immunol Immunopathol 17:145, 1987.
7. Edelman R: Vaccine adjuvants. Rev Infect Dis 2:370, 1980.
8. Chedid L: Adjuvants for vaccines. In Hadden JW, Chedid L, Dukor P, Spreafico F, Willoughby D (eds): "Advances in Immunopharmacology." Oxford: Pergamon Press, 1983, Vol. 2, p 401.
9. Retzinger GS, Meredith SC, Takayama K, Hunter RL, Kezdy FJ: The role of surface in the biological activities of trehalose 6,6′-dimycolate. J Biol Chem 256:8208, 1981.
10. Retzinger GS, Meredith SC, Hunter RL, Takayama K, Kezdy FJ: Identification of the physiologically active state of the mycobacterial glycolipid trehalose 6,6′-dimycolate and the role of fibrinogen in the biologic activities of trehalose 6,6′-dimycolate monolayers. J Immunol 129:735, 1982.
11. Hunter R, Strickland F, Kezdy F: The adjuvant activity of nonionic block polymer surfactants. I. The role of hydrophile-lipophile balance. J Immunol 127:1244, 1981.
12. Hunter RL, Bennett B: The adjuvant activity of nonionic block polymer surfactants. II. Antibody formation and inflammation related to the structure of triblock and octablock copolymers. J Immunol 133:3167, 1984.
13. Hunter RL, Bennett B: The adjuvant activity of nonionic block polymer surfactants. III. Characterization of selected biologically active surfaces. Scand J Immunol 23:287, 1986.
14. Woodard LF: Surfactants and surface formation in vaccine adjuvants and vehicles. In Stark DM (ed): "Immunological Adjuvants: The Old and the New." New York: American Association of Laboratory Animal Scientists Monograph Series VI, 1989 (in press).
15. Thein P: Adjuvants: Immunomodulation and immunostimulation. Vet Med Rev 59:3, 1988.
16. Allison AC: Mode of action of immunological adjuvants. J Reticuloendothel Soc 26:619, 1979.

17. Osebold JW: Mechanisms of action by immunologic adjuvants. J Am Vet Med Assoc 181:983, 1982.
18. Frost P, Lance EM: On the mechanism of action of adjuvants. Immunology 35:63, 1978.
19. Kaeberle ML: Function of carriers and adjuvants in induction of immune responses. In Nervig RM, Gough PM, Kaeberle ML, Whetstone CA (eds): "Advances in Carriers and Adjuvants for Veterinary Biologics." Ames: Iowa State University Press, 1986, p 11.
20. Garrett HE: "Surface Active Chemicals." Oxford: Pergamon Press, 1972, p 4.
21. Becher P: "Principles of Emulsion Technology." New York: Reinhold, 1955, p 3.
22. Becher P: "Emulsions: Theory and Practice." New York: Reinhold, 1965, p 209.
23. MacRitchie F: Proteins at interfaces. Adv Protein Chem 32:283, 1978.
24. Anonymous: "The HLB System: A Time Saving Guide to Emulsifier Selection." Wilmington, DE: ICI Americas, 1980.
25. Woodard LF, Jasman RL: Stable oil-in-water emulsions: Preparation and use as vaccine vehicles for lipophilic adjuvants. Vaccine 3:137, 1985.
26. Glenny AT, Buttle GAH, Stevens MF: Rate of disappearance of diphtheria toxoid injected into rabbits and guinea pigs: Toxoid precipitated with alum. J Pathol 34:267, 1931.
27. Aprile MA, Wardlaw AC: Aluminum compounds as adjuvants for vaccines and toxoids in man: A review. Can J Public Health 57:343, 1966.
28. Bunn TO, Nervig RM, Pemberton JR: Metallic salts as adjuvants for veterinary biologics. In Nervig RM, Gough PM, Kaeberle ML, Whetsone CA (eds): "Advances in Carriers and Adjuvants for Veterinary Biologics." Ames: Iowa State University Press, 1986, p 105.
29. Unanue GR, Askinas BA, Allison AC: A role of macrophages in the stimulation of immune responses by adjuvants. J Immunol 103:71, 1969.
30. Relyveld EH: Comparative studies on mineral adjuvants. Vaccine 3:153, 1985.
31. Relyveld EH: Preparation and use of calcium phosphate adsorbed vaccines. Dev Biol Stand 65:131, 1986.
32. Gupta RK, Sharma SB, Ahuja S, Saxena SN: The effect of aluminum phosphate adjuvant on the potency of pertussis vaccine. J Biol Stand 15:99, 1987.
33. Murdoch RD, Pepys J: Enhancement of antibody production by mercury and platinum group metal halide salts. Kinetics of total and ovalbumin-specific IgE synthesis. Int Arch Allergy Appl Immunol 80:405, 1986.
34. Hall JG: Studies on the adjuvant action of beryllium. I. Effects on individual lymph nodes. Immunology 53:105, 1984.
35. Hall JG, Spencer J: Studies on the adjuvant action of beryllium. II. Systemic effects with particular reference to secretory immunity. Immunology 53:115, 1984.
36. Behbehani K, Beller DI, Unanue ER: The effects of beryllium and other adjuvants on Ia expression by macrophages. J Immunol 134:2047, 1985.
37. Stalder K, Schrader R: Quartstaub als adjuvans beim entstehen einer zellvermi helten immunitat. Naturwissenschaften 67:569, 1980.
38. Mancino D, Buono G, Cusano M, Minucci M: Adjuvant effects of a crystalline silica on IgE and IgG1 antibody production in mice and their prevention by the macrophage stabilizer poly-2-vinylpyridine N-oxide. Int Arch Allergy Appl Immunol 71:279, 1983.
39. Gallily R, Garvey JS: Primary stimulation of rats and mice with hemocyanin in solution and adsorbed on bentonite. J Immunol 101:924, 1968.
40. Gelfand EW, Dosch HM, Hastings D, Shore A: Lithium: A modulator of cyclic AMP-dependent events in lymphocytes. Science 203:365, 1979.
41. Westman AP, McGregor AM, Lazarus JH, Rees Smith B, Hall R: The enhancement of immunoglobulin synthesis of human lymphocytes with lithium. Clin Immunol Immunopathol 22:400, 1982.

42. Desowitz RS, Barnwell JW: Effect of selenium and dimethyl dioctadecyl ammonium bromide on the vaccine-induced immunity of Swiss-Webster mice against malaria *(Plasmodium berghei)*. Infect Immun 27:87, 1980.
43. Koller LD, Exon JH: The two faces of selenium—deficiency and toxicity—are similar in animals and man. Can J Vet Res 50:297, 1986.
44. Lederer E: Cord factor and related synthetic trehalose diesters. Springer Semin Immunopathol 2:133, 1979.
45. Bekierkunst A, Yarkoni E, Flechner I, Morecki S, Vilkas E, Lederer E: Immune response to sheep red blood cells in mice pretreated with mycobacterial fractions. Infect Immun 4:256, 1971.
46. Granger DL, Yamamoto K, Ribi E: Delayed hypersensitivity and granulomatous response after immunization with protein antigens associated with a mycobacterial glycolipid and oil droplets. J Immunol 116:482, 1976.
47. Woodard LF, Renshaw HW, Burger D, McCain CS, Wilson RB: Cell-mediated immune responses of neonatal calves and adult cattle following inoculation with PPD of *Mycobacterium bovis* associated with a mycobacterial immunopotentiating glycolipid and oil droplets. Am J Vet Res 40:636, 1979.
48. Woodard LF, Toone NM, McLaughlin CA: Comparison of muramyl dipeptide, trehalose dimycolate and dimethyl dioctadecyl ammonium bromide as adjuvants in *Brucella abortus* 45/20 vaccines. Infect Immun 30:409, 1980.
49. Woodard LF, Jasman RL: Comparative efficacy of an experimental S45/20 bacterin and a reduced dose of strain 19 vaccine against bovine brucellosis. Am J Vet Res 44:907, 1983.
50. Winter AJ, Verstreate DR, Hall CE, Jacobson RH, Castleman WL, Meredith MP, McLaughlin CA: Immune response to porin in cattle immunized with whole cell, outer membrane, and outer membrane protein antigens of *Brucella abortus* combined with trehalose dimcolate and muramyl dipeptide adjuvants. Infect Immun 42:1159, 1983.
51. Ribi E: Structure-function relationship of bacterial adjuvants. In Nervig RM, Gough PM, Kaeberle ML, Whetsone CA (eds): "Advances in Carriers and Adjuvants for Veterinary Biologics." Ames: Iowa State University Press, 1986, p 35.
52. Saito R, Tanaka A, Sugiyama K, Azuma I, Yamamura Y, Kato M, Goren MB: Adjuvant effect of cord factor, a mycobacterial lipid. Infect Immun 13:776, 1976.
53. Parant M, Audibert F, Parant F, Chedid L, Soler E, Polonsky J, Lederer E: Nonspecific immunostimulant activities of synthetic trehalose -6,6′-diesters (lower homologs of cord factor). Infect Immun 20:12,1978.
54. Toubiana R, Ribi E, McLaughlin C, Strain SM: The effect of synthetic and naturally occurring trehalose fatty acid esters in tumor regression. Cancer Immunol Immunother 2:189, 1977.
55. McLaughlin CA, Ribi EE, Goren MB, Toubiana R: Tumor regression induced by defined microbial components in an oil-in-water emulsion is mediated through their binding to oil droplets. Cancer Immunol Immunother 4:109, 1978.
56. Ramanathan VD, Curtis J, Turk JL: Activation of the alternative pathway of complement by mycobacteria and cord factor. Infect Immun 29:30, 1980.
57. Johnson AJ, Gaines S, Landy M: Studies on the O antigen of *Salmonella typhosa*. V. Enhancement of the antibody response to protein antigens by the purified lipopolysaccharide. J Exp Med 103:225, 1956.
58. Rietschel ET, Galanos C, Luderitz O, Westphal O. Chemical structure, physiological function and biological activity of lipopolysaccharides and their lipid A component. In Webb D (ed): "Immunopharmacology and the Regulation of Leukocyte Function." New York: Marcel Dekker, 1982, p 183.

59. Gallanos C, Freudenberg A, Jay F, Nerkar D, Veleva K, Brade H, Strittmatter W: Immunogenic properties of lipid A. Rev Infect Dis 6:546, 1984.
60. Rietschel ET, Wollenweber HW, Russa R, Brade H, Zahringer U: Concepts of the chemical structure of lipid A. Rev Infect Dis 6:432, 1984.
61. Ukei S, Iida J, Shiba T, Kusumoto S, Azuma I: Adjuvant and antitumor activities of synthetic lipid A analogues. Vaccine 4:21, 1986.
62. Kotani S, Takada H, Tsujimoto M, Ogawa T, Mori Y, Sakuta M, Kawasaki A, Inage M, Kusumoto S, Shiba T, Kasai N: Immunobiological activities of synthetic lipid A and related compounds with those of bacterial lipopolysaccharides, re-glycolipid, lipid A, and muramyl dipeptide. Infect Immun 41:758, 1983.
63. Ribi E, Amano K, Cantrell J, Schwartzmann S, Parker R, Takayama K: Preparation and antitumor activity of nontoxic lipid A. Cancer Immunol Immunother 12:91, 1982.
64. Weiss DW, Dubos RJ: Antituberculosis immunity induced by methanol extracts of tubercle bacilli—Its enhancement by adjuvants. J Exp Med 103:73, 1956.
65. Munder PG, Modolell M: Adjuvant induced formation of lysophosphatides and their role in the immune response. Int Arch Allergy 45:133, 1973.
66. Strannegard O, Roupe G: Adjuvant effect of lysolecithin analogues on the development of contact sensitivity in mice. Int Arch Allergy Appl Immunol 51:198, 1976.
67. Arnold B, Miller JFAP, Weltzien HU: Lysolecithin analogs as adjuvants in delayed-type hypersensitivity in mice. I. Characterization of the adjuvant effect. Eur J Immunol 9:363, 1979.
68. Munder PG, Weltzien HU, Modolell M: Lysolecithin analogs: A new class of immunopotentiators. In Miescher PA (ed): ''Immunopathology.'' Basel: Schwabe & Co., 1976, p 411.
69. Ramon G, Richon R, Staub A: L'immunisation au moyen de microbes vivants—bacille diphtherique, staphylocoque, bacteridie charbonneuse—additionnes de substances adjuvants. Rev Immunol 3:389, 1937.
70. Rivenson S: Determination of the adjuvant capacity of aluminum hydroxide, carboxymethylcellulose and saponin in the production of foot-and-mouth disease complement fixing antibodies in guinea pigs. Gac Vet 20:209, 1959.
71. Amies CR: The use of topically formed calcium alginate as a depot substance in active immunization. J Pathol Bacteriol 77:435, 1959.
72. Dalsgaard K: Saponin adjuvants: Isolation of a substance from *Quillaja aaponaria molina* with adjuvant activity in foot-and-mouth disease vaccine. Arch Virusforschung 44:243, 1974.
73. James SL, Pearce EJ: The influence of adjuvant on induction of protective immunity by a non-living vaccine against schistosomiasis. J Immunol 140:2753, 1988.
74. Playfair JHL, De Souza JB: Vaccination of mice against malaria with soluble antigens. I. The effect of detergent, route of injection, and adjuvant. Parasite Immunol 8:409, 1986.
75. Wells PW, Emergy DL, Hinson CA, Morrison WI, Murray M: Immunization of cattle with a varient-specific surface antigen of *Trypanosoma brucei:* Influence of different adjuvants. Infect Immun 36:1, 1982.
76. Gall D: Observations on the properties of adjuvants. In Regamey RH, Hennesen W, Ikic D, Ungar J (eds): ''Symposia Series in Immunobiological Standardization.'' Basel: S. Karger, 1967, Vol. 6, p 39.
77. Morein B, Sundquist B, Hoglund S, Dalsgaard K, Osterhaus A: Iscom, a novel structure for antigenic presentation of membrane proteins from enveloped viruses. Nature 308:457, 1984.
78. Morein B: The iscom antigen-presenting system. Nature 332:287, 1988.

79. Morein B: Potentiation of the immune response by immunization with antigens in defined multimeric physical forms. Vet Immunol Immunopathol 17:153, 1987.
80. Soppi E, Lehtonen O: Selective effect of different retinoids on the primary antibody response in normal chickens. Immunopharmacology 8:91, 1984.
81. Tengerdy RP, Meyer DL, Lauerman LH, Lueker DC, Nockels CF: Vitamin E enhances humoral antibody response to *Clostridium perfringens*, type D, in sheep. Br Vet J 139: 147, 1983.
82. Afzal M, Tengerdy PR, Ellis RP, Kimberling CV, Morris CJ: Protection of rams against epididymitis by a *Brucella ovis*-vitamin E adjuvant vaccine. Vet Immunol Immunopathol 7:293, 1984.
83. Dukor P, Tarcsay L, Baschang G: Immunostimulants. Annu Rep Med Chem 14:146, 1979.
84. Williams DL, DiLuzig NR: Glucan-induced modification of murine viral hepatitis. Science 208:67, 1980.
85. Hamuro J, Rollinghoff M, Wagner H: Induction of cytotoxic peritoneal exudate cells by T-cell immune adjuvants of the β (1→3) glucan-type lentinan and its analogs. Immunology 39:551, 1980.
86. Reynolds JA, Kastello MD, Harrington DG, Crabbs CL, Peters CJ, Jemski JV, Scott GH, DiLuzio NR: Glucan-induced enhancement of host resistance to selected infectious diseases. Infect Immun 30:51, 1980.
87. Holbrook TW, Cook JA, Parker BW: Glucan-enhanced immunogenicity of killed erythrocytic stages of *Plasmodium berghei*. Infect Immun 32:542, 1981.
88. Benach JL, Habicht GS, Holbrook TW, Cook JA: Glucan as an adjuvant for a murine *Babesia microti* immunization trial. Infect Immun 35:947, 1982.
89. Diamantstein T, Klos M, Osawa H, Chen ZC: Chitin: An immunological adjuvant and a polyclonal B-lymphocyte activator. Int Arch Allergy Appl Immunol 68:377, 1982.
90. Nishimura K, Nishimura S, Nishi N, Numata F, Tone Y, Tokura S, Azuma I: Adjuvant activity of chitin derivatives in mice and guinea pigs. Vaccine 3:379, 1985.
91. Bice ED, Brunwell DLG, Salvaggio UE, Hoffman EC: Suppression of primary immunization by carrageenan—A macrophage toxic agent. Immunol Commun 1:615, 1972.
92. Evelegh MJ, Clark DA, McCandless EL: Carrageenan stimulates the release of dialyzable helper factors. Immunol Lett 5:247, 1982.
93. Little JR, Abegg A, Plut E: The relationship between adjuvant and mitogenic effects of amphotericin methyl ester. Cell Immunol 78:224, 1983.
94. Pedersen B, Norrilk B, Krebs HJ: The effect of monensin on antibody-dependent cell-mediated killing. Cell Immunol 72:108, 1982.
95. Noll H, Youngner JS: Virus-lipid interactions. II. The mechanism of adsorption of lipophilic viruses to water-insoluble polar lipids. Virology 8:319, 1959.
96. Woodhour AF, Jensen KE, Warren J: Antibody response to influenza vaccines combined with hexadecylamine. Soc Exp Biol Med 103:200, 1960.
97. Youngner JS, Axelrod V: Antigenicity of lipid-adsorbed diphtheria and tetanus toxoids. J Immunol 92:879, 1964.
98. Woodard LF: Adjuvant activity of water-insoluble surfactants. Lab Anim Sci 39:222, 1989.
99. Prager MD: Dimethyldioctadecyl ammonium bromide (DDA) as an immunologic adjuvant. Kodak Lab Chem Bull 56:1, 1985.
100. Jensen KE: Synthetic adjuvants: Avridine and other interferon inducers. In Nervig RM, Gough PM, Kaeberle ML, Whetstone CA (eds): "Advances in Carriers and Adjuvants for Veterinary Biologics." Ames: Iowa State University Press, 1986, p 79.
101. Kraaijeveld CA, Snippe H, Harmsen T, Benaissa-Trouw B: Enhancement of delayed-

type hypersensitivity and induction of interferon by the lipophilic agents DDA and CP-20,961. Cell Immunol 74:277, 1982.

102. Penny CL, Landi S, Shah P, Leung KH, Archer MC: Analysis of the immunoadjuvant octadecyl tyrosine hydrochloride. J Biol Stand 14:345, 1986.
103. Hunter RL, Bennett B: Structural basis of the activity of surface-active adjuvants. In Nervig RM, Gough PM, Kaeberle ML, Whetstone CA (eds): ''Advances in Carriers and Adjuvants for Veterinary Biologics.'' Ames: Iowa State University Press, 1986, p 61.
104. Zigterman GJWJ, Snippe H, Jansze M, Ernste EBHW, DeReuver MJ, Willers JMN: Nonionic block polymer surfactants enhance immunogenicity of pneumococcal hexasaccharide-protein vaccines. Infect Immun 56:1391, 1988.
105. Allison AC, Byars NE: The development of an adjuvant formulation that elicits cell-mediated and humoral immune responses to virus subunit and other antigens. In Majde JA (ed): ''Progress in Leukocyte Biology.'' New York: Alan R. Liss, 1987, Vol. 6, p 191.
106. Zigterman GJWJ, Jansze M, Snippe H, Willers JMN: Stimulation of liposome-induced humoral immune responses by non-ionic block polymer surfactants in Xid mice. Immunology 65:37, 1988.
107. Zigterman GJWJ, Snippe H, Jansze M, Willers JMN: Adjuvant effects of nonionic block polymer surfactants on liposome-induced humoral immune reponses. J Immunol 138: 220, 1987.
108. Hilgers LAT, Snippe H, Jansze M, Willers JMN: Immunomodulating properties of two synthetic adjuvants: Dependence upon type of antigen, dose, and time of administration. Cell Immunol 86:393, 1984.
109. Kishima M, Ross RF, Kuniyasu C: Cell-mediated and humoral immune response to *Mycoplasma hyopneumoniae* in pigs enhanced by dextran sulfate. Am J Vet Res 46:456, 1985.
110. Joio I, Emoid J: Adjuvant effect of DEAE-dextran on cholera vaccines. Vaccine 6:233, 1988.
111. Van der Meer C, Hofhuis FMA, Willers JMN: Killed *Listeria monocytogenes* vaccine becomes protective on addition of polyanions. Nature 269:594, 1977.
112. Brandely M, LaGrange P, Hurtrel B, Motta I, Truffa-Bachi P: Effects of sumarin on the immune responses to sheep red blood cells in mice. I. In vivo studies. Cell Immunol 93:280, 1085.
113. Arora MP, Crowle AJ: Adjuvanticity of Evans blue dye: Preliminary characterization of its mechanisms and mode of application in mice. J Reticuloendothel Soc 24:271, 1978.
114. Markham RJF, Wilkie BN: Local adjuvants: The influence of sodium dodecylbenzene sulfonate on immunization with aerosolized antigen. Experientia 35:414, 1979.
115. Lesourd BM, Vincent-Falquet JC, Deslandes D, Musset M, Moulias R: Influenza vaccination in the elderly: Improved antibody response with Imuthiol (Na diethyldithiocarbamate) adjuvant therapy. Int J Immunopharmacol 10:135, 1988.
116. Wong MSF, Hoskinson RM, Cox RI, Hawthorne DG, Solomon DH, Hinks NT, Donnelly JB: Use of a styrene-malic anhydride copolymer as an adjuvant with testosterone-serum albumin immunogens and its application to increasing ovulation rate in the ewe. Aust J Biol Sci 39:99, 1986.
117. Schroder U, Stahl A: Crystallized dextran nanospheres with entrapped antigen and their use as adjuvants. J Immunol Methods 79:127, 1984.
118. Anderson DW, Crowle AJ: Evans blue dye adjuvant enhances delayed hypersensitivity while blocking immunity to *Mycobacterium tuberculosis* in mice. Infect Immun 31:413, 1981.

119. Levin RH: Tilorone, its analogs, and developments in chemical immunology. Aldrichimica Acta 12:77, 1979.
120. Merluzzi VJ, Walker MM, Williams N, Susskind B, Hadden JW, Faanes RB: Immunoenhancing activity of NPT 15392: A potential immune response modifier. Int J Immunopharmacol 4:219, 1982.
121. Thong YH, Ferrante A: Immunopotentiation by pyrimethamine in the mouse. Clin Exp Immunol 39:190, 1980.
122. Johnson AG, Mohrman M, Odean M, Petrequin P: Adjuvant action of synthetic polynucleotides. In Nervig RM, Gough PM, Kaeberle ML, Whetstone CA (eds): "Advances in Carriers and Adjuvants for Veterinary Biologics." Ames: Iowa State University Press, 1986, p 71.
123. Freund J: The effect of paraffin oil and mycobacteria on antibody formation and sensitization. A review. Am J Clin Pathol 21:645, 1951.
124. Freund J: The mode of action of immunologic adjuvants. Adv Tuberc Res 7:130, 1956.
125. Herbert WJ: The mode of action of mineral-oil emulsion adjuvants on antibody production in mice. Immunology 14:301, 1968.
126. McKercher PD: Oil adjuvants: Their use in veterinary biologics. In Nervig RM, Gough PM, Kaeberle ML, Whetstone CA (eds): "Advances in Carriers and Adjuvants for Veterinary Biologics." Ames: Iowa State University Press, 1986, p 11.
127. Bokhout BA, vanGaalen C, van der Heijden PJ: A selected water-in-oil emulsion: Composition and usefulness as an immunological adjuvant. Vet Immunol Immunopathol 2:491, 1981.
128. Wilner BI, Evers MA, Troutman HD, Trader FW, McLean IW: Vaccine potentiation by emulsification with pure hydrocarbon compounds. J Immunol 91:210, 1963.
129. Shaw C, Alvord EC, Kies MW: Straight chain hydrocarbons as substitutes for the oil in Freund's adjuvants in the production of experimental "allergic" encephalomyelitis in the guinea pig. J Immunol 92:24, 1964.
130. Hilleman MR, Woodhour A, Friedman A, Weibel RE, Stokes J: The clinical application of adjuvant 65. Ann Allergy 30:152, 1972.
131. Bomford R: The adjuvant activity of fatty acid esters. The role of acyl chain length and degree of saturation. Immunology 44:187, 1981.
132. Reynolds JA, Harrington DG, Crabbs CL, Peters CJ, DiLuzio NR: Adjuvant activity of a novel lipid emulsion with inactivated viral vaccines. Infect Immun 28:937, 1980.
133. Siddiqui WA, Kan S, Kramer K, Case S, Palmer K: Use of a synthetic adjuvant in an effective vaccination of monkeys against malaria. Nature 289:64, 1981.
134. Carelli C, Audibert F, Chedid L: Persistent enhancement of cell-mediated and antibody immune responses after administration of muramyl dipeptide derivatives with antigen in metabolizable oil. Infect Immun 33:312, 1981.
135. Dingle JT, Jacques PJ, Shaw IH: "Liposomes in Applied Biology and Therapeutics." Amsterdam: North Holland, 1979.
136. Tom BH, Six H: "Liposomes and Immunobiology." New York: Elsevier, 1980.
137. Rouse BT, Turtinen L, Correa-Freire M: Use of liposomes in immunology. In Nervig RM, Gough PM, Kaeberle ML, Whetstone CA (ed): "Advances in Carriers and Adjuvants for Veterinary Biologics." Ames: Iowa State University Press, 1986, p 121.
138. Gregoriadis G, Davis D, Davies A: Liposomes as immunological adjuvants: Antigen incorporation studies. Vaccine 5:145, 1987.
139. Alving CR, Banerji B, Shiba T, Kotani S, Clements JD, Richards RL: Liposomes as vehicles for vaccines. In Mizrahi A, Hertman I, Klingberg MA, Kohn A (eds): "New Developments With Human and Veterinary Vaccines." New York: Alan R. Liss, 1980, p 339.

140. Dancey GF, Yasuda T, Kinsky SC: Enhancement of liposomal model membrane immunogenicity by incorporation of lipid A. J Immunol 119:1868, 1977.
141. Schuster BG, Neidig M, Alving BM, Alving CR: Production of antibodies against phosphocholine, phosphatidylcholine, sphingomyelins and lipid A by injection of liposomes containing lipid A. J Immunol 122:900, 1979.
142. Lowell GH, Smith LF, Seid RC, Zollinger WD: Peptides bound to proteosomes via hydrophobic feet become highly immunogenic without adjuvants. J Exp Med 167:658, 1988.
143. Lowell GH, Ballou WR, Smith LF, Wirtz RA, Zollinger WD, Hockmeyer WT: Proteosome-lipopetide vaccines: Enhancement of immunogenicity for malaria CS peptides. Science 240:800, 1988.
144. Brown L, Edelman E, Siemer L, Langer R: Controlled-release systems for macromolecules: methods and application. In Nervig RM, Gough PM, Kaeberle ML, Whetstone CA (eds): "Advances in Carriers and Adjuvants for Veterinary Biologics." Ames: Iowa State Unviersity Press, 1986, p 133.
145. Prince LM: "Microemulsions. Theory and Practice." New York: Academic Press, 1977.
146. Johnson KA, Shah DO: Effect of oil chain length and electrolytes on water solubilization in alcohol-free pharmaceutical microemulsions. J Colloid Interface Sci 107:269, 1985.
147. Stewart-Tull DES: The immunological activities of bacterial peptidoglycans. In Starr MP, Ingraham JL, Balows A (eds): "Annual Review of Microbiology." Palo Alto, CA: Annual Reviews Inc., 1980, p 311.
148. Ellouz F, Adam A, Ciorbaru R, Lederer E: Minimal structural requirement for adjuvant activity of bacterial peptidoglycan derivatives. Biochem Biophys Res Commun 59:1317, 1974.
149. Lederer E: New developments in the field of synthetic muramyl peptides, especially as adjuvants for synthetic vaccines. Drugs Exp Clin Res 12:429, 1986.
150. Chedid L, Audibert F, Jolivet M: Role of muramyl peptides for the enhancement of synthetic vaccines. Dev Biol Stand 63:133, 1986.
151. Warren HS, Chedid LA: Future prospects for vaccine adjuvants. CRC Crit Rev Immunol 8:83, 1988.
152. Woodard LF, Toone NM, McLaughlin CA: Immunogenic properties of soluble antigens or whole cells of *Brucella abortus* strain 45/20 associated with immunoadjuvants. II. Whole cells. Can J Comp Med 44:456, 1980.
153. Woodard LF, Toone NM, Jasman RL: *Brucella abortus* vaccines: Comparison of protection provided by immunopotentiated 45/20 bacterins and live strain 19 vaccine in guinea pigs. Am J Vet Res 42:1959, 1981.
154. Masek K, Zaoral M, Jezek J, Straka R: Immunoadjuvant activity of synthetic N-acetylmuramyl dipeptide. Experientia 34:1363, 1978.
155. Parant M, Audibert F, Chedid L, Level M, Lefrancier P, Choay J, Lederer E: Immunostimulant activities of a lipophilic muramyl dipeptide derivative and of desmuramyl peptidolipid analogs. Infect Immun 27:826, 1980.
156. Yamamura Y, Azuma I, Sugimura K, Yamawaki M, Uemiya M, Kusumoto S, Okada S, Shiba T: Immunological and antitumor activities of synthetic 6-0-mycoloyl-N-acetylmuramyl dipeptides. Proc Jpn Acad 53:63, 1977.
157. Allison AC, Byars NE, Waters RV: Immunologic adjuvants: Efficacy and safety considerations. In Nervig RM, Gough PM, Kaeberle ML, Whetstone CA (eds): "Advances in Carriers and Adjuvants for Veterinary Biologics." Ames: Iowa State University Press, 1986, p 91.
158. Staruch MJ, Wood DD: The adjuvanticity of interleukin 1 in vivo. J Immunol 130:2191, 1983.

159. Dinarello CA: An update on human interleukin-1: From molecular biology to clinical relevance. J Clin Immunol 5:287, 1985.
160. Duram SK, Schmidt JA, Oppenheim JJ: Interleukin 1: An immunological perspective. Annu Rev Immunol 3:263, 1985.
161. Kampschmidt RF: The numerous postulated biological manifestations of interleukin-1. J Leukocyte Biol 36:341, 1984.
162. Nencioni L, Villa L, Tagliabue A, Antoni G, Presentini R, Perin F, Silvestri S, Boraschi D: In vivo immunostimulating activity of the 163-171 peptide of human IL-1β. J Immunol 139:800, 1987.
163. Kishter S, Hoffman FA, Pizzo PA: Production of and response to interleukin-2 by cultured T cells: Effects of lithium chloride and other putative modulators. J Biol Respir Modif 4:185, 1985.
164. Kawamura H, Rosenberg SA, Berzofsky JA: Immunization with antigen and interleukin 2 in vivo overcomes Ir gene low responsiveness. J Exp Med 162:381, 1985.
165. Weinberg A, Merigan TC: Recombinant interleukin 2 as an adjuvant for vaccine induced protection. J Immunol 140:294, 1988.
166. Playfair JHL, DeSouza JB: Recombinant gamma interferon is a potent adjuvant for a malaria vaccine in mice. Clin Exp Immunol 67:5, 1987.
167. Kaiser ET, Kezdy FJ: Amphiphilic secondary structure: Design of peptide hormones. Science 223:249, 1984.
168. DeLisi C, Berzofsky JA: T-cell antigenic sites tend to be amphipathic structures. Proc Natl Acad Sci USA 82:7048, 1985.
169. Spouge JL, Guy HR, Cornette JL, Margalit H, Cease K, Berzofsky JA, DeLisi C: Strong conformational propensities enhance T-cell antigenicity. J Immunol 138:204, 1987.
170. Berzofsky JA: Features of T-cell recognition and antigen structure useful in the design of vaccines to ellicit T-cell immunity. Vaccine 6:89, 1988.

# Index